Lecture Notes in Networks and Systems

Volume 1244

The series "Lecture Notes in Networks and Systems" publishes the latest developments in Networks and Systems—quickly, informally and with high quality. Original research reported in proceedings and post-proceedings represents the core of LNNS.

Volumes published in LNNS embrace all aspects and subfields of, as well as new challenges in, Networks and Systems.

The series contains proceedings and edited volumes in systems and networks, spanning the areas of Cyber-Physical Systems, Autonomous Systems, Sensor Networks, Control Systems, Energy Systems, Automotive Systems, Biological Systems, Vehicular Networking and Connected Vehicles, Aerospace Systems, Automation, Manufacturing, Smart Grids, Nonlinear Systems, Power Systems, Robotics, Social Systems, Economic Systems and other. Of particular value to both the contributors and the readership are the short publication timeframe and the world-wide distribution and exposure which enable both a wide and rapid dissemination of research output.

The series covers the theory, applications, and perspectives on the state of the art and future developments relevant to systems and networks, decision making, control, complex processes and related areas, as embedded in the fields of interdisciplinary and applied sciences, engineering, computer science, physics, economics, social, and life sciences, as well as the paradigms and methodologies behind them.

Indexed by SCOPUS, EI Compendex, INSPEC, WTI Frankfurt eG, zbMATH, SCImago.

All books published in the series are submitted for consideration in Web of Science.

For proposals from Asia please contact Aninda Bose (aninda.bose@springer.com).

Igor Ilin · Ma Youzhong
Editors

Digital Systems and Information Technologies in the Energy Sector

Springer

Editors
Igor Ilin
Graduate School of Business Engineering
Peter the Great St. Petersburg Polytechnic
University
St. Petersburg, Russia

Ma Youzhong
School of Information Technology
Luoyang Normal University
Luoyang, Henan, China

ISSN 2367-3370 ISSN 2367-3389 (electronic)
Lecture Notes in Networks and Systems
ISBN 978-3-031-80709-1 ISBN 978-3-031-80710-7 (eBook)
https://doi.org/10.1007/978-3-031-80710-7

This Springer imprint is published by the registered company Springer Nature Switzerland AG
The registered company address is: Gewerbestrasse 11, 6330 Cham, Switzerland

Preface

International associations (IEA, IRENA, UN, etc.), large consulting agencies (McKinsey), companies from the energy and near-energy sector (Gazprom, Gazpromneft, EATON, General Electric, British Petroleum, etc.), opinion leaders in technology (I. Musk, S. Altman, etc.) have been actively talking about the relationship and possible intersections between digital transformation and energy transition in various aspects since 2017. The term "Twin Transition", which refers to the combined phenomenon of energy and digital transition, has become widespread.

At the same time, the issue has not yet received a proper response in the research community and is not widely publicized.

The book aims to gather researchers' views on each of the trends separately and on their possible synergetic effect and/or double challenge. The book is characterized by an original, interdisciplinary problem statement: description of various aspects of effective interaction between elements of energy and information technology/digital infrastructures, jointly providing information needs and compliance with energy efficiency requirements of modern economy. Both theoretical foundations of design and modeling of digital solutions for the energy sector, combining the requirements and capabilities of modern energy systems and digital technologies, and examples of practical implementation of such symbiotic solutions are addressed. There is a tangible demand from the practicing community to describe the possibilities of synergistic interaction between the two key trends of our days.

The book presents a variety of perspectives on the described phenomenon. There are four parts in the book. Energy specialists in the first part *Technical, Management and IT Aspects of Nowadays Energy Sector* describe their results in modeling, calculations and analysis of particular aspects of nowadays energy challenges. In the second part *Digital and IT Solutions for Industrial, Functional and Regional Challenges*, IT and digital technology experts present their views on nowadays IT and digital solutions for different aspects of business management, industry challenges and even regional management. The third part *Digital Aspects of Energy Sector* assess the possibilities of digital technologies for creating qualitatively new energy management approaches and their contribution to increasing the performance of energy systems, describe the possibilities of renewable energy sources and assess their real “alternativeness”, form the requirements for digital support of energy systems in view of the Twin Transition. Finally, the fourth part *Mathematical Models for Digital Solutions* contains a number of interesting mathematical models and approaches that can be realized by IT systems to enhance the efficiency of energy enterprises and other industries.

The special issues are devoted to the unique research task—development of the complex solution for Twin Transition challenge. The book represents the integrated experience and expertise of the researchers from different countries, universities and companies. This fact increases its attractiveness for international society. Based on the real data and case analysis, the articles of the issue can be widely distributed among the students, researchers and industrial partners.

St. Petersburg, Russia — Igor Ilin
Luoyang, China — Ma Youzhong

Contents

Mathematical Models for Digital Solutions

Technical, Management and IT Aspects of Nowadays Energy Sector

Development of a Methodology for Assessing the Investment Needs of the Low-Carbon Economy

Igor Ilin, Denis Bezruchko, and Polina Rukina

Abstract The article analyses the relevance of the problem of changing the structure of electricity and heat generation, taking into account the goal of increasing the share of low-carbon generation. The existing approaches to forecasting macroeconomic indicators, including the energy sector, are analysed. A new use of economic and mathematical modelling methods to build models of the sectoral balance is proposed. It has been established that a balanced economic and mathematical model that takes into account changes in the structure of energy generation and consumption can be an information base for calculating the need for investment in additional capacity. In this case, the investment capacity of the energy sector is calculated both on the basis of the first-order effects, expressed in the cost of energy capacities, and the second-order effects, expressed in the need to invest in enterprises for the production of energy capacities. The same model contains a calculation of the economic efficiency of investment, as well as a sensitivity analysis to changes in the input data for the calculation. The article concludes that the model developed according to the methodology being created can be a tool for justifying the targets for limiting carbon dioxide emissions under conditions of limited investment resources.

Keywords Low-carbon economy · Investments in power generation · Economic efficiency · Intersectoral balance

1 Introduction

The power generation industry is one of the basic branches of the national economy. It is well known that electricity and heat energy are included in the cost of any manufactured product [1]. It is also obvious that the production of gross domestic

I. Ilin · D. Bezruchko · P. Rukina (✉)
Peter the Great St. Petersburg Polytechnic University, Polytechnicheskaya, 29 B, St. Petersburg 195251, Russia
e-mail: rukina.polina.a@yandex.ru

I. Ilin and M. Youzhong (eds.), *Digital Systems and Information Technologies in the Energy Sector*, Lecture Notes in Networks and Systems 1244,
https://doi.org/10.1007/978-3-031-80710-7_1

product (GDP) is impossible without the power generation industry. Consequently, the importance of investment in the energy sector can hardly be overestimated, which determines the relevance of this study.

In addition to maintaining the industry in a workable state, the Russian energy sector faces additional challenges:

1. The need to implement the transition towards low-carbon energy generation due to international agreements (Paris Agreement [2, 3]) and domestic legislation.
2. Redistribution of generation in favour of renewable energy sources [4, 5] due to the gradual exhaustion of readily available non-renewable sources.
3. Ensuring energy security of the country, which is expressed in ensuring technological sovereignty both in terms of raw material sources and power equipment.
4. Ensuring sufficient economic efficiency of the industry to remain its investment attractiveness [6], on the one hand, and to prevent unreasonable growth of energy prices on the other hand (in order not to make all other industries that are consumers of energy economically inefficient) [7].

Energy availability is one of the factors of a favourable investment climate of the country. According to generally accepted international rules, the levelised cost of electricity (LCOE) is one of the key indicators of competitiveness in the world markets [8–10].

The levelised cost of electricity (LCOE) reflects the average unit selling price of electricity produced, which is sufficient to cover the cost of construction and maintenance of generating equipment over its useful life. LCOE is used as a tool to compare the integral competitiveness of electric power generation technologies (see Fig. 1).

The solution of the set tasks is carried out by investing in the energy sector [11]. The purpose of investing in the energy sector (from the point of view of the state) is to achieve a long-term trend of GDP growth in the country by ensuring the availability of energy for the needs of the economy.

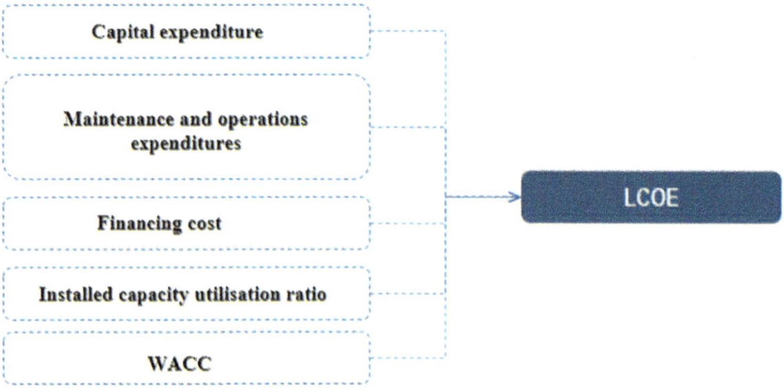

Fig. 1 Key factors for calculating LCOE

The object of study is the production of electric and thermal energy on the scale of the national economy.

To assess the investment capacity of the energy sector, we consider it necessary to identify the following areas of investment:

1. Investments in energy capacity to compensate for the loss of energy capacity (due to reaching depreciation limits or exhaustion of raw materials).
2. Investments in new energy capacity to meet growing energy demand.
3. Investments in the means of production of energy capacities.

The peculiarity of the energy sector in its current state is the practical impossibility of creating statistically significant reserves of electric and thermal energy. This implies that the same amount of energy is always produced and consumed, which makes it possible to build energy balances.

Forecasting of investments directly into energy capacities is connected with calculation of the difference between planned consumption and production of energy with a breakdown by types of energy. This difference gives the value of unsatisfied energy demand, which is satisfied by commissioning new energy capacities. In turn, a steady excess of the planned rate of commissioning of energy capacities over the available capacity for the production of energy equipment is the basis for investment in these production capacities. Such investment is determined by the derivative of the energy capacity demand function and is a second-order effect.

The policy of the Russian government to regulate investments in the power generation sector is implemented in the following ways:

1. Forecasting and guaranteeing demand for power equipment by concluding capacity supply agreements (CSA).
2. Defining requirements for the level of localisation of power equipment. Imported equipment is not allowed to participate in CSA programs for reasons of energy sovereignty of the country.
3. Application of state support measures for investments in the production of energy equipment: creation of a register of the best available technologies, application of special economic zones, conclusion of special investment contracts, etc.

When assessing the investment capacity in the power generation industry [12], the following factors should be taken into account:

1. Forecast of the sectoral structure of power generation.
2. Forecast of energy intensity of energy consuming industries.
3. General forecast of the sectoral energy consumption.
4. The full value of investment costs in additional capacities.
5. The full value of expenditures in construction of power equipment production facilities.

The economic and mathematical model for calculating investment capacity should provide not only the calculation of investment value, but also the calculation of the expected economic efficiency of investments, including the payback period of investments. Also, the model should provide an opportunity to calculate the sensitivity of investment efficiency to changes in the scenario conditions of the initial data.

2 Discussion

As noted earlier, the energy sector is fundamental to many other sectors of the national economy. Leontief's intersectoral balance models allow us to take into account the interconnection of different sectors of the economy. For the task of determining the investment capacity of the energy sector, we can take Leontief's models as a basis, but with a number of significant modifications.

We propose to solve the problem of assessing the investment capacity of the energy sector using the methodology of economic and mathematical modelling [13]. This methodology is based on the construction of a unified information-agnostic model, which is a time series of indicators linked to each other according to a certain algorithm. Like any other economic and mathematical model of investment orientation, our model should transform the initial data through intermediate calculations into conclusions about the economic efficiency of investment. Leontiev's [14] and his followers' [15–19] interindustry balance models should be such intermediate calculations. The data obtained in the course of calculations will be processed by methods of quantitative risk analysis to draw conclusions about the sustainability of the effectiveness of investments in the industry.

The logical scheme of the economic and mathematical model of the energy sector is presented in Fig. 2.

This logical framework needs to be explained. As mentioned above, the way to build the model is to create mutually linked time series. When building the series of energy cost and energy price forecasts, we focused on retrospective data for 2014–2022, taking into account trend extrapolation. The forecast of energy production and consumption is also based on trend extrapolation, but the generation structure has been adjusted to take into account the plans for advanced commissioning of generation from renewable sources. All other energy balance indicators are calculated on the basis of logical relationships and technological coefficients. The authors of this study assume that the method of setting the baseline forecasts for prices and generation volumes can be any, our methodology does not restrict the model user in the way of obtaining input data. We can use this fact when performing scenario analyses.

In addition to its direct task, the proposed scheme for calculating investment capacity also ensures the verification of a number of important internal relations:

1. Equality of production and consumption of heat and electricity.
2. Fulfilment of the carbon dioxide emission reduction plan.
3. Control of energy capacity availability over the entire planning horizon.

Practical implementation of the proposed methodology was carried out by building an economic and mathematical model in the Excel spreadsheet processor. In this model, we grouped the entire array of information into four logical blocks:

1. Block with initial data.
2. Block with tables of intersectoral balance.
3. Block with the results of calculations.

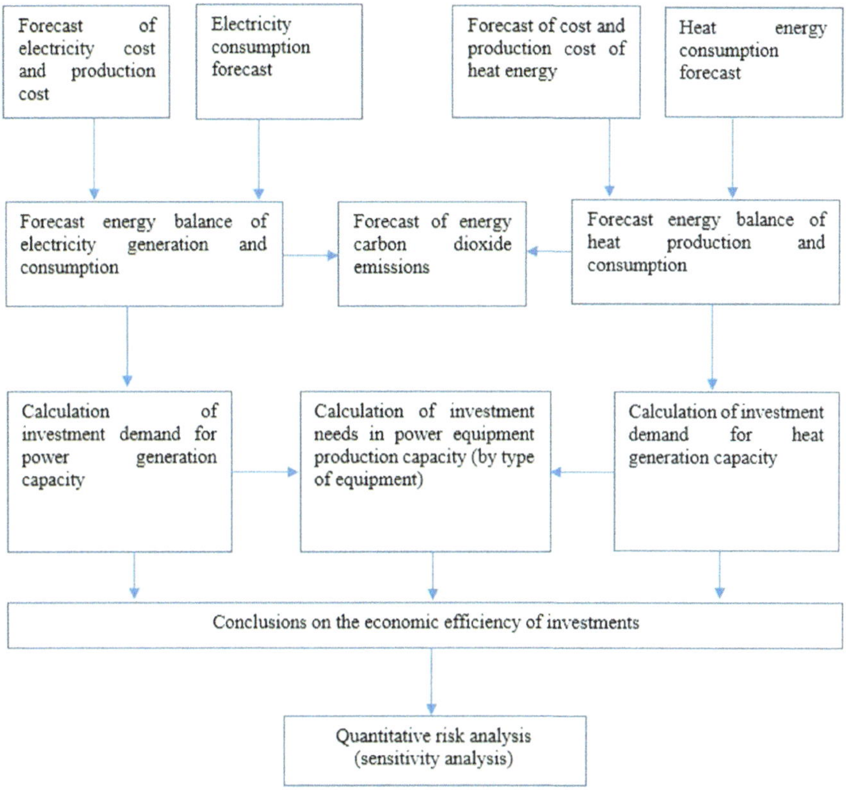

Fig. 2 Logical scheme of the economic and mathematical model of the energy sector

4. Block with quantitative risk analysis—sensitivity analysis [20].

Data interrelation in the model is carried out according to the scheme in Fig. 2. Sensitivity analysis is performed using an executable macro written in Visual Basic.

As our experience [20] shows, the economic efficiency of energy projects is determined by the following factors:

1. Intensity of equipment utilisation.
2. The cost of generating equipment.
3. The cost of supplying energy to the grid.
4. Cost of maintenance of power equipment.
5. The cost of financing.

Taking these factors as variables, we can perform a sensitivity analysis. As in classical investment projects, it is performed by alternately changing one group of parameters with a certain step while keeping the others unchanged. The results of sensitivity analysis allow us to identify the factors that have the greatest impact on investment efficiency.

3 Results of the Research

The economic and mathematical model developed by us was tested on a real data set of the Russian energy industry.

As input data we took the following data:

1. Data on the structure of production and consumption of thermal energy for 2017–2022.
2. Data on the structure of production and consumption of electricity for 2012–2022.
3. Data on the structure of the installed capacity of electric power for 2017–2022.
4. Actual data on integral indicators of cost and investment parameters of the energy sector.

The analysis of actual data on the energy balance allowed to identify the following key ratios in the energy sector (see Table 1).

The average utilisation of power generating capacities has also been determined (see Table 2).

Further, a forecast of electricity (Fig. 3) and heat (Fig. 4) consumption up to 2040 is made. The method used is extrapolation of the trend according to the exponential trend.

Further, the forecast of the structure of electricity (Fig. 5) and heat (Fig. 6) generation was made.

When calculating electricity generation, a forecast was made about the outstripping growth of wind generation (in the first place) and solar generation (in the second place) in order to increase the share of renewable energy sources in the energy balance structure. At the same time, it is planned to gradually reduce energy generation by means of TPPs in the volume ensuring the energy balance. It should be noted that the reduction of power generation by TPPs also reduces heat generation by TPPs. These volumes will have to be replenished by gas-fired boiler houses.

The growth of energy production volumes necessitates the introduction of new energy capacities (see Fig. 7).

Taking into account the manufacturing period of power equipment from 1 (SPP and HPP) to 5 years (NPP and HPP), investments in power capacities should be made in advance, before the equipment is put into operation. In addition, it can be seen that the production of a number of energy capacities will be carried out with acceleration. This means that it is necessary to make investments in the production

Table 1 Ratio of electricity and heat production by thermal power plant

	2017	2018	2019	2020	2021	2022	Mean
Heat energy (mln Gcal)	578	595	574	553	597	581	580
Electric power (mln kW)	702,832	716,151	713,946	655,642	714,777	738,279	706,938
Ratio	1217	1203	1244	1186	1197	1271	119

Table 2 Average utilisation of power generating capacities

Capacity utilisation ratio	2017 (%)	2018 (%)	2019 (%)	2020 (%)	2021 (%)	2022 (%)	Mean (%)
Nuclear power plant (NPP)	83	77	79	84	83	86	82
Thermal power plant (TPP)	49	50	49	45	50	51	49
Hydroelectric power plant (HPP)	44	45	45	49	49	45	46
Wind power plant (WPP)	11	13	19	14	18	31	22
Solar power plant (SPP)	12	10	10	13	13	14	12
Others	0	0	0	0	0	0	0
Total	48	48	47	46	49	50	48

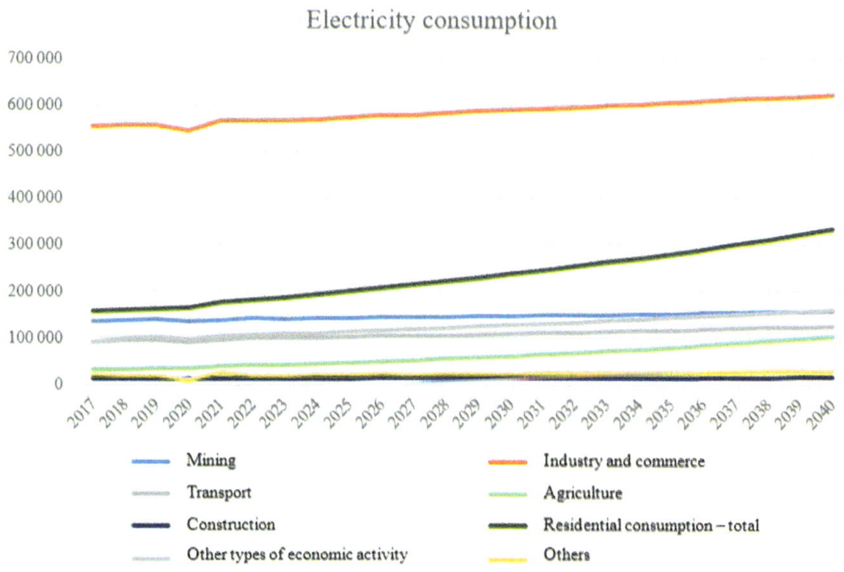

Fig. 3 Electricity consumption forecast

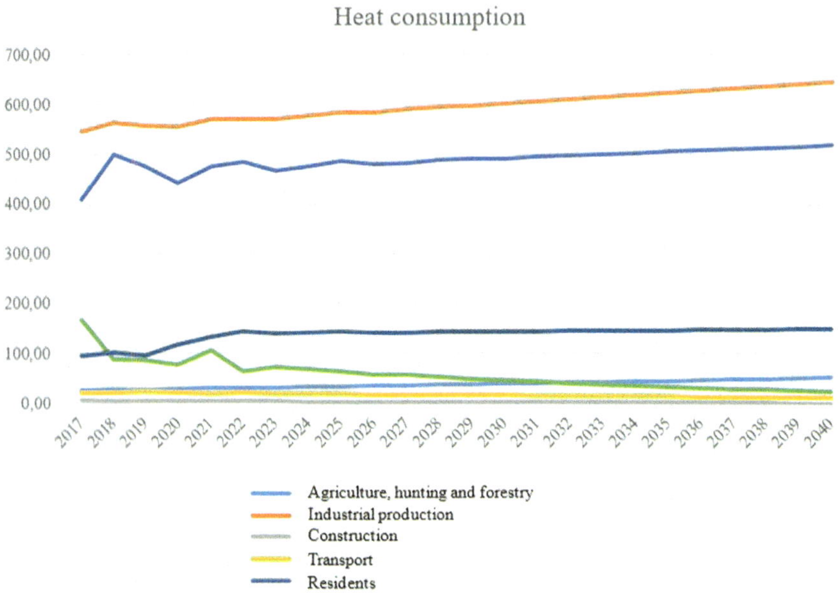

Fig. 4 Heat energy consumption forecast

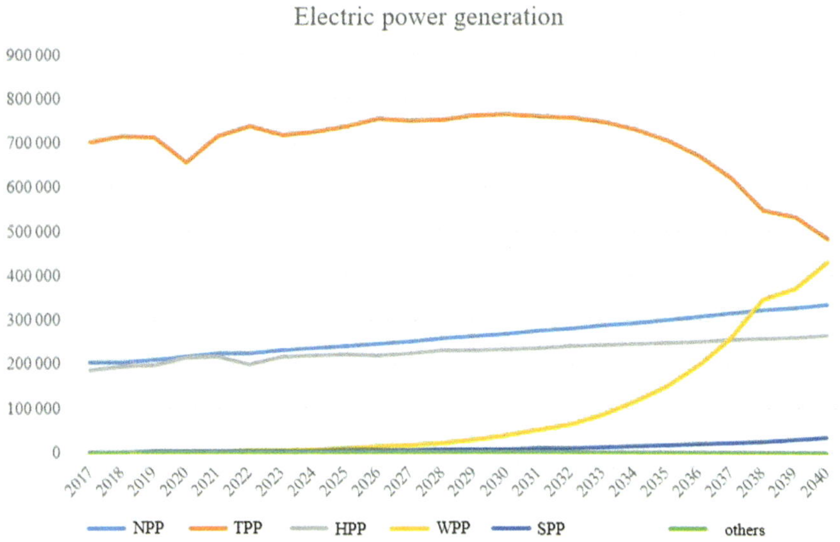

Fig. 5 Forecast of electricity generation structure

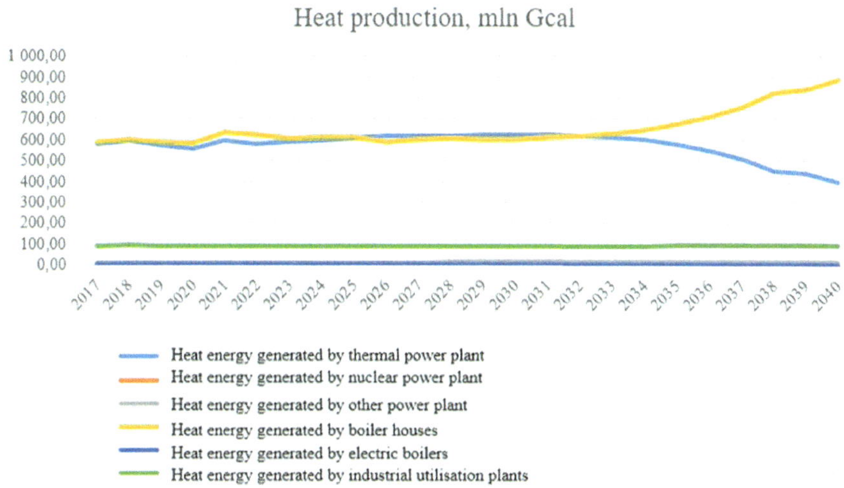

Fig. 6 Forecast of the heat generation structure

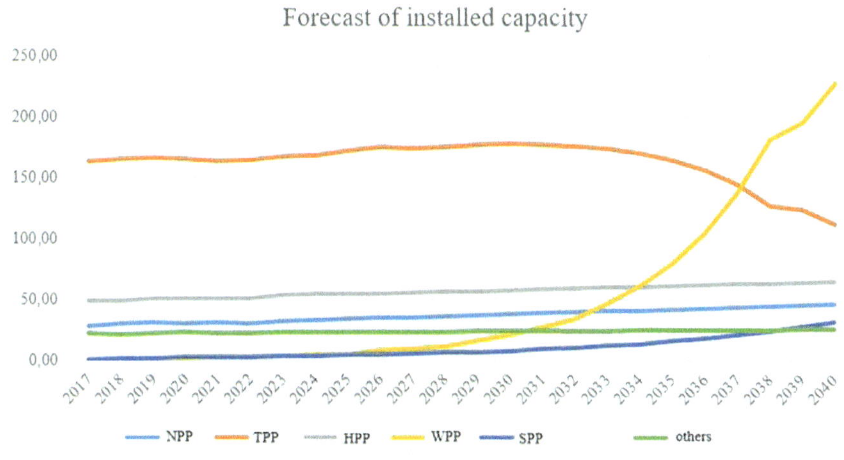

Fig. 7 Forecast of installed capacity demand

of power equipment. These investments are the second order derivative of the energy generation function. According to our calculations, it will be necessary to create new production facilities for the production of wind turbines (investment amount of 615.9 billion rubles until 2040, taking into account the production of components), SPP equipment (investment amount of 45.5 billion rubles), as well as equipment for the nuclear industry and HPPs (see Fig. 8 and Table 3).

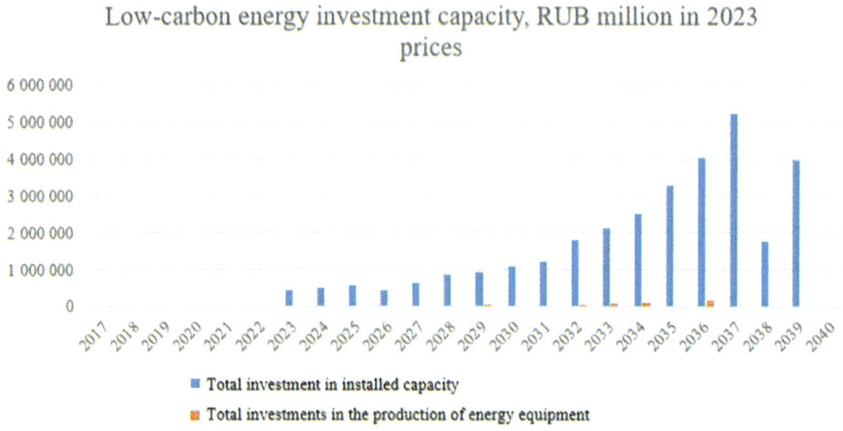

Fig. 8 Investment capacity of low-carbon energy by year, RUB mln

Table 3 Total investment capacity of low-carbon energy sector up to 2040, million RUB

Denomination	Investment in installed capacity	Investments in the production of energy equipment
Nuclear power plant (NPP)	2,364,239	6337
Thermal power plant (TPP)	0	0
Hydroelectric power plant (HPP)	1,622,576	11,709
Wind power plant (WPP)	24,632,360	615,903
Solar power plant (SPP)	2,891,554	45,794
Others	0	0
Total	31,510,730	679,742

4 Conclusion

The following conclusions have been drawn in the course of this study:

1. Changing the structure of energy generation is objectively necessary both to achieve a reduction in carbon dioxide emissions into the atmosphere and to reduce the consumption of non-renewable sources.
2. Changing the generation structure will require investments in the amount of 32 trillion rubles until 2040.
3. In order to fulfil the plans on commissioning of energy capacities, it is necessary to invest in the creation of new (or expansion of existing) production facilities for the production of energy equipment in the amount of 680 billion rubles.
4. When planning the reduction of electricity generation volumes by means of TPPs, it is necessary to take into account the replacement of the lost volumes of heat generation by means of boiler houses.

5. The methodology of investment capacity calculation based on the intersectoral balance presented in this article can also be used to calculate the investment capacity of individual regions, including isolated energy systems of the Far North.

Acknowledgements The research was supported by Russian Science Foundation grant No. 24-28-01796, https://rscf.ru/project/24-28-01796/

References

1. Ilin, I., Levina, A., Iliashenko, O.: Enterprise architecture analysis for energy efficiency of saint-petersburg underground. In: Energy Management of Municipal Transportation Facilities and Transport, pp. 1214–1223. Springer International Publishing, Cham (2017)
2. Paris Agreement. https://unfccc.int/sites/default/files/resource/parisagreement_publication.pdf. Last accessed 22 Sep 2024
3. Le, T.T., Sharma, P., Bora, B.J., Tran, V.D., Truong, T.H., Le, H.C., Nguyen, P.Q.P.: Fueling the future: a comprehensive review of hydrogen energy systems and their challenges. Int. J. Hydrogen Energy **54**, 791–816 (2024)
4. Shemyakina, A.A., Levina, A.I., Korablev, V.V., Lepekhin, A.A.: Architecture of the management system for hydrogen production at hydropplications. Int. J. Hydrogen Energy **69**, 1227–1235 (2024)
5. Gu, Y., Zhang, X., Myhren, J.A., Han, M., Chen, X., Yuan, Y.: Techno-economic analysis of a solar photovoltaic/thermal (PV/T) concentrator for building application in Sweden using Monte Carlo method. Energy Convers. Manage. **165**, 8–24 (2018)
6. Steffen, B.: The importance of project finance for renewable energy projects. Energy Econ. **69**, 280–294 (2018)
7. Poljanskihh, A., Levina, A., Dubgorn, A.: Investment in renewable energy: practical case in Estonia. In: MATEC Web of Conferences, vol. 193, p. 05065. EDP Sciences (2018)
8. Obi, M., Jensen, S.M., Ferris, J.B., Bass, R.B.: Calculation of levelized costs of electricity for various electrical energy storage systems. Renew. Sustain. Energy Rev. **67**, 908–920 (2017)
9. Castillo-Ramírez, A., Mejía-Giraldo, D., Muñoz-Galeano, N.: Large-scale solar PV LCOE comprehensive breakdown methodology. CT&F-Ciencia Tecnología y Futuro **7**(1), 117–126 (2017)
10. De Roo, G., Parsons, J.E.: A methodology for calculating the levelized cost of electricity in nuclear power systems with fuel recycling. Energy Econ. **33**(5), 826–839 (2011)
11. Ilin, I., Vasilenok, V., Marchenko, R.: Use of renewable energy and tax burden on CO_2 emissions in industrial enterprises. In: E3S Web of Conferences, vol. 110, p. 02102. EDP Sciences (2019)
12. Krömer, S., Gatzert, N.: Renewable energy investments with storage: a risk-return analysis. Int. J. Energy Sect. Manage. **12**(4), 714–736 (2018)
13. Svetunkov, S.G., Chanysheva, A.F.: Properties of complex dispersion and standard deviation. Revista Espacios **38**(33) (2017)
14. Leontief, W.W., Leontief, W.: Essays in economics: theories and theorizing, vol. 1. ME Sharpe (1976)
15. Fontela, E.: Leontief and the Future of the World Economy. Instituto LR Klein-Centro Stone (2002)
16. Leontief, W.: Structure of the world economy. Outline of a Simple Input-Output (1973)
17. Mykhailo, K.: Analysis and development of price models in the system MEANS of intersectoral balance. Econ. **10**(4), 125–138 (2021)
18. Arbex, M., Perobelli, F.S.: Solow meets Leontief: economic growth and energy consumption. Energy Econ. **32**(1), 43–53 (2010)

19. Dobos, I., Floriska, A.: A dynamic Leontief model with non-renewable resources. Econ. Syst. Res. **17**(3), 317–326 (2005)
20. Stroykov, G.A., Babyr, N.V., Ilin, I.V., Marchenko, R.S.: System of comprehensive assessment of project risks in energy industry. Int. J. Eng. Trans. A Basics **34**(7), 1778–1784 (2021)

Grid in Transition: Balancing Solar Growth and Energy Security in an Emerging Economy

Aghasi Tavadyan

Abstract This study examines the rapid transformation of a small, landlocked nation's electricity production landscape from 2015 to 2024, with projections to 2026. Leveraging comprehensive data analysis and advanced forecasting models, we explore the country's shifting energy mix, focusing on the exponential growth of solar power and its implications for grid management and energy security. Our findings reveal a dramatic increase in solar energy's contribution, from negligible levels to over 10% of total electricity generation, challenging traditional reliance on nuclear and thermal power. This transition presents both opportunities for enhanced energy independence and challenges for grid stability and infrastructure adaptation. The study offers critical insights into the technical, economic, and policy considerations that developing nations must navigate as they balance energy security, environmental sustainability, and economic development. By examining this case study, we contribute to the broader discourse on energy transitions in emerging economies and provide valuable lessons for policymakers, energy planners, and researchers worldwide grappling with similar challenges in the global shift towards sustainable energy systems.

Keywords Solar energy transition · Electricity grid modernization · Energy security · Developing economies · Sustainable development

1 Introduction

The global energy landscape is undergoing a profound transformation, driven by the imperative of sustainable development and the rapid advancement of renewable technologies. This transition is particularly challenging for developing nations,

A. Tavadyan (✉)
Russian-Armenian (Slavonic) University, 123 Hovsep Emin St, 0051 Yerevan, Armenia
e-mail: aghasi.tavadyan@rau.am

"AMBERD" Research Center, Armenian State University of Economics, Yerevan, Armenia

I. Ilin and M. Youzhong (eds.), *Digital Systems and Information Technologies in the Energy Sector*, Lecture Notes in Networks and Systems 1244,
https://doi.org/10.1007/978-3-031-80710-7_2

which must balance energy security, environmental sustainability, and economic development. Armenia, a small landlocked country in the South Caucasus, presents a compelling case study in this evolving energy paradigm.

Armenia's unique geopolitical position and historical reliance on imported fossil fuels make its journey towards energy security and sustainability especially noteworthy. The country's efforts align with the broader theoretical framework of sustainability transitions, which examines the long-term, multi-dimensional, and fundamental transformation processes through which established socio-technical systems shift to more sustainable modes of production and consumption [1].

This study examines Armenia's electricity production landscape from 2015 to 2024, with projections extending to 2026. By leveraging comprehensive data analysis and advanced forecasting models, we explore the country's shifting energy mix, focusing particularly on the rapid growth of solar power and its implications for grid management and energy security. This research is guided by the following questions:

1. How has Armenia's electricity production mix evolved from 2015 to 2024, and what are the projected trends up to 2026?
2. What factors are driving the growth of solar energy in Armenia, and how does this growth align with or challenge existing theories of energy transition in developing economies?
3. What are the implications of Armenia's changing energy mix for grid stability, infrastructure adaptation, and overall energy security?
4. How do Armenia's experiences inform our understanding of the challenges and opportunities facing developing nations in their energy transitions?

Our analysis reveals a complex interplay of factors shaping Armenia's energy future. The country's traditional reliance on nuclear and thermal power is being challenged by the exponential growth of solar energy, which accounted for 10.1% of electricity generation as of July 2024. This shift aligns with the multi-level perspective on socio-technical transitions [2–5], which emphasizes the interplay between niche innovations (like solar power), existing regimes, and broader landscape pressures.

The study is particularly timely, as Armenia grapples with frequent power outages and infrastructure vulnerabilities while simultaneously pursuing ambitious renewable energy targets. This context provides an opportunity to examine the practical challenges of implementing theoretical transition pathways in a real-world setting.

Through a rigorous examination of production data, forecasting models, and policy implications, this research contributes to the broader discourse on energy transitions in developing economies. It offers critical insights into the technical, economic, and policy considerations that must be navigated as countries like Armenia seek to balance energy security, environmental sustainability, and economic development.

As we delve into the intricacies of Armenia's electricity production and its future trajectory, this paper aims to provide advice for policymakers, energy planners, and researchers grappling with similar challenges in other regions. The findings presented here not only illuminate Armenia's specific context but also contribute to our understanding of the global energy transition, offering a nuanced perspective on the path towards a more sustainable and resilient energy future.

2 Literature Review

The integration of variable renewable energy sources (VRES) into existing power systems presents both opportunities and challenges for grid operators and policymakers. As Armenia aims to increase its share of renewable energy, particularly solar power, it faces a unique set of circumstances that require careful consideration of international experiences and best practices.

In [6] the authors provide a comprehensive overview of the challenges associated with VRES integration and potential solution technologies. They highlight that as the share of VRES increases, power systems face issues related to grid stability, reliability, and flexibility. For Armenia, which is seeking to rapidly expand its solar capacity, these challenges are particularly relevant. The authors emphasize the importance of developing a portfolio of solution technologies, ranging from improved forecasting methods to energy storage systems, to effectively manage high penetrations of VRES.

Forecasting plays a crucial role in the successful integration of VRES. Zamo et al. [7] presents a benchmark of statistical regression methods for short-term forecasting of photovoltaic (PV) electricity production. Their study, which compares various statistical methods without relying on technical information about the power plants, is particularly relevant to Armenia's context, where detailed technical data may not always be available. The authors' findings on the effectiveness of different forecasting methods could inform Armenia's approach to predicting solar power output, which is essential for grid stability and market operations.

The impact of VRES on power quality is another critical consideration. Liang [8] discusses the emerging power quality challenges due to the integration of renewable energy sources. Issues such as voltage and frequency fluctuations, which are caused by the variability of renewable resources, are highlighted. As Armenia increases its solar capacity, addressing these power quality concerns will be essential to maintain grid stability and reliability.

In [9] the authors apply complex network theory to assess the vulnerability of interdependent energy infrastructures. Their approach, which uses statistical indexes from graph theory as an alternative to power-flow techniques, could provide valuable insights for Armenia as it plans its grid expansion and modernization. The authors' methodology for analyzing cascading failures in coupled electric power and natural gas transmission systems may be particularly relevant if Armenia considers expanding its use of natural gas in conjunction with renewable energy.

The transition to a high-VRES system also has significant policy implications. Lehmann et al. [10] discusses the economic rationale for additional renewable energy targets and instruments in the context of EU climate and energy policy. While

Armenia is not part of the EU, the authors' analysis of the costs and benefits of various policy instruments could inform Armenia's own policy decisions as it sets targets for renewable energy integration.

In [11] the authors present a case study of Bangladesh, another developing country transitioning to renewable energy. Their analysis of different energy transition pathways, considering various policy scenarios and the impact of greenhouse gas emission costs, provides a useful framework for considering Armenia's own transition options.

Jonaitis et al. [12] offers valuable insights from Lithuania's experience in integrating wind power into their electric power system. Their study identifies two key technical objectives: determining the power transmission capacity of the electricity network and balancing the energy generated by wind power plants related to forecasting error control. While focused on wind energy, these challenges are equally applicable to solar integration in Armenia. The authors' analysis of forecast errors, balancing demands, and control reserves could provide a useful framework for Armenia to assess its own grid capabilities and requirements as it increases its solar capacity.

Kalehsar [13] examines Iran's transition to renewable energy, highlighting both challenges and opportunities. Although focused on Iran, this study offers relevant insights for Armenia, as both countries are rich in renewable energy sources but heavily dependent on fossil fuels. Kalehsar emphasizes the importance of appropriate policy formulations and strategies in leveraging Renewable Energy Technologies (RETs) for socio-economic development. This underscores the need for Armenia to develop comprehensive policies that not only promote renewable energy adoption but also ensure its integration contributes to broader economic and social goals.

This literature review underscores the multifaceted nature of the challenges and opportunities Armenia faces in its renewable energy transition. As we analyze Armenia's electricity production and potential for solar integration, we will draw on these insights to contextualize our findings and provide recommendations that are both technically sound and policy-relevant.

3 Methodology

This study employs a comprehensive approach to analyze and forecast Armenia's electricity production, utilizing both historical data analysis and advanced forecasting techniques.

3.1 Data Sources and Collection Methods

The primary data for this study was sourced from official Armenian government statistics, specifically from the Public Services Regulatory Commission of Armenia (psrc.am). The dataset encompasses monthly electricity production values from 2015 to 2024, disaggregated by energy source including thermal power plants, nuclear power, hydroelectric power, and solar energy. To ensure transparency and reproducibility, the complete dataset and analysis code have been made publicly available on GitHub.

3.2 *Analytical Techniques*

Our analysis employs a combination of descriptive statistics and time series forecasting models. The R programming language was used for all data processing, analysis, and visualization tasks.

For forecasting future electricity production, we implemented and compared four different time series models:

1. Auto-regressive Integrated Moving Average (ARIMA): We used an automatic ARIMA model selection process, which chooses the optimal model based on the Akaike Information Criterion (AIC) or Bayesian Information Criterion (BIC) [14].
2. Exponential Smoothing State Space Model (ETS): This model uses exponential smoothing methods to capture trend and seasonal components in the time series [15].
3. Bayesian Automatic Time Series Analysis (BATS): This model can handle complex seasonal patterns and automatically select the best model using a Bayesian approach [16].
4. Seasonal and Trend decomposition using Loess (STL) with ETS: This method first decomposes the time series into seasonal, trend, and remainder components using STL, then applies ETS forecasting to the seasonally adjusted data [17].

Each model was applied to the monthly historical data for each energy source, generating forecasts up to 24 months ahead. This approach allows us to capture both seasonality and trend in the data.

To assess forecast uncertainty, we generated prediction intervals at 50, 80, and 95% confidence levels [18]. The results from all four models were compared to provide a comprehensive view of potential future scenarios in Armenia's electricity production.

3.3 *Data Aggregation and Visualization*

While the forecasting was performed on monthly data to capture seasonal patterns accurately, we aggregated the results into year-over-year (YoY) values for visualization and interpretation purposes. This aggregation was done using efficient rolling window calculations. The YoY aggregation was performed after the forecasting process, ensuring that the seasonal information was fully utilized in the predictions.

This multi-model approach, combined with the YoY aggregation for visualization, allows us to capture different aspects of the time series data and provides a robust basis for analyzing trends and making projections about Armenia's future energy landscape. It balances the need for detailed forecasting with the requirement for easily interpretable results, crucial for informing policy decisions in the energy sector.

4 Current State of Armenia's Electricity Production

Armenia's electricity production landscape has undergone significant changes in recent years, characterized by a diversification of energy sources and a gradual shift towards renewable energy. This section analyzes the current state of electricity production by source and examines recent trends and challenges facing the sector.

4.1 Analysis of Production by Source

As of July 2024, Armenia's electricity production is derived from four primary sources: thermal power plants (TPPs), nuclear power, hydroelectric power plants (HPPs), and solar energy. The distribution of these sources has evolved over the past decade, reflecting both policy decisions and technological advancements (Fig. 1).

1. Thermal Power Plants (TPPs): TPPs continue to play a significant role in Armenia's electricity mix, accounting for 38.9% of total year-over-year (YoY) production in July 2024. This represents a slight decrease from previous years, likely due to the increasing share of renewable sources. TPPs primarily utilize natural gas imported from Russia, rendering this sector vulnerable to external price fluctuations and geopolitical tensions.
2. Nuclear Power: The Armenian Nuclear Power Plant (ANPP) remains a crucial baseload power source, contributing 31% of the country's YoY electricity in July 2024. This share has remained relatively stable over the years, underscoring the plant's importance in ensuring energy security. However, the aging infrastructure of the ANPP raises potential reliability and safety concerns.
3. Hydroelectric Power Plants (HPPs): HPPs have historically been a significant source of renewable energy for Armenia. However, their share has experienced a gradual decline, dropping from 32.2% in 2016 to 20% in July 2024. This decrease may be attributed to changing precipitation patterns and the increasing role of other renewable sources.
4. Solar Energy: The most notable trend in Armenia's electricity production is the rapid growth of solar energy. From a negligible share in 2016, solar power has grown to contribute 10.1% of the country's YoY electricity by July 2024. This

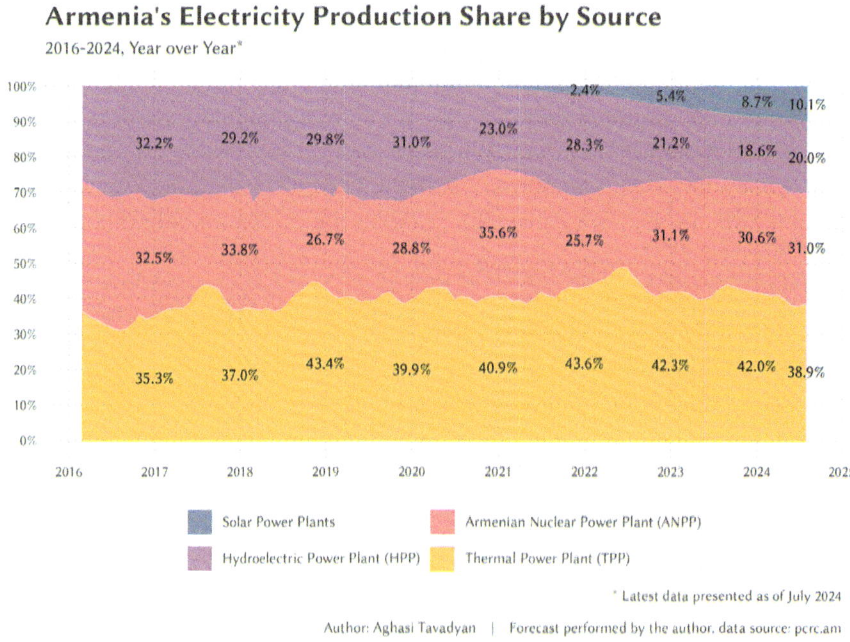

Fig. 1 Armenia's electricity production share by source, 2016–2024

exponential growth reflects both government initiatives to promote renewable energy and decreasing costs of solar technology.

4.2 *Recent Trends and Challenges*

1. Shift Towards Renewables: The most significant trend in Armenia's electricity sector is the rapid adoption of solar energy. This shift aligns with global trends towards cleaner energy sources and offers potential for increased energy independence.
2. Declining Hydroelectric Production: The decrease in hydroelectric power production is a concerning trend, potentially linked to climate change impacts on water resources. This decline underscores the need for diversification in renewable energy sources.
3. Infrastructure Vulnerabilities: Recent years have witnessed frequent power outages, particularly in urban areas such as Yerevan. These outages highlight the vulnerabilities in the existing power distribution infrastructure. Data from 2024 indicates that a significant portion of cable damages (69%) were caused by the electric company itself during maintenance or upgrade work, pointing to potential issues in infrastructure management and maintenance practices.

4. Regional Disparities: Analysis of power outages across different regions of Armenia reveals disparities in infrastructure reliability. Some areas experience a higher frequency of outages, suggesting uneven distribution of resources for maintenance and upgrades.
5. Aging Nuclear Infrastructure: While the ANPP continues to provide a stable baseload, its aging infrastructure presents ongoing challenges. Periodic shutdowns for maintenance and safety upgrades impact the overall stability of the electricity supply.
6. Dependence on Imported Fuel: The significant role of TPPs in electricity production means that Armenia remains dependent on imported natural gas, primarily from Russia. This dependence exposes the country to potential supply disruptions and price volatility.

These trends and challenges highlight the complex landscape of Armenia's electricity sector. While the growth in renewable energy, particularly solar, presents opportunities for increased energy independence and sustainability, issues such as infrastructure vulnerabilities and the decline in hydroelectric production pose significant challenges. Addressing these challenges will be crucial for ensuring a stable, efficient, and sustainable electricity supply for Armenia in the coming years.

5 Digital Forecasting of Electricity Production

The application of digital forecasting techniques to Armenia's electricity production provides valuable insights into future trends and potential challenges. This section presents the forecasting models employed and discusses their results, reliability, and implications for Armenia's energy sector.

5.1 Presentation of Forecasting Models and Results

To project future electricity production trends in Armenia, we employed multiple advanced time series forecasting models. These include Auto-Regressive Integrated Moving Average (ARIMA), Bayesian Structural Time Series (BATS), and Exponential Smoothing State Space (ETS) models. Each model was applied to historical data from 2015 to 2024 for the four primary sources of electricity production: thermal power plants (TPPs), nuclear power, hydroelectric power plants (HPPs), and solar energy.

Figure 2 illustrates the forecasted electricity production for each energy source up to 2026. The forecasts reveal several significant trends:

1. Thermal Power Plants (TPPs): The models predict a slight decline in TPP production, from 38.9% in July 2024 to approximately 38% by 2026. This trend suggests

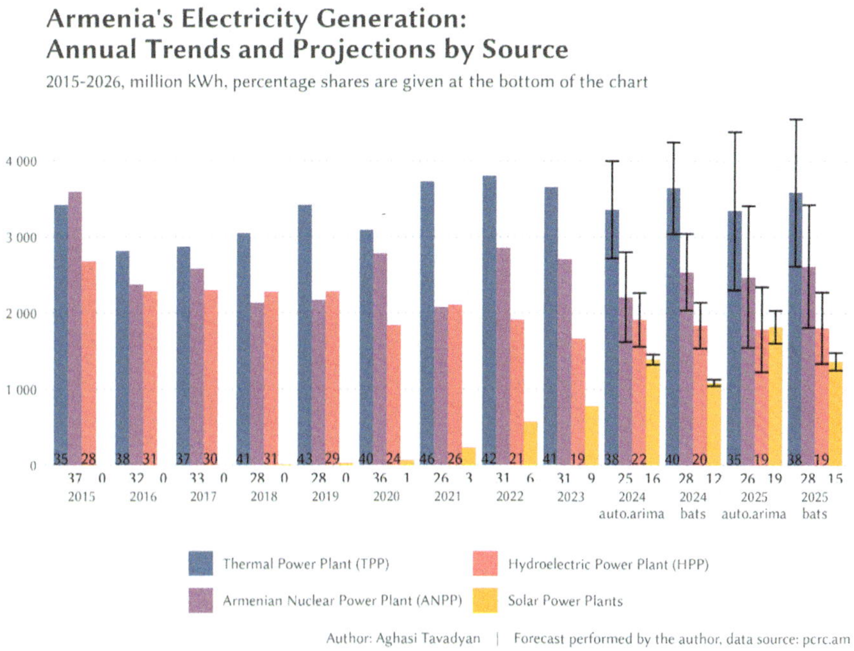

Fig. 2 Armenia's electricity generation: annual trends and projections by source, 2015–2026

a gradual reduction in reliance on fossil fuels, albeit at a slower pace than might be desired for rapid decarbonization.

2. Nuclear Power: Production from the Armenian Nuclear Power Plant (ANPP) is forecasted to remain relatively stable, maintaining its share at around 31% through 2026. This stability underscores the continued importance of nuclear energy in Armenia's electricity mix.
3. Hydroelectric Power Plants (HPPs): The models project a continued gradual decline in hydroelectric production, potentially dropping to around 19% by 2026. This forecast aligns with the observed historical trend and raises concerns about the long-term viability of hydropower in Armenia's changing climate.
4. Solar Energy: The most dramatic growth is predicted in solar energy production. The forecasts suggest that solar could contribute up to 15–16% of Armenia's electricity production by 2026, continuing its rapid upward trajectory.

5.2 *Discussion of the Reliability and Implications of Forecasts*

Reliability of Forecasts: The forecasting models employed in this study have demonstrated robust performance in capturing historical trends and seasonality. However, it

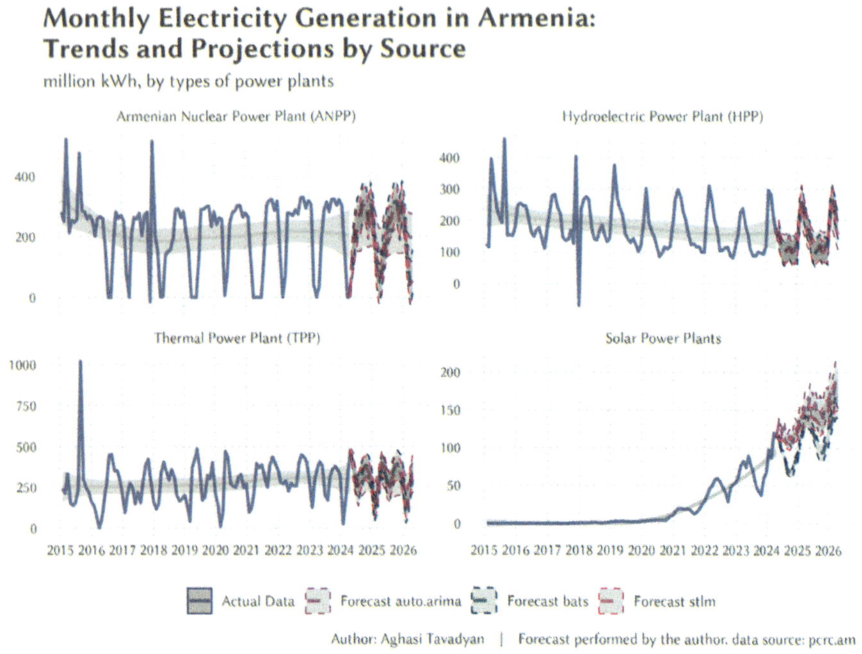

Fig. 3 Monthly electricity generation in Armenia: trends and projections by source

is crucial to acknowledge inherent uncertainties in long-term forecasts, particularly in a rapidly evolving sector such as energy.

Figure 3 presents a comparison of the reliability measures for each forecasting model across different energy sources. The shaded areas represent prediction intervals, providing a visual representation of forecast uncertainty. Notably, the forecasts for solar energy production show wider prediction intervals, reflecting the higher uncertainty associated with this rapidly growing sector.

Implications of Forecasts:

1. Transition to Renewables: The projected growth in solar energy production aligns with Armenia's goals for increasing renewable energy adoption. However, the pace of this transition may need to accelerate to meet ambitious climate targets.
2. Grid Stability Challenges: The forecasted increase in intermittent solar power, coupled with the decline in baseload hydroelectric power, suggests potential challenges for grid stability. This underscores the need for investment in energy storage and smart grid technologies.
3. Energy Security: The continued reliance on TPPs, albeit slightly decreasing, implies ongoing dependence on imported natural gas. This forecast highlights the importance of diversifying energy sources to enhance energy security.
4. Infrastructure Investment: The stable forecast for nuclear power production emphasizes the need for continued investment in maintaining and potentially upgrading the aging ANPP infrastructure.

5. Climate Adaptation: The projected decline in hydroelectric power production underscores the need for climate adaptation strategies in the energy sector, including diversification of renewable energy sources.
6. Policy Implications: These forecasts provide valuable input for policymakers, suggesting a need for policies that support the growth of solar energy while addressing the challenges of declining hydropower and the environmental impact of thermal power plants.

In conclusion, while these digital forecasts provide crucial insights into potential future scenarios for Armenia's electricity production, they should be interpreted with caution. Regular updates and refinements to these models, incorporating new data and evolving trends, will be essential for maintaining their relevance and reliability in guiding energy policy and investment decisions.

6 Challenges and Solutions in Armenia's Energy Transition

Armenia's energy sector is undergoing a significant transformation, characterized by a rapid increase in renewable energy production, particularly solar power, and the concurrent need for grid modernization. This section examines the challenges faced in this transition and explores potential solutions, with a focus on digital technologies and infrastructure improvements.

6.1 Renewable Energy Growth and Integration Challenges

As illustrated in Fig. 2, solar power's contribution to Armenia's national grid has grown exponentially, reaching 10.1% of total year-over-year production by July 2024. This growth, while promising for sustainability and energy independence, presents significant challenges:

1. Production-Consumption Mismatch: Solar energy production peaks at midday, while electricity consumption typically peaks in the evening, creating a temporal disparity.
2. Grid Capacity Limitations: The current power grid struggles to handle surplus electricity generated during peak solar production hours.
3. Regulatory Constraints: Existing regulations, such as the 150 kW limit on solar installations and reduced tariffs for excess electricity, may impede sector development.

6.2 *Infrastructure Vulnerabilities and Grid Management Issues*

Recent data from 2024 highlights critical vulnerabilities in Armenia's power distribution infrastructure, particularly in Yerevan.

Figure 4 illustrates the distribution of electrical network damage across Yerevan's districts in 2024. Key findings include:

1. From January to October 2024, 1569 cases of cable damage were recorded in Yerevan.
2. Surprisingly, 69% of these damages were caused by the electric company itself during maintenance or upgrade work.

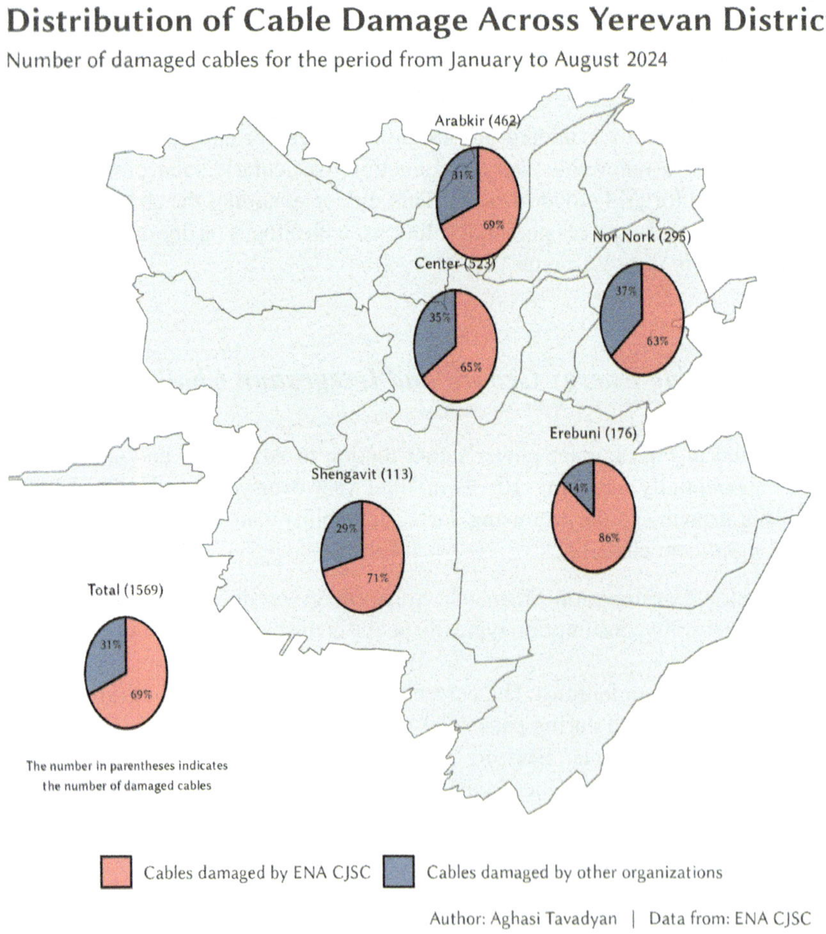

Fig. 4 Distribution of cable damage across Yerevan districts

3. Some districts, notably Kentron and Arabkir, experienced higher rates of damage.

These issues underscore the urgent need for infrastructure upgrades and improved maintenance practices.

Integrated Solutions: Digital Tools and Infrastructure Improvements.

Addressing these challenges requires a comprehensive approach combining digital technologies and infrastructure enhancements:

1. Advanced Forecasting and Grid Management Systems: Implementing machine learning algorithms for accurate solar output prediction and smart grid technologies for real-time monitoring and control.
2. Energy Storage Solutions: Developing capabilities such as pumped hydroelectric storage or battery systems to balance production-consumption mismatches.
3. Distribution Automation and Predictive Maintenance: Employing AI-powered analytics for proactive maintenance and automated systems for fault isolation and power rerouting.
4. Virtual Power Plants (VPPs) and Demand Response Programs: Digitally aggregating distributed resources and implementing programs to match consumption with production patterns.
5. Inter-state Grid Connections: Exploring connections to broader regional power grids to manage excess production and improve stability.
6. Regulatory Reform: Revising policies to encourage larger-scale solar installations and fair compensation for excess energy production.

The adoption of these integrated solutions is crucial for managing the increasing share of renewable energy while addressing infrastructure vulnerabilities. As Armenia continues its energy transition, the synergy between renewable integration and grid modernization will be key to ensuring a stable, efficient, and sustainable electricity supply.

This holistic approach to energy transition highlights the interconnected nature of renewable energy adoption and grid management challenges. It emphasizes the need for coordinated efforts in technological innovation, infrastructure investment, and policy reform to achieve Armenia's energy goals.

7 Discussion and Strategic Recommendations

Our analysis of Armenia's electricity production landscape from 2015 to 2024, with projections to 2026, reveals significant shifts in the country's energy mix and highlights several challenges and opportunities in its energy transition. This section discusses our findings in relation to our research questions and existing literature, and proposes strategic recommendations.

1. Evolution of Armenia's Electricity Production Mix Our findings show a dramatic increase in solar energy's contribution, from negligible levels to 10.1% by July 2024, with projections reaching 15–16% by 2026. This rapid growth aligns with

global trends in renewable energy adoption [19] but exceeds the pace observed in many developing countries. The decline in hydroelectric production, however, presents a challenge to Armenia's renewable energy goals.
2. Drivers of Solar Energy Growth The exponential growth of solar energy in Armenia can be attributed to falling technology costs, supportive policies, and increasing energy security concerns. This aligns with the multi-level perspective on socio-technical transitions [3], where niche innovations (solar power) are challenging the existing regime (traditional power sources).
3. Implications for Grid Stability and Energy Security The increasing share of intermittent renewable sources poses challenges for grid stability, as seen in other countries with high renewable penetration [20]. Armenia's experience underscores the need for grid modernization and energy storage solutions to manage these challenges effectively.
4. Lessons for Developing Nations in Energy Transition Armenia's case offers useful insights for other developing nations, particularly in balancing rapid renewable growth with grid stability and energy security. The country's experience highlights the importance of comprehensive planning and investment in infrastructure to support the energy transition.

Based on these findings, we propose the following strategic recommendations:

1. Accelerate Solar Energy Integration: Increase investment in solar infrastructure and revise regulations to encourage larger-scale installations.
2. Modernize Grid Infrastructure: Implement smart grid technologies and invest in energy storage solutions to manage intermittent renewable sources.
3. Diversify Renewable Energy Sources: Explore wind and geothermal energy to complement solar and address the decline in hydroelectric production.
4. Enhance Infrastructure Reliability: Implement a comprehensive upgrade program, prioritizing vulnerable areas and improving coordination with urban development initiatives.
5. Strengthen Energy Security: Develop a plan for reducing thermal power plant reliance and explore regional energy cooperation opportunities.
6. Prepare for Increased Electrification: Develop long-term plans to scale up production and distribution capacity, anticipating growth in demand from electric vehicles and heating.
7. Enhance Data-Driven Decision Making: Invest in advanced forecasting systems and develop a centralized energy information system.
8. Address Climate Change Impacts: Develop adaptation strategies for the energy sector, particularly for optimizing hydroelectric production.

These recommendations aim to build on progress of developing counties while addressing identified challenges. By implementing these strategies, Armenia can enhance its energy security, improve grid reliability, and move towards a more sustainable energy future, providing a model for other developing nations in their energy transitions.

8 Conclusion

This study has examined Armenia's electricity production landscape from 2015 to 2024, with projections extending to 2026, providing critical insights into the country's energy transition. Our analysis reveals several key findings:

1. Armenia has experienced a rapid growth in solar energy production, reaching 10.1% of total electricity generation by July 2024, with projections suggesting further growth to 15–16% by 2026.
2. This growth in solar energy is challenging the traditional reliance on nuclear and thermal power, presenting both opportunities and challenges for the country's energy security and grid stability.
3. The decline in hydroelectric production poses a challenge to Armenia's renewable energy goals and highlights the need for diversification of renewable sources.
4. Armenia's energy infrastructure faces significant vulnerabilities, as evidenced by frequent power outages, particularly in urban areas.

These findings directly address our research questions, demonstrating the evolution of Armenia's electricity production mix, identifying the factors driving solar energy growth, and highlighting the implications for grid stability and energy security. Moreover, Armenia's experience offers valuable lessons for other developing nations navigating the complexities of energy transition.

The significance of this study lies in its comprehensive analysis of a developing nation's rapid transition towards renewable energy, particularly solar power. It provides a nuanced understanding of the technical, economic, and policy challenges involved in such a transition, offering insights that can inform policy-making and energy planning in similar contexts.

However, this study has limitations. The projections are based on current trends and policies, and unforeseen technological advancements or policy changes could alter the trajectory. Additionally, the focus on electricity production does not fully capture the broader energy consumption landscape.

Future research could explore the social and economic impacts of Armenia's energy transition, examine the potential for other renewable sources such as wind and geothermal energy, and investigate the long-term implications of increased electrification in transportation and heating sectors.

In conclusion, Armenia's journey towards a more sustainable energy future, while challenging, offers a compelling case study in energy transition for developing nations. By addressing the identified challenges and leveraging its progress in solar energy adoption, Armenia has the potential to significantly enhance its energy security and contribute to global efforts in combating climate change.

References

1. Markard, J., Raven, R., Truffer, B.: Sustainability transitions: an emerging field of research and its prospects. Res. Policy **41**(6), 955–967 (2012)
2. Geels, F.W.: Technological transitions as evolutionary reconfiguration processes: a multi-level perspective and a case-study. Res. Policy **31**(8–9), 1257–1274 (2002)
3. Geels, F.W.: The multi-level perspective on sustainability transitions: responses to seven criticisms. Environ. Innov. Soc. Trans. **1**(1), 24–40 (2011)
4. Politaeva, N.A., Illin, I.V., Oparina, A.M., Donetskova, A.S.: New energy approaches to the use of waste biosorbents of microalgae Chlorella kessleri. Povolzhskii Ekologicheskii Zhurnal **2022**(3), 322–335 (2022)
5. Ilin, I., Vasilenok, V., Marchenko, R.: Use of renewable energy and tax burden on CO_2 emissions in industrial enterprises. E3S Web Conf. **110**, 02102 (2019)
6. Sinsel, S.R., Riemke, R.L., Hoffmann, V.H.: Challenges and solution technologies for the integration of variable renewable energy sources—a review. Renew. Energy **145**, 2271–2285 (2020)
7. Zamo, M., Mestre, O., Arbogast, P., Pannekoucke, O.: A benchmark of statistical regression methods for short-term forecasting of photovoltaic electricity production, part I: deterministic forecast of hourly production. Sol. Energy **105**, 792–803 (2014)
8. Liang, X.: Emerging power quality challenges due to integration of renewable energy sources. IEEE Trans. Ind. Appl. **53**(2), 855–866 (2016)
9. Beyza, J., Garcia-Paricio, E., Yusta, J.M.: Applying complex network theory to the vulnerability assessment of interdependent energy infrastructures. Energies **12**(3), 421 (2019)
10. Lehmann, P., Gawel, E., Strunz, S.: EU climate and energy policy beyond 2020: are additional targets and instruments for renewables economically reasonable? Eur. Dimens. Germany's Energy Transit. Oppor. Confl., 11–26 (2019)
11. Gulagi, A., Ram, M., Solomon, A.A., Khan, M., Breyer, C.: Current energy policies and possible transition scenarios adopting renewable energy: a case study for Bangladesh. Renew. Energy **155**, 899–920 (2020)
12. Jonaitis, A., Gudzius, S., Morkvenas, A., Azubalis, M., Konstantinaviciute, I., Baranauskas, A., Ticka, V.: Challenges of integrating wind power plants into the electric power system: Lithuanian case. Renew. Sustain. Energy Rev. **94**, 468–475 (2018)
13. Kalehsar, O.S.: Iran's transition to renewable energy: challenges and opportunities. Middle East Policy **26**(2), 62–71 (2019)
14. Hyndman, R.J., Khandakar, Y.: Automatic time series forecasting: the forecast package for R. J. Stat. Softw. **27**, 1–22 (2008)
15. Hyndman, R., Koehler, A.B., Ord, J.K., Snyder, R.D.: Forecasting with Exponential Smoothing: The State Space Approach. Springer Science & Business Media (2008)
16. De Livera, A.M., Hyndman, R.J., Snyder, R.D.: Forecasting time series with complex seasonal patterns using exponential smoothing. J. Am. Stat. Assoc. **106**(496), 1513–1527 (2011)
17. Cleveland, R.B., Cleveland, W.S., McRae, J.E., Terpenning, I.S.T.L.: A seasonal-trend decomposition. J. Off Stat **6**(1), 3–73 (1990)
18. Tavadyan, A.: Uncertainty Bands: A Guide to Predicting and Regulating Economic Processes. Anthem Press (2022)
19. Scientific Research Institute of Energy: Energy Balance of the Republic of Armenia for 2022. Yerevan (2024)
20. Kroposki, B., Johnson, B., Zhang, Y., Gevorgian, V., Denholm, P., Hodge, B.M., Hannegan, B.: Achieving a 100% renewable grid: operating electric power systems with extremely high levels of variable renewable energy. IEEE Power Energy Mag. **15**(2), 61–73 (2017)

Redirection of Russian Oil Exports: Analyzing the Impact of Western Sanctions

Ashot Tavadyan and Aghasi Tavadyan

Abstract This study illuminates how Western sanctions against Russia in 2022 catalyzed a dramatic restructuring of global oil trade patterns. Through innovative application of mirror statistics, we investigate the causes and consequences of this radical shift in Russian oil export dynamics, revealing transformative impacts on world economic trajectories. By 2023, our estimates indicate that 38.6% of Russian oil exports were redirected through India, accompanied by significant price anomalies. This sanctions-induced trade shift led to a sharp increase of oil imports by India, playing a crucial role in mitigating the impact on Russian oil exports while potentially boosting India's GDP. We provide causal evidence of how this phenomenon led to substantial GDP growth revisions in both Russia and oil-importing countries. Our findings extend beyond oil trade, offering vital insights into the broader economic consequences of sanctions and the resilience of export-dependent growth. This research not only expands our understanding of contemporary sanctions evasion techniques but also provides critical guidance for policymakers grappling with the complex interplay of geopolitical pressures and global trade patterns.

Keywords Economic sanctions · Russian oil exports · Mirror statistics · Trade redirection · Energy geopolitics

1 Introduction

The global oil market has undergone significant transformations in recent years, with geopolitical tensions and international sanctions reshaping long-established trade patterns. At the epicenter of this shift is Russia, a key player in the world oil market, consistently ranking among the top global producers and exporters of crude oil. The

A. Tavadyan · A. Tavadyan (✉)
Russian-Armenian (Slavonic) University, 123 Hovsep Emin St, 0051 Yerevan, Armenia
e-mail: aghasi.tavadyan@rau.am

I. Ilin and M. Youzhong (eds.), *Digital Systems and Information Technologies in the Energy Sector*, Lecture Notes in Networks and Systems 1244,
https://doi.org/10.1007/978-3-031-80710-7_3

events of 2022, particularly Western sanctions, have catalyzed a dramatic reconfiguration of Russian oil export routes, sending ripples through the global energy landscape and economic systems.

Prior to 2022, Russian oil exports followed well-established patterns, with a substantial portion flowing to European and Asian markets. In 2021, Russia exported a total of 231.6 million tons of crude oil, with China being the largest importer at 30.7% (71.02 million tons), followed by the Netherlands at 16.1% (37.36 million tons) and Germany at 8.3% (19.17 million tons). This stable trade relationship formed a cornerstone of both Russian economic strategy and the energy security of importing nations.

However, the imposition of wide-ranging sanctions by Western nations in response to the Ukraine conflict precipitated a seismic shift in this status quo. The sanctions, designed to exert economic pressure on Russia, included restrictions on oil imports, financial transactions, and technology transfers. These measures, coupled with corporate withdrawals from Russian markets, effectively closed off much of the Western market to Russian oil exports.

In response to these challenges, Russia has pivoted eastward, dramatically increasing its oil exports to Asian markets, particularly India. This shift is starkly illustrated by the change in Russian oil exports to India: from a mere 1.92 million tons (0.8% of Russian exports) in 2021, it surged to 37.0 million tons (15.2% of India's imports) in 2022, and further to 88.9 million tons (38.6% of India's imports) in 2023. This remarkable increase has not only altered global oil trade routes but has also raised questions about the effectiveness of sanctions, the resilience of the Russian economy, and the changing dynamics of global energy geopolitics.

This study aims to provide a comprehensive analysis of these changes in Russian oil exports, focusing on the period from 2022 to early 2024. A significant challenge in this analysis is that following the commencement of its military operation in Ukraine in February 2022, Russia ceased publishing official trade statistics. To overcome this data limitation, we employ mirror statistics, using import data from Russia's trading partners to estimate Russian oil exports.

Our research objectives are:

1. To quantify and analyze the changes in volume and structure of Russian oil exports following the imposition of sanctions in 2022, with a particular focus on:
    a. The redirection of trade flows from European to Asian markets, especially India.
    b. The year-over-year changes in export volumes to key countries, as reflected in mirror statistics.
2. To examine the economic implications of these shifts, including:
    a. The estimated impact on Russia's oil export revenues and overall economic indicators, based on partner country import data.
    b. The effects on the economies of new primary importers, particularly India.
    c. The broader consequences for global oil market dynamics and pricing.

3. To explore the wider implications of this phenomenon, including:
   a. The effectiveness of economic sanctions in altering long-term trade patterns, as evidenced by the observed changes in trade flows.
   b. The resilience and adaptability of international trade networks under geopolitical pressure.
   c. The potential long-term impacts on global energy security and policy.

By utilizing mirror statistics and other available data sources, this study aims to provide a robust analysis of Russian oil export trends despite the lack of official Russian trade data. This approach allows us to offer insights into the rapidly evolving landscape of global oil trade in the context of significant geopolitical tensions.

2 Literature Review

The analysis of Russian oil export dynamics in the wake of Western sanctions requires a multifaceted examination of trade data discrepancies, the effects of economic sanctions, and the broader implications for oil trade and economic diversification. This literature review synthesizes key findings from these interconnected fields, providing a foundation for our subsequent analysis.

2.1 Trade Data Discrepancies and Mirror Statistics

The accuracy and reliability of international trade statistics have long been subjects of academic debate, particularly in the context of countries with transitional economies or those facing geopolitical challenges. The authors of [1] highlight what they term the "mirror problem" in international trade data, where bilateral trade flows reported by exporting and importing countries often show discrepancies. This issue is particularly relevant when studying trade flows between countries with complex economic relationships, such as Russia and its trading partners in the current geopolitical climate.

Benita and Urzua [2] offers a methodological framework for analyzing mirror trade statistics, demonstrating its application in the context of China-Latin America trade. Their approach provides valuable insights for examining potential discrepancies in trade data between Russia and its oil importers. The use of mirror statistics as a tool for uncovering hidden economic activity is further developed by Eberstadt [3] in his study of North Korea's capital goods trade. This methodology proves particularly relevant when official data from one country (in our case, Russia) may be incomplete or unreliable due to sanctions or other factors.

The authors of [4] delve deeper into the causes of trade data discrepancies, focusing on the evasive behaviors of exporters and importers. Their findings on US-China trade relations provide a useful framework for understanding potential motives for discrepancies in oil trade statistics. Phan Thi Thu and Hung [5] offers additional perspectives on the application of mirror statistics in countries with transitional economies, which can be particularly insightful for our analysis of Russian oil exports.

The work of Tavadyan and Tavadyan [6] on navigating uncertainty in export dynamics provides a crucial link between trade data discrepancies and the specific context of transitional economies like Armenia and Russia. Our interval method offers a novel approach to uncovering export dynamics that could be particularly useful in our analysis of Russian oil exports under sanctions. This builds upon work by Tavadyan [7] on the principle of minimum interval of uncertainty and intervals of uncertainty in the economy, which provides a theoretical foundation for dealing with data inconsistencies in transitional economies.

2.2 *Economic Sanctions and Their Effects*

The role of economic sanctions in shaping international trade patterns is critical to understanding the dynamics of Russian oil exports. Gurvich and Prilepskiy in [8] provide an analysis of how Western financial sanctions affected the Russian economy, offering a baseline for understanding the pre-2022 sanctions environment. However, the sanctions landscape has changed dramatically since 2022.

Manushin in [9, 10] offers a comprehensive analysis of Russia's anti-sanction and sanction economic policy from 2022 to 2025, highlighting the shift from reactive measures to a more strategic approach in countering sanctions. This work is crucial for understanding Russia's evolving strategies to mitigate the impact of sanctions, including the potential use of countries as trade intermediaries.

Astrov et al. in [11, 12] provide insights into monitoring the impact of sanctions on the Russian economy, particularly focusing on sanctions evasion through third countries. Their methodology for normalizing export values offers a useful tool for analyzing potential sanctions evasion through third countries.

Kholodilin and Netšunajev [13] and Gold et al. [14] offer additional perspectives on the impact of sanctions on the Russian economy and its trading partners, providing a broader context for understanding the current situation.

2.3 *Oil Trade Dynamics and Economic Diversification*

The specific dynamics of oil trade in the international economy form another critical pillar of our analysis. Mehrotra and Carbonnier [15] investigate abnormal pricing in international commodity trade, with a particular focus on Switzerland. Their methodologies and findings on price discrepancies in high-value commodities like oil provide valuable insights for examining Russian oil export flows.

Drobyshevsky and Pavlov [16] offer a decomposition of economic growth in the Russian Federation up to 2024, providing a baseline for expected economic trajectories prior to the latest round of sanctions. Their work allows for a comparison between projected and actual economic development post-2022 sanctions.

Skufina et al. [17] analyze forecast documents for the socio-economic development of the Russian Arctic, providing valuable insights into regional economic planning and potential changes in resource extraction and trade patterns that may affect oil flows. Jensen and Tarr [18] and Malkina and Balakin [19] offer additional perspectives on economic diversification and development in the context of transitional economies and under sanctions. Tavadyan's and Adilbayev's [20] analysis of prospects for economic relations between Armenia and Kazakhstan provides a regional perspective on economic diversification efforts that could influence oil trade patterns.

This comprehensive review of literature sets the stage for our analysis of Russian oil export dynamics in the face of Western sanctions. By synthesizing insights from trade data analysis, the impacts of economic sanctions, and the specific dynamics of oil trade and economic diversification, we establish a robust foundation for examining how Russia has adapted its oil export strategies in response to geopolitical pressures. Our study aims to build upon this existing body of knowledge by providing a detailed analysis of recent changes in Russian oil export patterns, with a particular focus on the redirection of trade flows to Asian markets and the economic implications of these shifts.

3 Methodology

This study employs a mixed-method approach, combining quantitative analysis of trade data with qualitative assessment of policy changes and market dynamics. Our methodology is designed to overcome the challenges associated with data limitations, particularly the absence of recent official Russian export data. To ensure transparency and reproducibility, the complete dataset and analysis code have been made publicly available on GitHub.

3.1 Data Sources and Time Period

Our primary data sources include:

1. UN Comtrade Database: We use this for historical trade data up to 2021 for Russia. For all other countries, we have HS6 trade data up to and including the first quarter of 2024.

2. National Statistical Offices: We utilize data from the Federal State Statistics Service (Rosstat) of the Russian Federation for available Russian economic indicators.
3. World Oil Council Reports: These provide insights into global oil market trends and prices (World Oil Council, 2024).
4. Supplementary Sources: We also draw from the OECD Bilateral Trade Database (OECD, 2023) and the World Bank's World Integrated Trade Solution (WITS) database (World Bank, 2021).

Our analysis primarily focuses on the period from 2019 to the first quarter of 2024. However, we include data from 2017 where relevant to establish longer-term trends and provide additional context. This extended time frame allows us to:

1. Establish pre-sanctions baselines and long-term trends (2017–2021)
2. Identify the immediate impact of sanctions (2022)
3. Track the evolution of trade patterns post-sanctions (2023).

3.2 Mirror Statistics Approach

To address the lack of recent Russian export data, we employ the "mirror statistics" or "reverse statistics" method, a widely recognized approach in international trade analysis (United Nations, 2011). This method involves using import data from partner countries to estimate the exports of the country under study. Our approach follows these steps:

1. Identify the latest available period of direct reporting for each country.
2. Use original export data up to this period.
3. Estimate subsequent export data using import reports from partner countries.

This approach aligns with the International Monetary Fund's (IMF) recommendations for addressing asymmetries in trade statistics (IMF, 2022) and is endorsed by the World Bank for filling data gaps in international trade analysis (World Bank, 2021).

3.3 Statistical Analysis Methods

Our analysis incorporates several statistical methods:

1. Trend Analysis: We examine the dynamics of oil export trends over time.
2. Comparative Analysis: We compare reported prices and volumes between countries to identify anomalies.
3. Economic Impact Assessment: While we don't conduct formal correlation analysis, we examine how changes in oil export dynamics affect GDP and other economic indicators for Russia and major oil importers. This assessment is based

on a comparative analysis of economic data before and after significant changes in oil exports, considering potential distortions in economic indicators caused by this phenomenon.

3.4 *Limitations and Considerations*

While mirror statistics provide valuable estimates, they are not without limitations. Discrepancies can arise due to differences in valuation methods, timing, and classification practices between countries (OECD, 2023). We acknowledge these potential discrepancies and interpret our results with due caution.

Moreover, the lack of recent data from key partners, may affect the accuracy of our estimates. To address this, we conduct sensitivity analysis to account for potential changes in the data.

Given the sensitive nature of the topic and its geopolitical implications, we maintain a strict commitment to objectivity and academic integrity throughout our analysis. We rely exclusively on official and publicly available data sources, and our interpretations are based on empirical evidence rather than political considerations.

This methodology enables us to conduct a comprehensive analysis of the changing dynamics of Russian oil exports, offering insights into the economic implications of these changes while acknowledging and addressing the inherent limitations in the available data.

4 Results and Analysis

The period from 2019 to 2023 witnessed significant shifts in the volume and structure of Russian oil exports, largely influenced by geopolitical events and subsequent economic sanctions. To illustrate these changes, we will examine several key visualizations.

Figure 1 presents the annual volume of Russian oil exports from 2019 to 2023. The data reveals several notable trends:

1. Pre-sanction stability (2019–2021): Russian oil exports remained relatively stable, fluctuating between 231.6 and 266.6 million tons annually. This period was characterized by established trade relationships, primarily with European countries and China.
2. Post-sanction surge (2022): Following the imposition of Western sanctions in 2022, Russian oil exports surprisingly increased to 279.0 million tons. This unexpected rise can be attributed to several factors, including: (a) Stockpiling by traditional buyers before sanctions took full effect (b) Increased purchases by non-Western countries, particularly India and China (c) Possible data discrepancies due to the challenges in tracking Russian exports post-sanctions

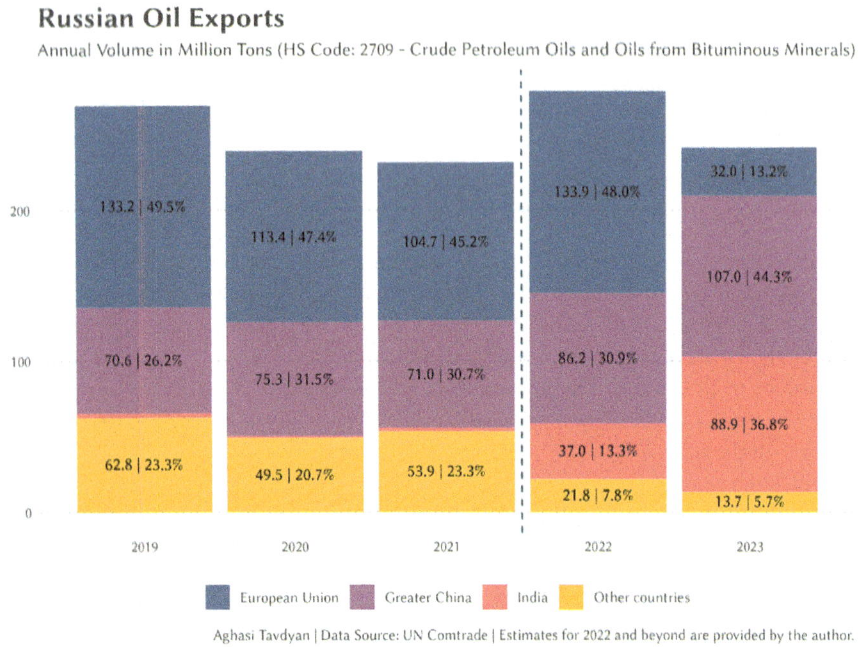

Fig. 1 Annual volume of Russian oil exports from 2019 to 2023

3. Adjustment and decline (2023): The volume of exports decreased to 241.6 million tons in 2023, reflecting the impact of sanctions and the restructuring of Russia's oil trade partnerships.

To further understand these changes, we'll examine the shifting structure of Russian oil exports.

Figure 2 illustrates the distribution of Russian oil exports in 2021, before the imposition of sanctions. Key observations include:

1. China was the largest importer, accounting for 30.7% (71.02 million tons) of Russian oil exports.
2. European countries collectively represented a significant portion of exports, with the Netherlands (16.1%), Germany (8.3%), and Poland (4.8%) being major importers.
3. India's share was minimal at 0.8% (1.92 million tons), highlighting its limited role in Russian oil imports pre-sanctions.

Figure 3 shows the estimated distribution of Russian oil exports in 2022, reflecting the immediate impact of sanctions:

1. China maintained its position as the largest importer, slightly increasing its share to 30.9% (86.24 million tons).

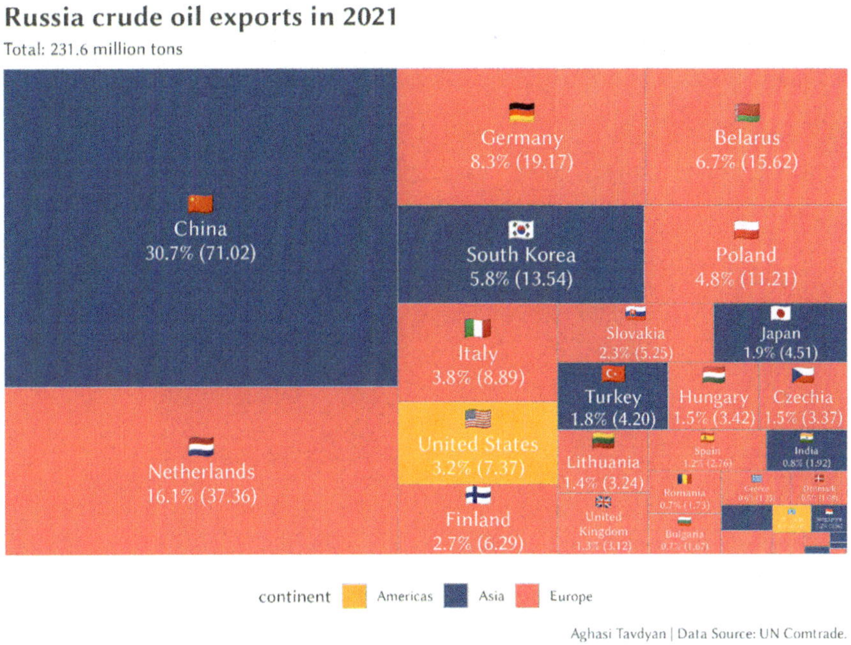

Fig. 2 Distribution of Russian oil exports in 2021

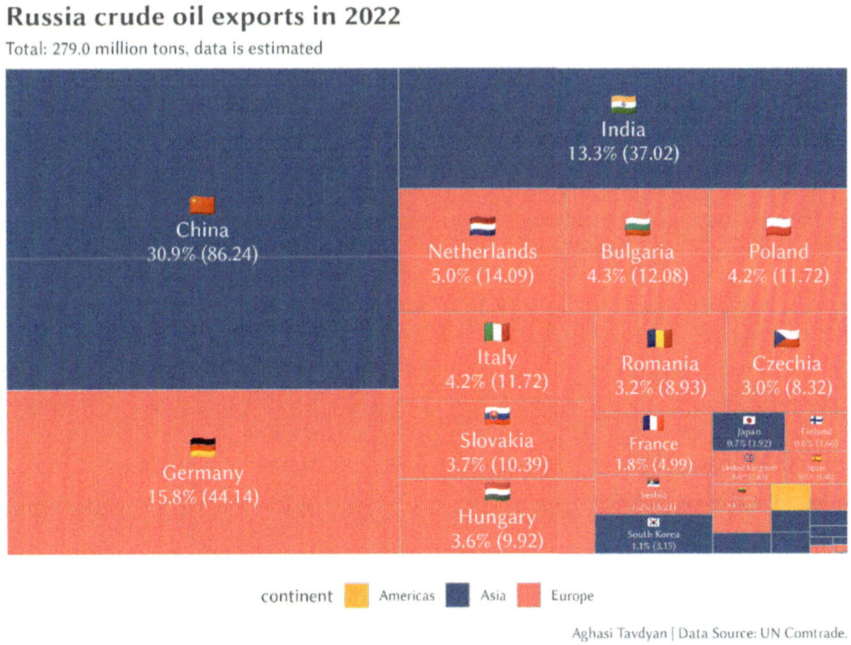

Fig. 3 Estimated distribution of Russian oil exports in 2022

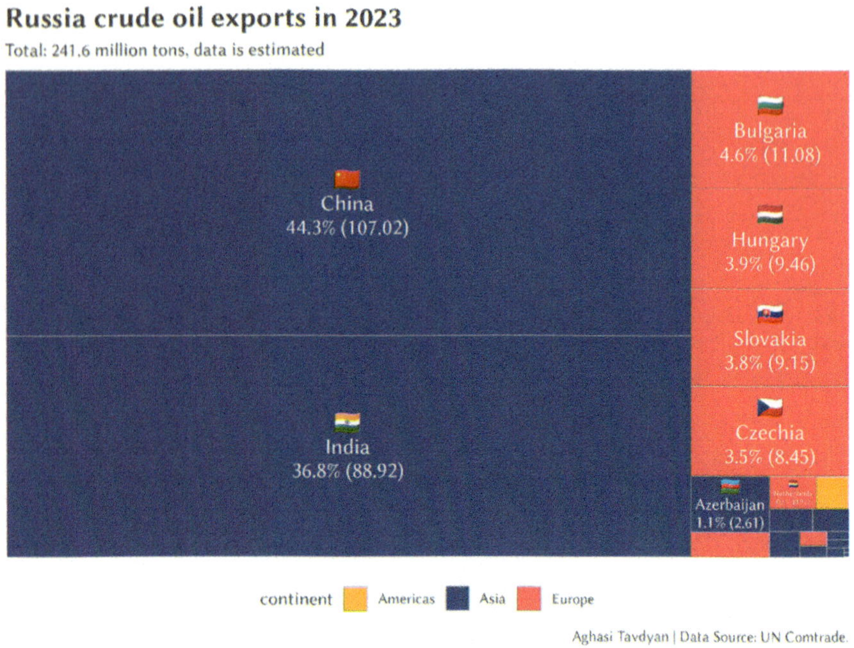

Fig. 4 Estimated structure of Russian oil exports in 2023

2. India emerged as a significant new buyer, with its share jumping to 13.3% (37.02 million tons), a dramatic increase from the previous year.
3. European countries, while still present, showed reduced imports. Germany's share, for instance, increased to 15.8% (44.14 million tons), possibly due to stockpiling before stricter sanctions were implemented.

Figure 4 depicts the estimated structure of Russian oil exports in 2023, revealing a dramatic shift in trade patterns:

1. China's dominance grew substantially, accounting for 44.3% (107.02 million tons) of Russian oil exports.
2. India solidified its position as the second-largest importer, with its share rising to 36.8% (88.92 million tons).
3. European countries' shares diminished significantly, with only a few countries like Bulgaria (4.6%) and Hungary (3.9%) maintaining notable imports.

These visualizations clearly demonstrate the profound impact of Western sanctions on Russian oil exports. The most striking changes include:

1. A significant eastward shift in export destinations, with China and India becoming the dominant buyers.
2. A sharp decline in exports to traditional European markets.

3. An overall reduction in export volume from the peak in 2022 to 2023, indicating the challenges Russia faces in redirecting its oil exports.

This restructuring of Russian oil exports has far-reaching implications for global energy markets, geopolitical relationships, and the effectiveness of economic sanctions as a policy tool. The next sections will delve into the broader economic impacts on Russia and its new primary importing countries.

4.1 Economic Impact on Russia and Importing Countries

The redirection of Russian oil exports following Western sanctions has had significant economic implications for both Russia and its new primary importing countries, particularly India. This shift in trade patterns has reshaped the global oil market and affected the economies of involved nations.

As Fig. 5 illustrates, India's oil import landscape has undergone a dramatic transformation since 2021. The most striking change is the surge in Russian oil imports:

1. Pre-sanctions (2019–2021): Russian oil imports to India were minimal, accounting for only 1.9 million tons (0.8% of India's total oil imports) in 2021.

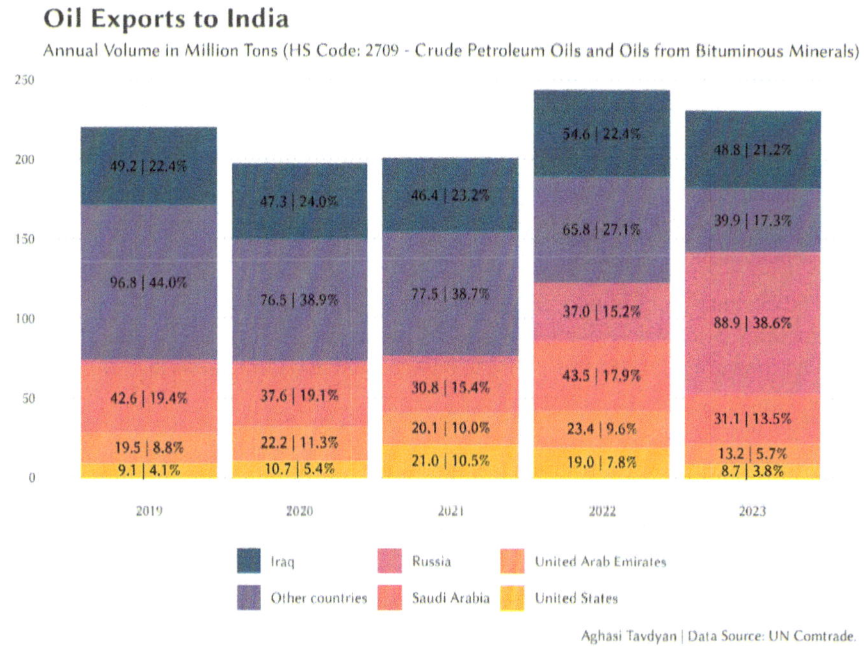

Fig. 5 Annual oil imports to India by country (2019–2023)

2. Post-sanctions (2022–2023): Russian oil imports to India skyrocketed to 37.0 million tons (15.2% of India's imports) in 2022, further increasing to 88.9 million tons (38.6% of India's imports) in 2023.

This rapid increase in Russian oil imports has several implications:

1. For Russia:
   a. Mitigation of sanctions impact: The pivot to Asian markets, particularly India, has allowed Russia to partially offset the loss of European buyers.
   b. Potential price concessions: To attract new buyers, Russia may have offered discounts, potentially affecting its oil revenues.
2. For India:
   a. Energy security: Increased imports from Russia have diversified India's oil sources, potentially enhancing its energy security.
   b. Economic opportunity: As a major importer of crude oil, India has capitalized on potentially discounted Russian oil, which could benefit its national economy.
   c. Refining industry boost: India's significant crude oil imports, coupled with its lack of crude oil exports, suggest a thriving refining industry. The country processes imported crude oil into various petroleum products.
3. Global market impact:
   a. Shift in trade flows: The redirection of Russian oil to Asian markets has reshaped global oil trade routes.
   b. Price dynamics: Changes in supply patterns may have influenced global oil prices and regional price differentials.

It's important to note that while India has dramatically increased its imports of Russian crude oil, it does not export crude oil itself. Instead, India processes the imported crude in its refineries, potentially exporting refined petroleum products. This aspect of India's oil trade warrants further investigation, particularly regarding:

1. The destination of India's refined petroleum product exports.
2. Potential price discrepancies in these exports.
3. The overall impact on India's trade balance and economic growth.

5 Discussion

The dramatic shift in Russian oil export patterns following Western sanctions in 2022 has far-reaching implications for global energy markets, economic sustainability, and international relations. This section discusses the key implications of this export redirection, the sustainability of economic growth based on oil exports, and the regulatory and policy considerations that arise from these changes.

5.1 Implications of Export Redirection

The redirection of Russian oil exports from Western markets to Asian countries, particularly India and China, has several significant implications:

1. Geopolitical Realignment: The shift has strengthened economic ties between Russia and major Asian economies, potentially leading to new geopolitical alliances. This realignment could have long-term effects on global power dynamics and international cooperation.
2. Sanctions Effectiveness: The ability of Russia to redirect its oil exports raises questions about the effectiveness of Western sanctions. While the sanctions have disrupted traditional trade patterns, they have partially isolated Russia from the global oil market.
3. Global Energy Market Dynamics: The redirection has altered global oil supply chains, potentially affecting price stability and energy security for various regions. Countries that have increased their reliance on Russian oil may face new vulnerabilities or opportunities.
4. Economic Interdependence: The increased oil trade between Russia and countries like India creates new economic interdependencies. This could influence future diplomatic relations and policy decisions among these nations.

5.2 Sustainability of Economic Growth Based on Oil Exports

The heavy reliance on oil exports for economic growth, as exemplified by Russia's pivot to new markets, raises concerns about long-term economic sustainability:

1. Volatility Risk: Economies heavily dependent on oil exports are vulnerable to price fluctuations in the global oil market. This volatility can lead to unpredictable government revenues and economic stability development.
2. Resource Depletion: Over-reliance on non-renewable resources like oil is inherently unsustainable in the long term. Countries focused on oil exports may neglect the development of other economic sectors.
3. Diversification Challenges: For both exporting and importing countries, the current situation underscores the need for economic diversification. However, the short-term benefits of oil trade may disincentivize investments in alternative industries or energy sources.
4. Technological Disruption: The global push towards renewable energy and electric vehicles poses a long-term threat to oil-dependent economies. Failure to adapt to these trends could result in significant economic challenges in the future.

The redirection of Russian oil exports represents a significant shift in global energy trade with multifaceted implications. While it has allowed Russia to mitigate some effects of Western sanctions, it also highlights the vulnerabilities of oil-dependent economies and the complex interplay between energy trade, geopolitics, and economic sustainability. Both oil-exporting and importing countries should implement policies that promote economic diversification to reduce vulnerability to oil market fluctuations. As the global energy landscape continues to evolve, adaptive and forward-looking policies will be essential to navigate these challenges and opportunities.

6 Limitations and Future Research

While this study provides valuable insights into the dynamics of Russian oil exports and their economic implications, it's important to acknowledge several limitations that also point to directions for future research.

A primary limitation of this study is the lack of recent trade data from key partners, particularly China and India, which have become major importers of Russian oil. The absence of data from these countries in UN Comtrade for 2022, 2023, and 2024 may affect the accuracy of our estimates of Russian oil exports.

The reliability of mirror statistics, while a valuable tool in the absence of direct export data. However, we acknowledge that using mirror statistics may give approximate estimates, especially when data from key trading partners is missing.

Our study primarily focuses on the Russian oil export sector.

Future research should:

1. Include a more detailed analysis of how other sectors of the Russian economy are functioning in relation to the changed direction of oil exports and its price dynamics.
2. Explore potential spillover effects of changes in oil trade on other areas of the economy.
3. Analyze the long-term sustainability of economic growth dependent on the oil factor.

7 Conclusions

This study has examined the significant shifts in Russian oil exports following the imposition of Western sanctions in 2022, revealing substantial changes in global oil trade patterns and their economic implications.

7.1 Summary of Key Findings

1. Dramatic Shift in Export Destinations: Our analysis reveals a marked redirection of Russian oil exports from Western markets to Asian countries, particularly India and China. This shift represents a fundamental change in global oil trade dynamics.

2. Surge in India's Russian Oil Imports: The most striking finding is the exponential growth in India's imports of Russian oil. From a mere 1.92 million tons (0.8% of India's imports) in 2021, Russian oil imports to India surged to 37.0 million tons (15.2%) in 2022 and further to 88.9 million tons (38.6%) in 2023.
3. Resilience of Russian Oil Exports: Despite Western sanctions, Russia has managed to maintain significant oil export volumes by pivoting to new markets, demonstrating the adaptability of global oil trade networks.
4. Economic Implications: The redirection of oil exports has had substantial economic impacts, not only for Russia but also for importing countries like India, potentially influencing their economic growth and energy policies.
5. Data Challenges: The study highlighted some challenges in accurately tracking oil trade flows due to the absence of official Russian data and inconsistencies in international trade statistics.

7.2 *Future Outlook for Russian Oil Exports*

The future of Russian oil exports remains subject to several key factors:

1. Geopolitical Dynamics: The ongoing geopolitical situation, including the duration and intensity of Western sanctions, will continue to play a crucial role in shaping Russian oil export patterns.
2. Global Energy Transition: The pace of global transition towards renewable energy sources will impact long-term demand for oil, potentially affecting Russia's export strategies.
3. Economic Adaptations: Russia's ability to maintain and potentially expand its new export markets, particularly in Asia, will be critical for the sustainability of its oil-dependent economy.
4. Technological Developments: Advancements in oil extraction, transportation, and alternative energy technologies could influence the competitiveness and viability of Russian oil exports.
5. Market Competition: The response of other major oil producers to Russia's market shifts could lead to increased competition in Asian markets, potentially affecting prices and trade volumes.

The redirection of Russian oil exports represents a significant shift in global energy trade with far-reaching implications. While it has allowed Russia to mitigate some effects of Western sanctions, it also highlights the interconnectedness of global energy markets and the complex interplay between energy trade, geopolitics, and economic policy. As the global energy landscape continues to evolve, adaptive strategies and forward-looking policies will be essential for all stakeholders in navigating these challenges and opportunities. Future research, particularly focusing on the role of major importers like India and China, will be crucial in fully understanding the long-term implications of these shifts in global oil trade patterns.

References

1. Linsi, L., Burgoon, B., Mügge, D.K.: The problem with trade measurement in international relations. Int. Stud. Q. **67**(2) (2023)
2. Benita, F., Urzua, C.M.: Mirror trade statistics between China and Latin America. J. Chinese Econ. Foreign Trade Stud. **9**(3), 177–189 (2016)
3. Eberstadt, N.: The DPRK's international trade in capital goods, 1970–1995: indications from mirror statistics. J. East Asian Aff. **12**(1), 165–223 (1998)
4. Ferrantino, M.J., Liu, X., Wang, Z.: Evasion behaviors of exporters and importers: evidence from the US–China trade data discrepancy. J. Int. Econ. **86**(1), 141–157 (2012)
5. Phan Thi Thu, H., Hung, N.V.: Mirror study of Vietnam's international merchandise trade: findings and implications on cross-border trade administration. SECO/WTI Academic Cooperation Project Working Paper Series, 26 (2017)
6. Tavadyan, A., Tavadyan, A.: Navigating uncertainty: an interval method to uncover export dynamics–insights from the Republic of Armenia. Int. J. Prof. Bus. Rev. **8**(10), 377 (2023)
7. Tavadyan, A.: Uncertainty Bands: A Guide to Predicting and Regulating Economic Processes. Anthem Press (2022)
8. Gurvich, E., Prilepskiy, I.: The impact of financial sanctions on the Russian economy. Russ. J. Econ. **1**(4), 359–385 (2015)
9. Manushin, D.V.: Anti-sanction and sanction economic policy of Russia in 2022–2025. Part 2: aftermaths of sanctions war, management of changes, clarification of research concepts. Russ. J. Econ. Law **18**, 37 (2024)
10. Manushin, D.V.: Anti-sanction and sanction economic policy of Russia in 2022–2025. Part 3: analysis of anti-crisis measures, managing Russian sanctions and anti-sanctions. Russ. J. Econ. Law, 332 (2024)
11. Astrov, V., Scheckenhofer, L., Semelet, C., Teti, F.: Are western trade sanctions effective? (No. 6). Vienna Inst. Int. Econ. Stud. (2024)
12. Astrov, V., Kochnev, A., Scheckenhofer, L., Stamer, V., Teti, F.: Monitoring the impact of sanctions on the Russian economy (No. 46). EconPol Policy Report (2024)
13. Kholodilin, K.A., Netšunajev, A.: Crimea and punishment: the impact of sanctions on Russian economy and economies of the euro area. Baltic J. Econ. **19**(1), 39–51 (2019)
14. Gold, R., Hinz, J., Valsecchi, M.: To Russia with love? The impact of sanctions on regime support (2024)
15. Mehrotra, R., Carbonnier, G.: Abnormal pricing in international commodity trade: empirical evidence from Switzerland. Resour. Policy **74**, 102352 (2021)
16. Drobyshevsky, S., Pavlov, P.: Decomposition of economic growth in the Russian federation through 2024. Russ. Econ., 305–310 (2018)
17. Skufina, T.P., Baranov, S.V., Samarina, V.P.: Analysis of forecasting documents for the socio-economic development of the Russian Arctic. Arctic **48**, 49–63 (2022)
18. Jensen, J., Tarr, D.: Deep trade policy options for Armenia: the importance of trade facilitation, services and standards liberalization. Econ. **6**(1), 20120001 (2012)
19. Malkina, M.Y., Balakin, R.V.: Trends in the Russian economy development during the new anti-Russian sanctions. Russ. J. Econ. Law, 287 (2024)
20. Tavadyan, A., Adilbayev, J.: Prospects for the development of economic relations between Armenia and Kazakhstan. 21st Century **2**(43), 81–90 (2017)

Methodology for Assessing the Economic Efficiency of Internal Investment Projects in Energy Sector Companies

Denis Bezruchko, Igor Ilin, Arsenii Sluzhaev, and Manfred Esser

Abstract The study addresses the challenge of evaluating the economic efficiency of internal investment projects in energy companies, which are often focused on innovations that do not directly generate revenue but require significant investments. The main objective of the study is to develop a universal methodology for assessing these internal projects based on economic and mathematical modeling. The research utilizes statistical and analytical methods, specifically key performance indicator (KPI) trees, along with investment and sensitivity analysis to assess the effectiveness of the proposed measures. The results include the creation of a new methodology for evaluating internal projects, along with a refined understanding of the concept of internal investment projects, which enhances decision-making in the energy sector.

Keywords Innovations in the energy sector · Economic efficiency · Internal investment project

1 Introduction

The continuous adoption of new technologies in the energy sector is an ongoing process driven by the need to maintain the industry's competitiveness in the global market. As the energy demands of modern society continue to grow, energy companies face increasing pressure to innovate. This pressure is not only driven by competition but is also compounded by regulatory demands and calls from governments and environmentally conscious organizations to adopt cleaner and more efficient

D. Bezruchko · I. Ilin · A. Sluzhaev (✉)
Peter the Great St. Petersburg Polytechnic University, Polytechnicheskaya, 29 B, St. Petersburg 195251, Russia
e-mail: sluzhaevars@outlook.com

M. Esser
GetIT, Dresden, Germany

I. Ilin and M. Youzhong (eds.), *Digital Systems and Information Technologies in the Energy Sector*, Lecture Notes in Networks and Systems 1244,
https://doi.org/10.1007/978-3-031-80710-7_4

technologies. Internal investment projects, aimed at enhancing the efficient use of a company's resources, have become a crucial component in achieving these goals.

Moreover, the need to adapt to evolving regulatory requirements and minimize the environmental impact of energy companies has provided additional impetus for innovation. This has led to an intensification of research and development (R&D), which is not only focused on improving performance and reducing costs but also on justifying the economic efficiency of implementing these technologies [1–4]. As a result, there is a growing demand for methodologies that allow for the accurate assessment of the economic efficiency of internal projects, ensuring that investments generate adequate returns while complying with environmental and regulatory standards.

The energy sector is characterized by the high capital intensity of large investment projects, such as the construction of new energy generation and transmission facilities. While the evaluation of the economic efficiency of such large-scale projects occurs relatively infrequently, internal investment projects that focus on process optimization and equipment modernization are much more common within companies. These projects are designed to improve efficiency without directly impacting the company's revenue. However, they can significantly reduce both capital and operational costs, improve environmental performance, and enhance overall productivity.

Since internal investment projects do not typically result in direct revenue increases, the primary focus is on cost reduction and efficiency improvements. In this context, it is essential to develop robust methodologies for evaluating economic efficiency, as traditional investment analysis methods, which rely on revenue growth, may be inadequate. Assessing the economic efficiency of internal projects requires approaches that account for changes in the company's cash flow resulting from the implementation of these projects. Therefore, evaluating the effectiveness of internal investment projects in the energy sector is a crucial task, requiring both theoretical and practical substantiation to support decision-making processes and ensure the sustainable growth of the company in a highly competitive and regulated environment.

The object of this study is the projects of structural subdivisions of energy companies aimed at implementing innovations into the companies' operations.

The subject of the study is the methodology for evaluating internal investment projects of energy companies.

The purpose of the study is to develop a universal methodology for assessing the economic efficiency of internal investment projects of companies based on economic and mathematical modeling.

The data for this study were based on internal projects of energy companies.

Thus, the **objectives of this study** are:

- Analyzing the dataset of internal investment projects of energy companies,
- Developing economic and mathematical models for these projects,
- Developing a methodology for analyzing the economic efficiency of projects,
- Analyzing the economic efficiency of projects according to the developed methodology,

- Evaluating the analysis results,
- Preparing conclusions on the applicability of the methodology.

Statistical and analytical methods were used in this study. To evaluate the effectiveness of the implementation of proposed measures, methods for assessing economic efficiency based on key performance indicator (KPI) trees, enhanced by investment analysis and sensitivity analysis methods, were applied.

The following contributions to scientific knowledge in the subject area were made in addressing the stated objective:

- A new and improved methodology for assessing the economic efficiency of internal company projects using investment analysis tools was developed, described, and applied,
- The concept of internal investment projects was refined.

2 Discussion

The fundamental difference between an internal investment project and a classical ("external") project is that such a project does not lead to changes in the quantity or quality of the enterprise's products. At the same time, the project is associated with initial investments, meaning it possesses all the characteristics of an investment project [5–7]. The focus of an internal investment project is aimed at improving the efficiency of the company's resource utilization and is generally related to the introduction of innovations in the technological sphere or in business processes, which ultimately leads to a reduction in company costs. Therefore, in this context, an internal investment project is defined as an investment project aimed at reducing the company's expenses.

Since any implementation of innovations requires investments, the justification of their economic efficiency is carried out using investment analysis methods. It is widely known that investment analysis is based on the discounting of a project's cash flow, which involves bringing future cash flows to the present moment. This discounting is done by dividing the cash flow amounts by the discount factor raised to the power corresponding to the year of project implementation. At the same time, the simple (undiscounted) cash flow of the project is calculated as the difference between the inflows and outflows of cash from current activities, minus the investment cash flow, and is expressed by the following formula (in matrix form):

$$CF = (TR - TC) - I,$$

where:

CF Cash flow of the project (vector),
TR Total revenue from current activities,
TC Total costs from current activities,

I Investments [8-10].

As noted earlier, the focus of the research is internal investment projects involving the introduction of innovations in the energy sector. In these projects, the revenue part of the project, TR, does not change, but the project impacts costs (TC) and investment capital intensity (I). Additionally, it should be noted that TR is generally an unknown value within the studied projects since the project-initiating division does not possess information about the sales structure of the entire company.

In this case, to correctly evaluate the economic efficiency of the project, it is necessary to shift from classical investment analysis methods to comparative analysis methods.

Comparative analysis does not examine the project's or the company's overall cash flow, but rather its change due to the implementation of the project introducing new technologies (see formula):

$$\Delta CF = (\Delta TR - \Delta TC) - \Delta I$$

As noted earlier, the change in the company's revenue resulting from the internal project is not calculated and is assumed to be zero. Therefore, the change in cash flow will equal the change in the total costs of the project's implementation:

$$\Delta CF = -(\Delta TC + \Delta I)$$

Thus, to accurately calculate the economic effect of the project (equal to the change in cash flow), it is necessary to calculate the total project costs "as it was" under the old technology and "as it will be" under the new technology.

The obtained value of the change in cash flow (ΔCF) provides the magnitude of the economic effect in absolute terms, in the currency of the calculation. If the effect extends beyond the first year, the cash flow must be discounted by a vector of discount factor values (i) and the discounted cash flow (ΔDCF) analyzed. By applying classical investment analysis methods, the values of NPV, IRR, and the simple and discounted payback periods are derived (see Table 1).

The main challenge in evaluating the change in costs (ΔTC) lies in identifying the influencing factors of the investment project and establishing the logical relationship between these factors and ΔTC. The quality of this logical relationship determines the accuracy of the project's efficiency assessment.

The methodology for evaluating the economic efficiency of internal projects is based on functional-cost analysis (FCA). FCA methods are widely used in business analysis due to their strong potential for conducting techno-economic assessments of project efficiency. A specific case of FCA is the methodology for investment analysis of internal projects based on constructing a key performance indicator (KPI) tree. As recommended in, all economic effects are calculated for the project's end user (the company) and take into account both the costs of implementing innovations and the savings achieved through their use [11–14].

Table 1 Efficiency indicators of an internal investment project

Notation	Name	Calculation formula	Economic content
ΔCF	Change in cash flow	$-(\Delta TC + \Delta I)$	The amount of savings over the project life cycle in absolute terms
NPV	Net present value	$-(\Delta CF/i)$	The present value of the project's economic effect in absolute terms
IRR	Internal rate of return	Discount rate at which NPV = 0	The relative value of the economic effect, expressed as the project's average annual return
PBP	Payback period	Period when $\Delta CF = 0$	The payback period (calculated using the simple cash flow)
DPBP	Discounted payback period	Period when $\Delta DCF = 0$	The payback period (calculated using the discounted cash flow)

The evaluation of the implementation effects and the efficiency indicators of a project is conducted at four levels of key performance indicators (KPIs):

1. Level 4 ("from bottom to top")—Project influencing factors: At this level, the direct project influence on measurable "physical" indicators is determined (e.g., time for a technological operation, variable cost norms per unit of production, etc.). Here, it is essential to reflect the "project"—"influencing factor" relationship.
2. Level 3—Physical parameters: At this level, the effect of the project implementation is determined in physical terms over a period of time (e.g., labor intensity measured in person-years, or variable costs in physical units per year). The indicators for levels 3 and 4 must be presented in two versions: "as is" (before the project) and "to be" (after the project). This is necessary to fix KPIs for monitoring purposes, as it is crucial to understand the absolute physical indicators achieved by the project's implementation and the baseline for comparison.
3. Level 2—Operational KPIs: At this level, the effects of the project are recorded in monetary terms, such as labor cost savings (million rubles per year) or reduced expenses (million rubles per year).
4. Level 1—Investment KPIs (economic efficiency indicators): The key efficiency indicators are the following four metrics: NPV (Net Present Value), IRR (Internal Rate of Return), PBP (Payback Period), and DPBP (Discounted Payback Period). These economic indicators are calculated based on changes in the company's cash flow due to the project implementation, taking into account both direct effects and the total of direct and indirect effects. Notably, accounting for indirect effects enhances the economic efficiency of the project.

Investment KPIs are determined by forming cash flows based on operational KPIs by year and full implementation costs, including both capital (CapEx) and operational (OpEx) expenses.

The primary goal of KPI levels is to establish a clear causal relationship between project influencing factors and the project's economic performance indicators—"investment KPIs." In the economic-mathematical model of the project, it is essential to ensure full consistency between the indicators at all KPI levels.

Further investment analysis methods can be employed if the economic calculations are performed using the methodology of economic-mathematical modeling of investment projects. Economic-mathematical modeling is understood as the process of constructing an abstract representation (model) of a real or hypothetical financial situation. Nowadays, it is assumed that economic-mathematical models are built using software, with all input data for the model being variable.

The logic of building a model is related to converting the initial data into standard financial reports for future periods, which, in turn, serve as the input for any type of economic analysis. In described case, the intermediate goal of building the model is to obtain the project's cash flow line ΔCF for subsequent risk analysis using sensitivity analysis and conducting a stress test using the Monte Carlo method. Let us take a closer look at these types of analysis.

Sensitivity analysis is one of the quantitative methods for project risk analysis. Technically, sensitivity analysis is performed by changing one of the project's main influencing factors incrementally, while keeping the other factors unchanged. Each such change is followed by recalculating the project's efficiency indicators and recording the new values in a table. Afterward, the altered factor is returned to its baseline value, and the same process is repeated for other influencing factors.

The recalculated efficiency indicators of the project are compiled into a table, from which the following conclusions are drawn:

- Influencing factors are ranked by the degree of their impact on the project's efficiency indicators,
- The ranges of influencing factors are identified, within which the project's efficiency remains in an acceptable range,
- Project scenarios are developed based on the identified characteristics.

Sensitivity analysis is recommended when the project's payback period exceeds the calculation interval of the financial model (e.g., 1 year). This type of analysis is particularly appropriate for projects with medium or low efficiency, where ensuring the financial stability of the project over time becomes crucial. By testing various assumptions and altering key influencing factors, sensitivity analysis helps identify how changes in these variables impact the project's outcomes. For such projects, the goal is to ensure that the project remains financially viable under a range of different scenarios, especially when uncertainty surrounds key cost or revenue drivers. If the payback period is achieved within the first year, however, sensitivity analysis may be deemed unnecessary. In such cases, indicators like the Internal Rate of Return (IRR), Payback Period (PBP), and Discounted Payback Period (DPBP) cannot be accurately calculated, and the project's economic efficiency is typically self-evident without further testing of assumptions.

Monte Carlo analysis, on the other hand, takes a probabilistic approach to project evaluation by constructing a financial model that simulates numerous possible outcomes. This method is particularly useful for projects where the variables influencing the outcomes are highly uncertain. To conduct this type of analysis, the first step is to define the range of probable values for each influencing factor. It is also necessary to establish a probability distribution function for the values within this range; if this function is unknown, a normal distribution is typically used as a standard. The financial model is then recalculated multiple times—often in the thousands—with each recalculation incorporating random values within the predefined ranges for the key factors. The efficiency indicators, such as NPV or IRR, are recorded based on these values in each simulation. After multiple iterations, the resulting data set is analyzed using statistical methods to determine trends, averages, and risks. Ultimately, probability distribution graphs are constructed to visually represent the likelihood of various outcomes, offering decision-makers a clearer understanding of the project's economic efficiency under different conditions of uncertainty.

The conclusions that can be drawn from Monte Carlo analysis include:

- The most likely values of the project's economic efficiency indicators (which often do not match the efficiency indicators calculated based on the baseline scenario),
- The probability of unfavorable scenarios for the project,
- The recommended reserve amounts to cover key risks (e.g., the risk of a cash shortfall).

Monte Carlo analysis is also applied only to projects with medium and low economic efficiency, primarily to identify risks of unfavorable scenarios. For projects with high economic efficiency, the use of this method is not justified and provides no significant information for decision-making.

3 Results

Next, the application of the methodology for assessing the efficiency of an internal project will be demonstrated using the example of implementing a digital twin in an energy facility. The use of digital twin technology in energy management has become increasingly important, particularly in complex systems like power generation. In 2017, the departmental project "Digital Energy" acted as a catalyst, accelerating the digitalization process of the Russian energy sector. This project not only highlighted the importance of digital transformation in energy companies but also underscored the need for more efficient resource management. Special attention has been given to Digital Twin technology, which is now recognized as one of the key strategic technological trends in the energy sector due to its wide-ranging applications and benefits.

A digital twin is a virtual model of a real object or system, allowing for real-time monitoring of its condition, the prediction of potential deviations, and virtual testing of various operational scenarios. This technology enables companies to gain deeper insights into system performance by simulating different outcomes without any risk to the actual infrastructure. The continuous feedback loop between the real-world object and its digital counterpart allows for constant optimization of operations, reducing inefficiencies and improving overall system reliability.

In the context of a thermal power plant (TPP), the digital twin can be deployed to monitor the operational status of equipment, optimize management protocols, and streamline maintenance processes. Furthermore, it can play a pivotal role in the development and implementation of innovative technologies by simulating their impact before physical installation. This predictive capability not only enhances operational efficiency but also reduces maintenance and repair costs. Additionally, it leads to significant reductions in energy consumption, as processes can be optimized to avoid waste. By constantly monitoring equipment performance and forecasting potential breakdowns, the digital twin also helps minimize the risk of accidents, thereby improving both safety and operational uptime. This integration of digital twin technology into thermal power plants represents a fundamental shift in how energy facilities are managed and maintained, offering substantial economic and environmental benefits.

The expected effects of project implementation are:

- Reduction in fuel costs;
- Reduction in unplanned repair costs;
- Reduction in labor costs.

In general terms, the economic effect represents a useful result that is reflected in cost assessments. To calculate fuel costs, it is necessary to know the annual power output of the TPP and the price of natural gas [15–17].

In this case, the annual power output is 1 million kWh, and the gas price per 1000 m^3 ranges from 5550 to 5700 rubles, with an average price of 5625 rubles per 1000 m^3, or approximately 6 rubles per 1 m^3. According to rough estimates, 0.25–0.3 m^3 of gas is required per 1 kWh. Using 0.3 m^3 as a basis, the cost per 1 kWh is 1.8 rubles, and for 1 million kWh, the cost is 1,800,000 rubles.

The change in the number of personnel is expected to reduce by 20 people. The predicted effects of implementing the digital twin at the TPP are presented in Table 2.

The calculation of the annual economic effect for each factor is shown in Table 3.

Based on the calculated change in cash flow resulting from project implementation, the key performance indicators (KPIs) of the project were determined (see Table 4).

All economic calculations were performed within an economic-mathematical model, allowing for recalculations when input data changes, as well as conducting

Table 2 Predicted project effects

Indicators (rub/year)	Before implementation	After implementation	Change
Unplanned repair costs	8,000,000	6,500,000	– 1,500,000
Labor costs	19,600,000	18,200,000	– 1,400,000
Fuel costs	1,800,000	1,728,000	– 72,000

Table 3 Calculation of cash flow from project implementation

Period	2025	2026	2027	2028 and beyond
Base year	Operation	Operation	Operation	
Investments (thousand rubles)	15,000,000			
Unplanned repair costs		1,500,000	1,500,000	1,500,000
Labor costs		1,400,000	1,400,000	1,400,000
Fuel costs		72,000	72,000	72,000
Cash flow	– 15,000,000	2,972,000	2,972,000	

Table 4 Key performance indicators of the project

Indicator	Value
NPV (rubles)	14,720,055
IRR (%)	15
Payback period (years)	6
Discounted payback period (years)	9

sensitivity analysis. As it seen, sensitivity analysis provides a more objective quantitative risk assessment compared to traditional scenario analysis, as previously recommended by authors such as [18–20].

Additionally, when working on the project, it is identified social efficiency indicators—measurable project values that provide societal benefits. In this case, these include the overall reduction in technological risks (including accident risks) as a result of the timely implementation of this investment project.

Unlike the work of most researchers, such as [15], the use of sensitivity analysis to assess the impact of project risks is prioritized (for the first time in internal projects). Sensitivity analysis helped identify the degree of influence of various factors on project efficiency.

It was found that the most important factors affecting project efficiency are as follows:

- Changes in cost parameters,
- Changes in the magnitude of effects,
- Changes in costs,
- Changes in the planned budget of the project's investment costs.

The results of the analysis calculations are shown in Fig. 1.

As unplanned repair costs and labor costs (decreased savings) increase, the discounted payback period also extends. There is also a direct correlation with the amount of investment: with a large capital expenditure, the discounted payback period may not be reached within the next 10 years, the duration planned for the project.

Conclusions: The project is highly sensitive to changes in investment volume and least sensitive to changes in fuel costs.

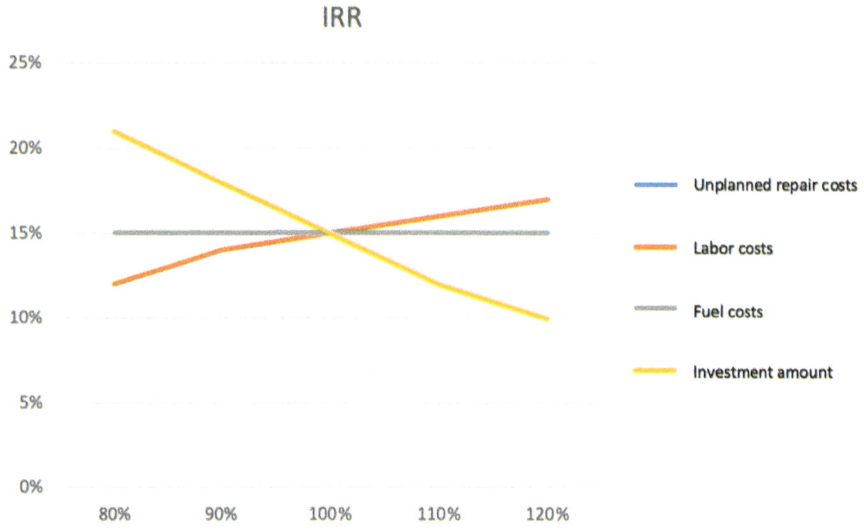

Fig. 1 Sensitivity analysis of IRR versus effect sizes (for direct effects only)

4 Conclusion

The majority of innovation-related projects in the energy sector are categorized as internal investment projects, where the focus lies predominantly on improving operational efficiency rather than generating additional revenue. These projects typically aim to reduce costs in production, maintenance, or energy use, while making the company more efficient overall. As such, they do not directly influence sales or income from product sales but instead impact the cost structure of processes, particularly by lowering operational and capital expenditures. This makes the assessment of economic efficiency a particularly challenging but necessary task for energy companies, as the benefits of such projects are often difficult to quantify using traditional investment analysis methods.

The most reliable results in evaluating the economic efficiency of these projects are obtained through comparative analysis, which focuses on investment and operating costs. This approach allows for a more accurate assessment of the project's impact on cash flows, where the true benefits of internal projects—such as cost savings and process improvements—can be captured. Unlike classical methods that focus primarily on revenue generation, internal investment projects rely on the project's ability to enhance operational efficiency, leading to improvements in cash flow rather than direct revenue increases. The comparative method thus provides a more suitable framework for evaluating the financial viability of these types of projects.

To ensure a logical and consistent link between project drivers and the key performance indicators (KPIs) of the company, a four-level KPI tree was developed and utilized. This tree provided a structured approach to map the project's impact on cost drivers and operational metrics, ensuring that changes in cash flow could be tracked systematically. The changes in cash flow brought about by the implementation of the project were used as the primary data source for calculating key economic indicators, such as Net Present Value (NPV), Internal Rate of Return (IRR), Payback Period (PBP), and Discounted Payback Period (DPBP). Sensitivity analysis was then applied to test the robustness of the project's economic efficiency under different scenarios by incrementally adjusting the influencing factors.

For projects with medium to low economic efficiency, the further application of sensitivity analysis and Monte Carlo analysis for the better understanding of potential risks is recommended. These methods allow companies to quantify project risks more effectively and develop strategies to mitigate them. In particular, Monte Carlo simulations offer a deeper insight into how uncertainties in key variables, such as fluctuating costs or varying implementation timelines, can impact the overall success of the project. This is especially relevant for projects with uncertain or long-term benefits, as it provides a probabilistic understanding of the project's financial outcomes.

It is important to note that this study marks the first time that a sensitivity analysis of KPIs has been conducted specifically for internal investment projects in the energy sector. The findings show that the magnitude of the cost-saving effect, rather than the initial investment, plays a more significant role in determining the project's overall efficiency. This insight shifts the focus of internal project evaluations from simply minimizing capital investments to maximizing operational efficiency, offering a more practical and results-oriented approach to investment decision-making within energy companies.

By understanding these dynamics and applying a more nuanced investment analysis approach, energy companies can better prioritize internal investment projects, allocate resources more efficiently, and improve decision-making processes in the context of innovation and operational enhancements.

Acknowledgements The research was supported by Russian Science Foundation grant No. 24-28-01796, https://rscf.ru/project/24-28-01796/.

References

1. Iliashenko, O., Bikkulova, Z., Dubgorn, A.: Opportunities and challenges of artificial intelligence in healthcare. E3S Web Conf. **110**, 02028 (2019)
2. Ahram, T.Z.: Advances in Artificial Intelligence, Software and Systems Engineering Joint Proceedings of the AHFE 2018 International Conference on Human Factors in Artificial Intelligence and Social Computing, Software and Systems Engineering. Loews Sapphire Falls Resort, Universal Studios, Orlando, USA (2018)
3. Bril, A., Kalinina, O., Ilin, I.: Small innovative company's valuation within venture capital financing of projects in the construction industry. MATEC Web Conf. **106**, 08010 (2017)
4. Anisiforov, A., Dubgorn, A., Lepekhin, A.: Organizational and economic changes in the development of enterprise architecture. E3S Web Conf. **110**, 02051 (2019)

5. Ilin, I.V., Levina, A.I., Dubgorn, A.S., Abran, A.: Investment models for enterprise architecture (EA) and IT architecture projects within the open innovation concept. J. Open Innov. Techn. Market, Complex. **7**(1), 69, 1–18 (2021)
6. Guerard, J., Saxena, A., Gultekin, M.: Investing in Assets: Theory of Investment Decision-Making (2021). https://doi.org/10.1007/978-3-030-43547-9_11
7. Brusco, S., Panunzi, F.: Internal financing, managerial compensation, and multiple tasks. Ann. Finance (2020). https://doi.org/10.1007/s10436-020-00368-9
8. Robinson, J.: Dual rate discounted cash flow analysis. J. Valuation **2**, 143–157 (1986). https://doi.org/10.1108/eb007990
9. Parker, D., Robinson, J.: A brief history of the Australian discounted cash flow practice standard. J. Property Investment Finan. **18**(2), 196–211 (2000). https://doi.org/10.1108/14635780010324349
10. Schill, M.J.: Discounted cash flow analysis. Darden Case No. UVA-F-172810.2139/ssrn.2974530
11. Mineeva, M., Ogloblin, S., Sofrankova, Y.: Development of functional cost analysis in machine-building enterprises. Eur. Proc. Soc. Behav. Sci. **9**(3), 145–162 (2020)
12. Fabianová, J., Janeková, J., Fedorko, G., Molnár, V.: A comprehensive methodology for investment project assessment based on Monte Carlo simulation. Appl. Sci. **13**(10), 6103 (2023)
13. Gigli, S., Landi, D., Germani, M.: Cost-benefit analysis of a circular economy project: a study on a recycling system for end-of-life tyres. J. Clean. Prod. **231**, 258–271 (2019)
14. Svetunkov, S.G., Chanysheva, A.F.: Properties of complex dispersion and standard deviation. Espacios **38**(33), 25 (2017)
15. Guan, S., Hu, W., Zhou, H.: Implementation of online digital twin framework for thermal power plant. In: Sun, Z., Das, P.K. (eds.) Proceedings of the 10th International Conference on Energy Engineering and Environmental Engineering, pp. 1–7. Springer (2024). https://doi.org/10.1007/978-3-031-48204-5_16
16. Zhao, R., Yan, D., Liu, Q., Leng, J., Zhang, X.: Digital twin-driven cyber-physical system for autonomously controlling of micro punching system. IEEE Access **7**, 9459–9469 (2019). https://doi.org/10.1109/ACCESS.2019.2891060
17. Wang, C., Xu, J., Wang, K.: SoC-based digital twin of power system simulation. In: Proceedings of 2020 5th Asia Conference on Power and Electrical Engineering (ACPEE 2020), pp. 185–191 (2020). https://doi.org/10.1109/ACPEE48638.2020.9136196
18. Morgan, D.L.: Integrating Qualitative and Quantitative Methods: A Pragmatic Approach. Sage Publications, Thousand Oaks (2013)
19. Morse, J.M.: Principles of mixed methods and multi-method research design. In: Teddlie, C., Tashakkori, A. (eds.) Handbook of Mixed Methods in Social and Behavioral Research, pp. 189–208. Sage Publications, Thousand Oaks, CA (2003)
20. Czernich, N., Falck, O., Kretschmer, T., Woessmann, L.: Broadband infrastructure and economic growth. Econ. J., 505–532 (2011)

Arctic Zone Energy Sector Ontology

Nina Trifonova, Alexandra Borremans, Anastasia Levina, and Alissa Dubgorn

Abstract The Arctic region has enormous potential for extracting mineral and energy resources, especially oil and gas. However, efficient development requires careful and balanced decisions to mitigate environmental impacts and ensure sustainable resource management. To simplify the decision-making process for stakeholders, this paper proposes a top-level ontology for the Arctic oil and gas industry. This ontology will systematize information across multiple domains including geological studies, environmental considerations and social impacts. By creating a structured knowledge base, this ontology will enable stakeholders to analyse complex interdependencies, develop effective management strategies and promote resource development in this delicate ecosystem.

Keywords Regional ontology · Arctic · Energy resources · Oil and gas sector

1 Introduction

As the world's fossil fuel reserves dwindle and there is an urgent need to find sustainable energy solutions, the Arctic's hydrocarbon wealth is attracting significant attention from researchers, companies and nations. But unlocking this potential requires a delicate balance that ensures responsible development while preserving the region's fragile ecosystem and the livelihoods of indigenous peoples.

The harsh climate, unpredictable weather conditions and permafrost create significant technical and logistical obstacles to the development of energy resources [1–5]. These challenges, combined with growing concerns about environmental impacts and

---

The original version of the chapter has been revised. A correction to this chapter can be found at https://doi.org/10.1007/978-3-031-80710-7_33

N. Trifonova · A. Borremans · A. Levina (✉) · A. Dubgorn
Peter the Great St. Petersburg Polytechnic University, Polytechnicheskaya 29, 195351 St. Petersburg, Russia
e-mail: levina_ai@spbstu.ru

I. Ilin and M. Youzhong (eds.), *Digital Systems and Information Technologies in the Energy Sector*, Lecture Notes in Networks and Systems 1244,
https://doi.org/10.1007/978-3-031-80710-7_5

the need to prioritise the rights of indigenous peoples, require a robust and adaptable governance structure.

Ontologies are powerful tools for structuring and organising knowledge domains. They provide a formal representation of concepts and their relationships, facilitating information sharing, knowledge integration and informed decision making. In the context of the Arctic oil and gas industry, ontology serves as a critical roadmap to guide sustainable development [6]. The creation of an Arctic Zone ontology will establish a clear understanding of the multifaceted aspects of the Arctic oil and gas industry, covering resource potential, environmental issues, socio-economic impacts and regulatory frameworks. The ontology will create a common language for information exchange between different stakeholders-industry players, governments, researchers, and indigenous communities-that enables effective communication and shared decision making.

The integration of disparate data sources, ranging from geological studies and environmental monitoring to socio-economic indicators and indigenous knowledge, will make it possible to support comprehensive assessments and policy development in the region.

The objective of this paper is to propose a top-level ontology for the Arctic oil and gas industry, i.e. a framework for navigating the complexities of this critical sector. This framework will serve as a step to develop more detailed and specific ontologies for specific aspects of the industry, such as resource exploration, extraction technologies, environmental monitoring and stakeholder engagement.

In order to cover this topic, three research questions have been formulated, which will form the basis of this article:

- RQ1: Which stakeholders can be identified for the energy sector in the Arctic zone?
- RQ2: What data do stakeholders need to make good management decisions?
- RQ3: What elements should the top-level ontology that will hold the data needed by stakeholders consist of?

This study consists of the following parts: method—defining the methodological framework of the research; results—reviewing the existing literature in the field of ontology creation and describing the proposed top-level ontology; discussion—interpreting the results and answering the research questions, describing further research activities; conclusion—summarizing the results of the study.

By creating a comprehensive ontology, we aim to contribute to a common understanding of the Arctic oil and gas sector, enabling stakeholders to navigate a complex web of challenges and opportunities. This knowledge infrastructure will pave the way for informed decision-making, ensuring that the vast energy potential of the Arctic is utilised in a sustainable, responsible and harmonious manner, taking into account the region's unique environment and the rights of indigenous peoples.

2 Materials and Methods

The ontology lifecycle, an optimised process, revolves around the refinement of the prototype. It progresses through individual stages, each of which corresponds to a specific action envisaged by the chosen approach. Eventually, the ontology enters a state of persistence in which knowledge is collected, evaluated and documented throughout the lifecycle [7].

Existing studies describe mostly narrow domains for the application of ontologies: the Arctic ice edge [8, 9], indigenous people [10], Arctic shipping [11], seascape Grenoble et al. [12]. However, for the energy domain it is necessary to use a different approach for ontology construction, as this domain is difficult to formalise.

In our study, we use Enterprise model approach—an approach based on ontology construction using knowledge about the ontology of an organisation [13–16]. This approach provides a well-developed and clear ontology of the Arctic region that achieves stakeholder goals and describes the energy sector in the most comprehensive way possible.

In our work, we used the following algorithm:

1. Define the scope and purpose. Clearly define the specific area that the ontology will cover and define the intended use based on stakeholder needs.
2. Identify key objects and relationships.
3. Define the classes and properties of the ontology.

Subsequently, it will be necessary to populate the instance data and refine the ontology.

Ontology evaluation criteria can be used generic, as described in [17, 18], or be defined for a particular ontology. Evaluation of an ontology for the Arctic oil and gas sector requires consideration of various aspects covering its structure, content and relevance to specific applications. We propose to use the following criteria:

1. Coverage and completeness: Does the ontology integrate the scope of relevant concepts, from geological formations and extraction methods to environmental regulations and socio-economic impacts? Does it provide sufficient detail for specific applications? For example, are different types of oil and gas reservoirs, drilling methods and environmental risks represented? (Relevant for ontology with instances) Does the ontology cover the unique challenges and opportunities presented in the Arctic environment, such as permafrost, extreme weather conditions and indigenous communities?
2. Structure and organisation: Is the ontology structured logically and coherently, with clear definitions and relationships between concepts? Does the ontology use a well-defined hierarchy to organise concepts to allow efficient navigation? Is the ontology modular, allowing specific modules to be added or removed according to different applications?
3. Accuracy and validity: Are the definitions and relationships in the ontology based on scientific knowledge and industry standards? Does the ontology accurately reflect real-world concepts and relationships in the Arctic oil and gas sector?

4. Usability and applicability: Is the ontology easily accessible and easily understood by both technical and non-technical users? Does the ontology meet the specific requirements of different applications, such as exploration and production planning, environmental impact assessment or policy development?
5. Ethical and social considerations: Does the ontology respect the rights and interests of indigenous communities? Does the ontology include conceptions and relationships relevant to minimising the environmental impact of the energy sector in the Arctic?
6. Maintenance and evolution: Is the ontology designed to accommodate future changes and updates in the oil and gas sector and the Arctic environment? Are the stakeholders in the oil and gas sector academia and relevant communities involved in the development and maintenance of the ontology?

Evaluating the ontology against these criteria will provide a comprehensive assessment of its quality, suitability and potential to support decision-making in the Arctic oil and gas sector.

3 Results

The Arctic oil and gas sector is a complex ecosystem in which multiple stakeholders compete for influence and benefit. The Arctic zone offers various potential benefits to stakeholders, but it is important to recognise the complex and sensitive nature of the region. The ecosystem is a sensitive environment with many areas that affect both the natural world and communities. Key areas include environmental impacts (climate change, biodiversity loss, pollution, habitat disturbance), socio-economic impacts (indigenous peoples, economic development, political tensions, community traditions), technological and logistical challenges (extreme conditions, limited infrastructure, safety and risk management, environmental monitoring), governance and regulation (international co-operation, national and international cooperation, local and regional co-operation, local and regional cooperation), and the environment.

The balance between economic opportunities, environmental and social considerations is key to ensuring a sustainable future for the region [19, 20] including in the energy sector.

In order to navigate this space, it is necessary to understand the different interests and priorities of stakeholders. In response to RQ1, let us identify the main stakeholders of the Arctic oil and gas sector and consider their drivers.

Arctic nations such as Russia, Canada, the United States, Norway and Denmark (Greenland) have significant territorial interests in the Arctic and view oil and gas extraction as a strategic economic opportunity, potentially affecting their national economies and energy security. International organisations should also be highlighted: bodies such as the Arctic Council, the UN and the International Maritime Organisation play a role in shaping environmental norms, promoting responsible development and ensuring international cooperation in the region.

In the area of energy resource development, international oil and gas companies such as ExxonMobil, Shell and Rosneft are actively exploring and producing oil and gas resources in the Arctic, driven by the potential for massive reserves and growing energy demand. Banks, investment funds and insurance companies are providing financial support for oil and gas projects and development. They increasingly consider environmental and social factors in their investment decisions, influencing industry practices. Also, one stakeholder is the region's indigenous peoples: the Inuit, Saami, Nenets and other indigenous groups. These communities have lived in the Arctic for thousands of years and rely on the region's natural resources for their livelihoods. They are concerned about the potential environmental impact of oil and gas development on their traditional way of life, hunting and cultural heritage. Communities often have a direct impact on the economic opportunities and social changes brought about by energy development. They are affected by job creation, increased infrastructure and economic growth, as well as potential environmental problems and social disruption.

And, of course, environmental organisations such as Greenpeace, WWF and other conservation groups. These organisations advocate for the protection of the fragile Arctic environment and ecosystems, highlighting the potential risks of oil spills, climate change and habitat destruction associated with oil and gas operations. To ensure responsible and sustainable development of Arctic oil and gas resources, it is important to understand the different perspectives and interests of these stakeholders. To this end, it is necessary to:

- Open and transparent dialogue. Facilitating communication and co-operation between all stakeholders to solve problems.
- Prioritising environmental protection. Implementing strict regulations and best practices to minimise environmental risks and protect sensitive Arctic ecosystems.
- Respect for the rights of indigenous peoples. Respecting the traditions of indigenous communities in meaningful consultation, making their voices heard and respecting their traditional knowledge and rights.
- Promoting economic diversification. Supporting the development of alternative industries and sustainable livelihoods to reduce dependence on oil and gas production.
- Making balanced system decisions. Effective management of the region's oil and gas complex is difficult without the use of a systematised information space.

Once the interests of the parties have been identified, it is necessary to describe the data that are necessary for stakeholders to fulfil their objectives.

This table serves as a response to RQ2.

For the purposes of our study, it is appropriate to group all the data listed into 4 groups:

1. Socio-economic
2. Geology, geophysics and energy
3. Infrastructure
4. Ecology

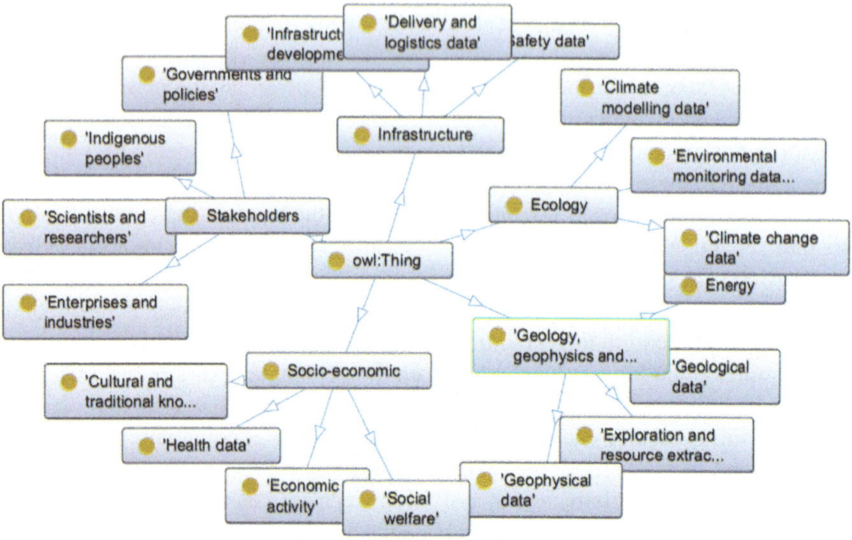

Fig. 1 Classes and subclasses of the ontology (developed by the author)

Based on these groups and stakeholder information, a top-level ontology was constructed (Fig. 1).

According to RQ3 we define that Ontology of the oil and gas sectors of the Arctic zone contains 5 classes and 18 subclasses, including:

1. Socio-economic

   a. Cultural and traditional knowledge
   b. Economic activity
   c. Health data
   d. Social welfare

2. Geology, geophysics and energy

   e. Energy
   f. Exploration and resource extraction data
   g. Geological data
   h. Geophysical data

3. Infrastructure

   i. Delivery and logistics data
   j. Infrastructure development data
   k. Safety data

4. Ecology

   l. Climate change data

   m. Climate modelling data
   n. Environmental monitoring data

5. Stakeholders

   o. Indigenous peoples
   p. Governments and policies
   q. Enterprises and industries
   r. Scientists and researchers

Each subclass corresponds to several other subclasses detailing the Arctic oil and gas sector. The subclasses of the 4 classes are described in more detail, except for the stakeholder class. For this class it was decided to conduct further research for a more detailed description.

Thus the ontology will take the following form (Fig. 2).

The ontology contains 53 Logical axioms and describes the data required by the stakeholders. The figures below show separately each of the classes of the created ontology (Fig. 3, 4, 5, 6, 7, 8).

Thus, by extending the ontology, a comprehensive ontology of the energy industry can be created that will have the necessary information for oil and gas stakeholders in the Arctic zone. The proposed ontology integrates the scope of relevant concepts, from geological formations and extraction methods to environmental regulations and socio-economic impacts. It also provides sufficient detail for specific stakeholder applications. The ontology encompasses the unique challenges and opportunities presented in the Arctic environment, such as permafrost, extreme weather conditions and indigenous communities.

The ontology is structured logically and coherently, with clear definitions and relationships between concepts, and uses a well-defined hierarchy to organise concepts for efficient navigation. The ontology is modular, allowing specific modules to be added or removed according to different stakeholder needs. The definitions and relationships in the proposed ontology are based on scientific knowledge and industry standards. The ontology accurately reflects real-world concepts and relationships in the Arctic oil and gas sector. The ontology is easily accessible and understandable to both technical and non-technical users. It respects the rights and interests of indigenous communities and includes concepts and relationships relevant to minimising the environmental impacts of the energy sector in the Arctic. This ontology, designed to accommodate future developments and updates in the Arctic oil and gas sector.

4 Discussion

Arctic's oil and gas sector is a complex system of competing interests. Balancing between two spheres—economic development and environmental protection—determines the need to contain the competition for influence in the region of different stakeholders.

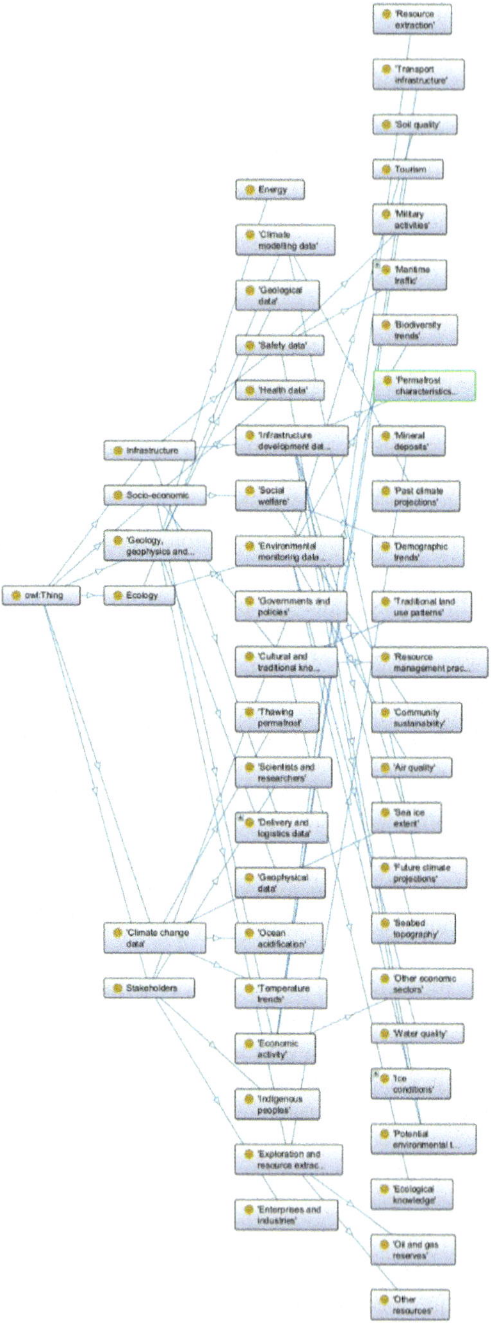

Fig. 2 Final ontology (developed by the author)

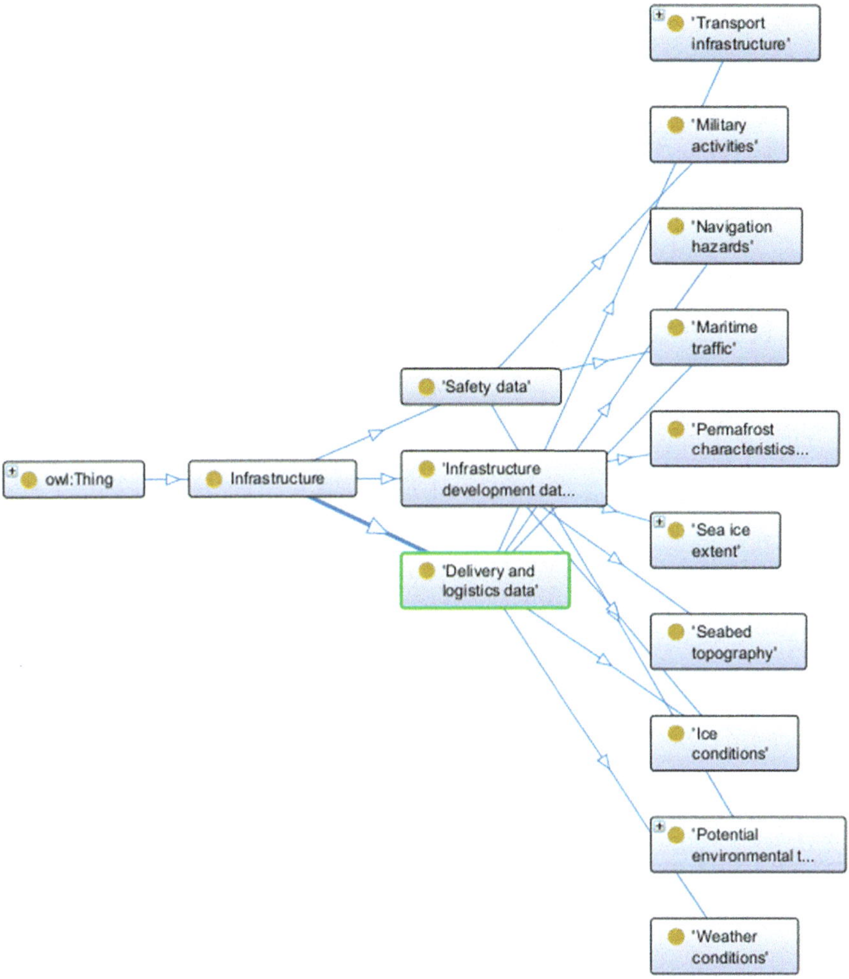

Fig. 3 Neighborhood of the class "Infrastructure" (developed by the author)

Key issues include:

1. Environmental impacts: Climate change, biodiversity loss, pollution and habitat destruction are significant challenges.
2. Socio-economic impacts: The impact on indigenous peoples, economic development, political tensions and social traditions must be carefully considered.
3. Technological and logistical challenges: Extreme conditions, limited infrastructure and security risks pose unique challenges.
4. Governance and regulation: International, national and local co-operation.

Countries such as Russia, Canada, the United States, Norway and Denmark (Greenland) see oil and gas as strategic economic opportunities potentially impacting

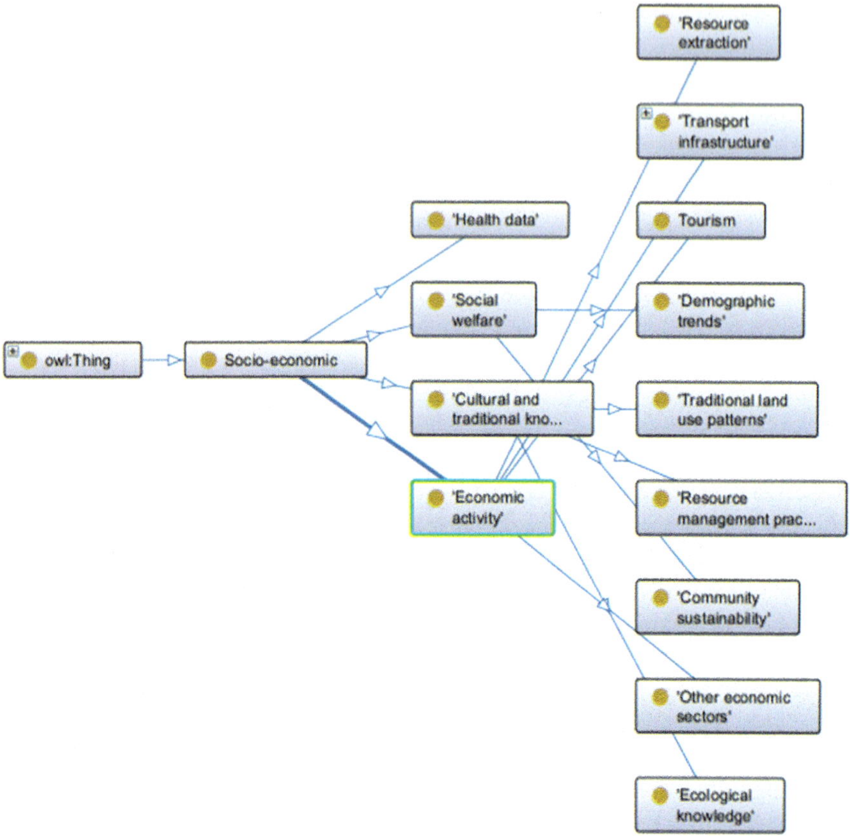

Fig. 4 Neighborhood of the class "Socio-economic" (developed by the author)

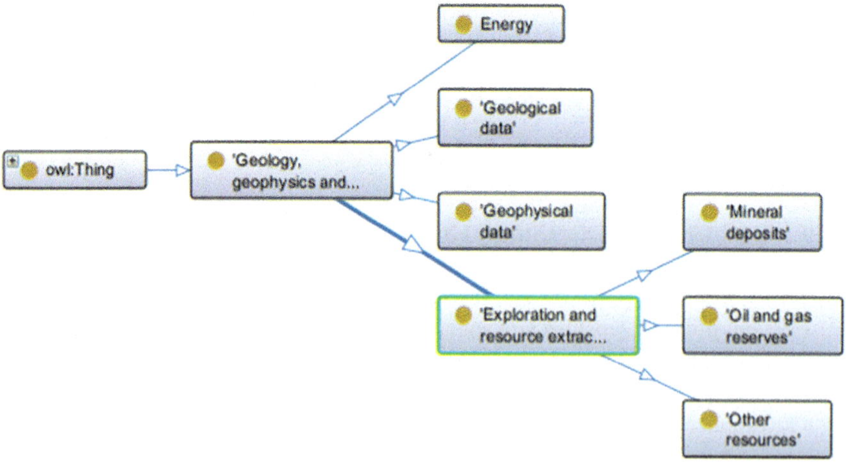

Fig. 5 Neighborhood of the class "Geology, geophysics and energy" (developed by the author)

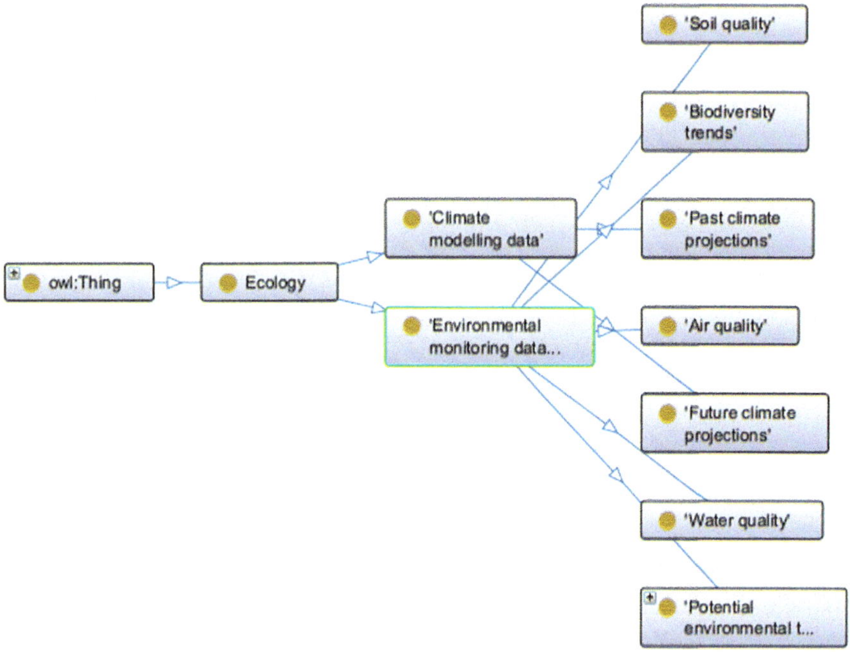

Fig. 6 Neighborhood of the class "Ecology" (developed by the author)

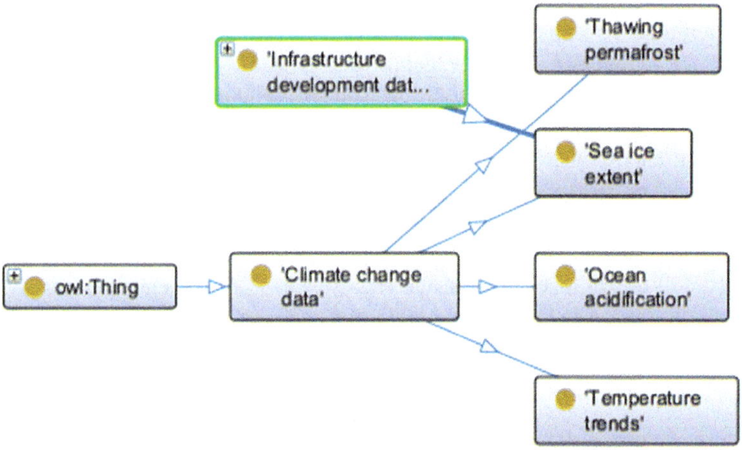

Fig. 7 Neighborhood of the class "Climate change data" (developed by the author)

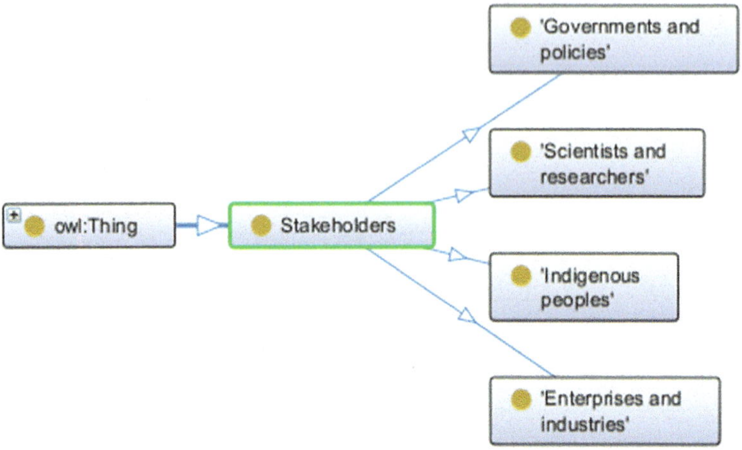

Fig. 8 Neighborhood of the class "Stakeholders" (developed by the author)

their national economies and energy security. Exxonmobil, Shell and Rosneft are drawn to the Arctic's vast reserves and growing global energy needs. In turn groups such as Greenpeace and WWF protect the fragile Arctic environment and ecosystems by highlighting the risks of oil spills, climate change and habitat destruction.

The key to a sustainable future lies in finding a balance between economic development, environmental considerations and social responsibility.

We identified 4 groups of data needed by stakeholders to achieve their goals in the region: Socio-economic, Geology, geophysics and energy, Infrastructure, Ecology.

In turn, the proposed ontology of the Arctic oil and gas sector contains 5 classes and 18 subclasses that contain the above data.

This ontology, containing 53 logical axioms, provides a comprehensive framework for the Arctic oil and gas industry. It covers a wide range of concepts including geological formations, extraction methods, environmental regulations and socio-economic impacts, allowing stakeholders to access relevant data. Expansion of this ontology will create a comprehensive representation of the energy industry in the Arctic, facilitating data sharing and informed decision making.

Future research plans to delve deeper into the ontology and look at specific cases for its use.

5 Conclusion

The development of the oil and gas sector in the Arctic is a complex task requiring a balanced approach that takes into account economic, environmental and social aspects. The development of a robust ontology of the Arctic zone is critical to advancing scientific understanding and facilitating informed decision-making.

Current research struggles with inconsistent terminology and fragmented data across disciplines, hindering interaction and comprehensive analyses. The ontology will establish a common understanding of Arctic concepts, including physical geography, ecosystems, human activities, and the effects of climate change. It provides a unified framework for stakeholders to access relevant information, facilitating more informed decision-making.

The development of this ontology contributes to a comprehensive understanding of the Arctic energy industry, facilitates data sharing, and promotes sustainable development in the region. This standardised framework will ensure efficient data integration, facilitate the discovery of automated knowledge and promote collaboration between scientific communities. In addition, it will provide a framework for developing predictive models and supporting policy interventions for sustainable development in this rapidly changing environment. Balancing economic growth, environmental protection, and social well-being, it paves the way for collaborative efforts among stakeholders to ensure responsible and sustainable development of Arctic oil and gas resources.

Table 1 Data required for stakeholders (compiled by the author)

Stackholder	Data	Description
Indigenous peoples	Cultural and traditional knowledge	Data on traditional land use patterns, resource management practices and ecological knowledge specific to their communities
	Health data	Information on Arctic-specific environmental and health risks, including exposure to pollutants and climate-related health effects
	Socio-economic data	Understanding community livelihoods, economic activities and social well-being for sustainable development
Governments and policies	Climate change data	Information on temperature trends, sea ice extent, thawing permafrost and ocean acidification to inform adaptation and mitigation strategies
	Environmental monitoring data	Air, water and soil quality data and biodiversity trends to assess environmental impacts on human activities and implementation of environmental regulations
	Economic data	Information on resource extraction, transport infrastructure, tourism and other economic sectors for sustainable development plans
	Safety data	Data on maritime traffic, military activities and potential environmental threats to national security and regional stability
Enterprises and industries	Exploration and resource extraction data	Information on mineral deposits, oil and gas reserves and other resources to inform investment decisions and optimise resource management
	Infrastructure development data	Data on ice conditions, seabed topography and permafrost characteristics to guide infrastructure projects such as ports and pipelines
	Delivery and logistics data	Information on sea ice conditions, navigational hazards and weather conditions to optimise delivery routes and ensure safe transportation
Scientists and researchers	Climate modelling data	Information on past and future climate projections to understand the impacts of climate change and develop mitigation strategies
	Environmental monitoring data	Data on air, water and soil quality, biodiversity and ecosystem processes to assess environmental change and its impacts
	Social and economic data	Information on demographic trends, community resilience and economic development to understand the social and economic impacts of environmental change
	Geological and geophysical data	Information on subsurface geology, ocean floor topography and permafrost characterisation to understand the physical environment of the Arctic

Acknowledgements The research was supported by Russian Science Foundation grant No. 23-78-10190, https://rscf.ru/project/23-78-10190/.

References

1. Dmitrieva, D., Romasheva, N.: Sustainable development of oil and gas potential of the arctic and its shelf zone: the role of innovations. J. Mar. Sci. Eng. **8**(12), 1003 (2020)
2. Katysheva, E.: Analysis of the interconnected development potential of the oil, gas and transport industries in the Russian Arctic. Energies **16**(7), 3124 (2023)
3. Stroykov, G., Vasilev, Y.N., Zhukov, O.V.: Basic principles (indicators) for assessing the technical and economic potential of developing Arctic offshore oil and gas fields. J. Mar. Sci. Eng. **9**(12), 1400 (2021)
4. Ilin, I., Vasilenok, V., Marchenko, R.: Use of renewable energy and tax burden on CO_2 emissions in industrial enterprises. In: E3S Web of Conferences, vol. 110, pp. 02102 (2019)
5. Poljanskihh, A., Levina, A., Dubgorn, A.: Investment in renewable energy: practical case in Estonia. In: MATEC Web of Conferences, vol. 193, pp. 05065 (2018)
6. Campos, J.G. et al.: INSIDE: an ontology-based data integration system applied to the oil and gas sector. In: Proceedings of the XIX Brazilian Symposium on Information Systems SBSI '23: XIX Brazilian Symposium on Information Systems, pp. 94–101. ACM, Maceió Brazil (2023)
7. Jones, D., Bench-Capon, T., Visser, P.: Methodologies for ontology development. University of Liverpool (1998). https://cgi.csc.liv.ac.uk/
8. Veland, S., Lynch, A.H.: Arctic ice edge narratives: scale, discourse and ontological security. Arct. Ice Edge. Narrat. **49**(1), 9–17 (2017)
9. Fadeev, A.M.: Technological independence and import substitution in the implementation of energy projects in the Arctic. Technoeconomics 2 **2**(5), 66–75 (2023). https://doi.org/10.57809/2023.2.2.5.6
10. Castro, D., Hossain, K., Tytelman, C.: Arctic ontologies: reframing the relationship between humans and rangifer. Polar geography 39. Arct. Ontol. **2**, 98–112 (2016)
11. Vicentiy, A.V.: Digitalization of Arctic shipping along the northern sea route. In: IOP Conference Series: Earth and Environmental Science, vol. 816(1), pp. 012023 (2021)
12. Grenoble, L.A., McMahan, H., Petrussen, A.K.: An ontology of landscape and seascape in Greenland: the linguistic encoding of land in Kalaallisut. Int. J. Am. Linguist. **85**(1), 1–43 (2019)
13. Antunes, G. et al.: Ontology-based enterprise architecture model analysis. In: Proceedings of the 29th Annual ACM Symposium on Applied Computing : SAC '14, pp. 1420–1422. Association for Computing Machinery, New York, NY, USA (2014)
14. Dietz, J.L.G., Hoogervorst, J.A.P.: Enterprise ontology in enterprise engineering. In: Proceedings of the 2008 ACM Symposium on Applied Computing : SAC '08, pp. 572–579. Association for Computing Machinery, New York, NY, USA (2008)
15. Sunkle, S., Kulkarni, V., Roychoudhury, S.: Analyzing enterprise models using enterprise architecture-based ontology. In: Hutchison, D., Moreira, A. et al. (eds.) Model-Driven Engineering Languages and Systems: Lecture Notes in Computer Science, Coll, vol. 8107, pp. 622–638. Springer Berlin Heidelberg, Berlin, Heidelberg (2013)
16. Gomez-Perez, A.: Some ideas and examples to evaluate ontologies. In: Proceedings the 11th Conference on Artificial Intelligence for Applications, pp. 299–305. IEEE Comput. Soc. Press, Los Angeles, CA, USA (1995)
17. Maydanova, S.A., Ilin, I.V.: Digital transformation strategy development in an enterprise architecture context. Technoeconomics 2. **1**(4), 64–75 (2023). https://doi.org/10.57809/2023.2.1.4.6

18. Gómez-Pérez, A., Fernández, M., de Vicente, A.: Towards a method to conceptualize domain ontologies. In: Proceedings of the Ontological Engineering 12th European Conference on Artificial Intelligence (ECAI'96), pp. 41–51. Budapest, Hungary (1996)
19. Borremans, A. et al.: Arctic sustainable development: digital and logistics infrastructure in the region. In: Devezas, T.C. et al. (eds.) Understanding the Digital Transformation of Socio-Economic-Technological Systems: Lecture Notes in Networks and Systems, vol. 951, pp. 3–15. Springer Nature Switzerland, Cham (2024)
20. Levina, A. et al.: Effective management of multimodal logistics in Russia. In: Ilin, I., Jahn, C., Tick, A. (eds.) Digital Technologies in Logistics and Infrastructure: Lecture Notes on Data Engineering and Communications Technologies, vol.157, pp. 264–272. Springer International Publishing, Cham (2023)

Optimization of Functionality of Energy Management System in Network Trading Companies

Olga Voronova, Viktoria Sheleyko, and Vladimir Vasilyev

Abstract This research suggests a functional model of the energy management subsystem for the FMCG retail outlet. The topic proves to be highly relevant due to the digitalization of retail, development of ESG- and eco-friendly agenda in enterprise management. Throughout the research, the authors define the role of energy monitoring in energy management and preservation in retail. The main directions of energy monitoring in retail outlets have been identified and characterized. The most widespread monitoring tools in the modern retail market are also distinguished and specified. As a result, the authors define major disadvantages of their application and develop the range of promising solutions. Results of the research can be used in further development of IT subsystem for energy managemnet at retail enterprises. The authors also elaborate on the major specifics of energy monitoring at network FMCG companies, the structure of energy costs, and the typification of modern IT technologies associated with energy consumption.

Keywords FMCG · Retail outlet · Energy management · ESG agenda

1 Introduction

Today, retail is rapidly scaling up with the growing number of retail outlets and space per capita. At the same time, the volume of associated costs increases together with the general awareness over the ESG agenda [1]. In this environment, FMCG retail starts to be driven by the urge to reduce costs associated with the purchase and consumption of energy resources, or utility payments, to be precise. Efficient energy

O. Voronova (✉) · V. Sheleyko
Peter the Great Saint-Petersburg Polytechnic University, Saint-Petersburg 195251, Russia
e-mail: iliina_ov@spbstu.ru

V. Vasilyev
OOO Neva Exploitation, Saint-Petersburg 194354, Russia

I. Ilin and M. Youzhong (eds.), *Digital Systems and Information Technologies in the Energy Sector*, Lecture Notes in Networks and Systems 1244,
https://doi.org/10.1007/978-3-031-80710-7_6

use and wise application of energy-saving technologies is turning into a top priority that requires optimization of energy management tools.

The PDCA cycle shapes the foundation for the international standards in energy and environmental management (ISO 50001 and ISO 140001). According to these, the energy management process ("plan") should rely on data on the current use of energy resources, while in order to ensure continuous improvement ("act") it is necessary to provide grounds for consistent monitoring of their dynamics [2, 3].

2 Materials and Methods

This research employs such methods of theoretical research as analysis, synthesis, classification, deduction, and induction, which contribute to determining theoretical and practical techniques in energy consumption and monitoring in FMCG retail. What is more, a whole-scale assessment based on these methods shaped the authors' perspective on energy optimization.

The study is carried out on the grounds of existing research on the topic and the author's conclusions on energy-saving management perspectives. The theoretical basis for the research is shaped by the international standards on energy and environmental management, as well as studies on digitalization and automation. Another important facet under consideration is the development of green and ESG policies in retail.

Application of practical (empirical) research methods, in particular, description and modelling, allowed the authors to suggest a range of suggestions aimed at optimization of energy consumption management in network retail companies.

3 Results and Discussion

Traditionally, FMCG retail features the following areas of energy consumption and monitoring:

1. Power supply:
   a. Refrigeration equipment
   b. Lighting
   c. Air-conditioning and ventilation
   d. Electric heating and air curtains
2. Water supply
3. Central heatingSubsequent paragraphs, however, are indented.

Table 1 below presents a more comprehensive structure of energy consumption monitoring.

Table 1 Energy monitoring in retail outlets, FMCG-retail (designed by the authors)

Monitoring area feature	Power supply				Central heating	Water supply
	Refrigeration equipment	Lighting	Air-conditioning	Electric heating		
HCC facilities to be managed and monitored	Cooling chests; Refrigerated displays; Refrigeration units (including built-in lighting and electric heating devices)	Lighting units	Industrial air conditioners with a shared cooling machine (in Russia, including built-in heating module)	Autonomous heating from electric boilers	Central heating equipment, including pipe line, boiler facility (if present)	Sanitary equipment; Pipe line
Facility management tools	Individual temperature controllers; Individual lighting controllers (relays)	Individual or group switches (including remote control) installed in the electrical network; Built-in (including remote control) switches		Individual temperature controllers; Individual lighting controllers (relays)	Individual temperature controllers	Water pressure regulator
Built-in monitoring tools	Individual temperature sensors; Opening sensors (relay); Alarms	–		Electric power sensors; Individual temperature sensors; Alarms	Universal heat meters; Water meters; Alarms	Water meters; Alarms

(continued)

Table 1 (continued)

Monitoring area feature	Power supply				Central heating	Water supply
	Refrigeration equipment	Lighting	Air-conditioning	Electric heating		
Additional monitoring tools	Network cards; Point energy utilization smart sensors; Multi-channel energy meters; Universal energy meters				Pressure sensor; Leakage sensor; Temperature sensors	
Monitoring indicators	Operational mode; Operation failures	Operational mode; Operation failures	Maintained temperature; Operational mode; Operation failures		Maintained temperature; Operational mode; Operation failures	Operation failures (i.e. leakage)

(continued)

Table 1 (continued)

Monitoring area feature	Power supply				Central heating	Water supply
	Refrigeration equipment	Lighting	Air-conditioning	Electric heating		
Results of monitoring	Data on operational mode; Data from binary input sensors	Data on energy consumption of equipment, including intensity and operation failures	Data on energy consumption of equipment, including intensity and operation failures; Data on maintained temperature			Data on state of the water system
Monitoring tasks in energy management	Notifications on operation failures; Automation of technical maintenance of refregiration equipment; Control of operational mode; Control of staff's complience with rules on expoitation and maintenance; Control of energy use volume	Provision of zone control of lighting; Control of operational mode	Provision of automated control of heat mode; Provision of zone ligting control at retail areas in accordance with the planogram; Control of energy use volume; Provision of service for forecating the outside temperature; Provision of forecatinf service; Provision of service for heat loss analytics			Ensuring protection against leakage; Control of energy use volumes; Provision of service for forecasting water temperature

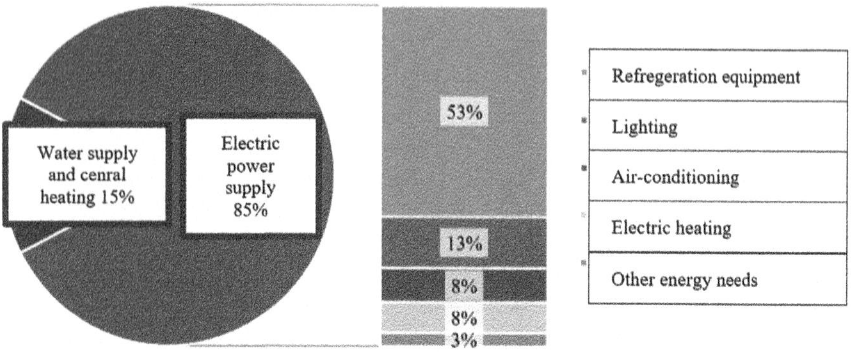

Fig. 1 Traditional structure of energy costs in FMCG-retail, the Russian Federation (designed by the authors)

According to the table, monitoring of energy consumption in retail outlets boils down to a set of control tools to address various types of equipment that make part of an infrastructure complex that consumes a certain amount of energy costs.

At the same time, tracking the cost dynamics of a retailer plays one of the major roles in the management of economic resources [4, 5]. In this regard, a number of energy management functions represent a matter of strategic importance for financial management.

Based on business practice, the data on the structure of energy costs in FMCG retail are summarized in Fig. 1.

Based on the data presented above, electricity supply accounts for the majority (85%) of all energy costs. At the same time, costs for refrigeration equipment amount to more than half (53%) of the overall cost structure, lighting equipment—13%, air conditioning and ventilation, heating equipment, and air curtains—8%. Such distribution explains why energy management traditionally prioritizes the monitoring of refrigeration equipment performance since it is expected to provide a broader picture of energy optimization.

However, it would be misleading to neglect the high level of costs associated with failure risks, direct costs of energy supply, regardless of a smaller share they account for.

For instance, water supply costs make only 15% of the total financial costs. Neverthelss, a retail outlet may undergo serious financial loss in case of watter supply failure, such as leaks. Similarly, in case of partial or complete failure of lighting, ventilation units, heating, and/or other infrastructural equipment, a retail outlet is likely to lose income because of reduced customer traffic. Or worse, a company may be suspended due to the violation of safety rules at the retail facility.

Apparently, energy monitoring is an important supporting business process that tracks all company processes associated with energy consumption. Using its output data, it is possible to achieve a significant reduction in energy costs as well as improve energy efficiency in all operations of production.

Ongoing digitalization of retail [6] requires a whole-scale range of IT solutions in monitoring to collect reliable and accurate data on the performance of each and every retail outlet as part of a single information system. It is the only way to encourage efficient decision-making in energy management [7–12].

The leading solution in retail would be a digital twin as a major data source [13]. Figure 2 depicts the essence of the digital twin.

The data that maintains and updates the digital twin is collected by various sensors and cameras located in the retail outlet (including the ones presented in Table 1), as well as direct databases on the operational activities.

In this regard, a generalized complex of tools for monitoring and management at the retail outlet rests on 3 types of modern innovative technologies (Fig. 3).

The most common monitoring tools in the complex of infrastructural support of the digital twin include sensors, CCTV cameras, and digital price tags.

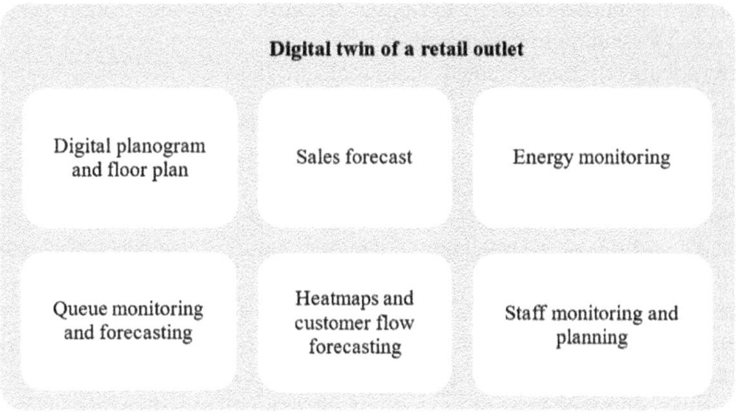

Fig. 2 Digital twin of a retail outlet (designed by the authors based on [14])

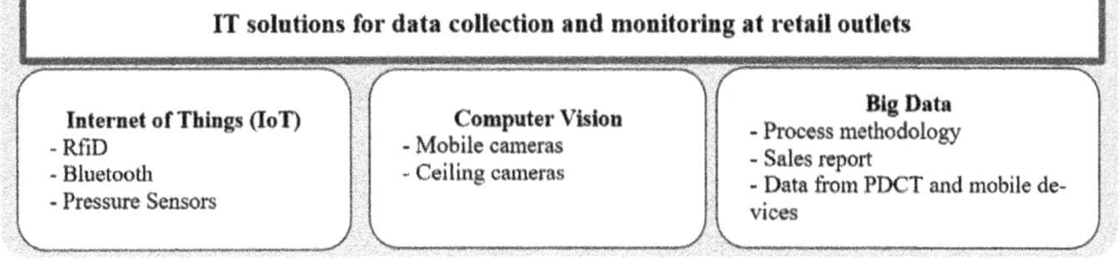

Fig. 3 Typification of modern IT technologies in development and maintenance of a digital twin of a retail outlet

4 Results and Discussion

The most common tools for monitoring the activities of retail outlets in chain trading companies are elaborated on below.

Sensors. Any piece of electrical equipment features either of the following sensor types:

- Integrated individual sensors and signalling systems;
- Additional network cards that ensure collection and transmission of data from internal sensors to a single server and allow remote control.
- Current transformer-based sensors that record the data on power consumption of an individual facility, retail space, and/or electric service panel and enable transfering data to a single server ("smart sensors").

Since digitalization makes integral part of tracking and control over engineering equipment, the introduction of IoT-based network cards and "smart sensors" is becoming more and more commonplace.

When adequately implemented, these tools provide:

1. Regular monitoring of energy consumption indicators at a retail outlet, including those facilities and appliances that are not equipped with integrated sensors (e.g., air conditioners and lighting units).
2. Visualization of energy data in accordance with the hierarchy of equipment, specific sections of the retail outlet, and/or electric service panels that make part of the engineering complex.
3. Operation mode control, including automation in accordance with the set parameters (models), as well as the zone control. For instance, zone control of air conditioners over retail stalls with vegetables can be performed via the set planogram; and zone control of lighting devices via the manually set data on working hours and customer flow at a particular retail outlet.

These and other functions of modern sensors that trace and control energy equipment are aimed at improving energy efficiency and optimizing maintenance costs.

However efficient these solutions may be, their mass implementation requires careful and wise approach to each and every facility of the retail space, including the installation of wireless posts and/or modems. These measures naturally call for additional funds and labour costs. In large FMCG retailers, this issue is especially relevant due to the acute need to track performance, determine peak loads, and the intensity of energy consumption in conditions of dynamic customer traffic.

Cameras. At present, photo and video cameras make one of the necessary prerequisites for managing retail outlets. Depending on the purpose and planning solutions, cameras can have different locations (ceiling cameras, wall cameras, and mobile cameras). For example, in day-to-day operations, staff may use mobile device cameras in order to manually monitor the external condition of equipment, planograms, and handle complaints, etc. While security management is more likely

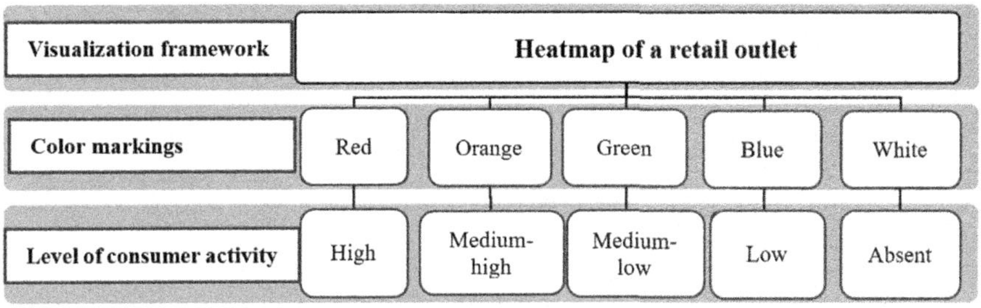

Fig. 4 Heatmap classification of customer flow at the retail outlet (designed by the authors)

to rely on ceiling cameras to trace activity at a retail outlet, marketing can rely on those too, in case they are equipped with computer vision technology and special software to generate heatmaps.

Heatmaps is a tool for visualizing the intensity of customer flows at a retail outlet for a certain period of time. The data is collected via computer vision technology and special software. As a result, videos from cameras are processed by algorithms in order to identify customers and track their movement in accordance with the zone technology presented in Fig. 4.

However, relying on visual information only does not reflect the comprehensive picture of business functions related to monitoring the customer flow. Therefore, the heatmap complex also collects and aggregates the following statistical information:

- Number of visitors during the day within the set counting intervals;
- Average number of visitors per hour;
- Peak load hours.

When applied as a complex, these data allow analyzing the following indicators of maintenance and merchandising at the retail outlet:

- Demand for goods;
- Efficiency of promotion and sales campaigns;
- Performance of specific products or product ranges;
- Ergonomics of the planogram;
- Main customer routes.

However, the attempts to implement heatmaps alone are likely to fail, since an entire combination of factors should also be considered. It is crucial to integrate services for monitoring the intensity of customer flows with the data on premises and retail outlets, together with the register of sales equipment.

Electronic price tags. Physically, electronic (digital) price tags are e-ink screens located on shelves in a special holder equipped with a battery compartment and a control module (individual or shared). Being a product of IoT technologies, the radio frequency identification, or RFID, ensures automatic price updates for products presented at a particular retail outlet via synchronizing data from the server, access points, and price tags in the retail space.

The major benefits of electronic price tags for FMCG retail include:

- Automated price update based on data from the recent price register;
- Development of an up-to-date digital planogram of a retail outlet based on joint consideration of a floor plan, a register of retail equipment, and data on the physical location of price tags in the retail space.

The digital planogram based on the data on electronic price tags makes one of the key elements of the digital twin [15–17] that ensures managing and monitoring the display of goods in real time. What is more, this tool is characterized by outstanding strategic potential in merchandising, thus defining the effciency of offline sales management.

FMCG retail outlets gain a number of operational impacts produced by electronic price tags. According to the authors' calculations, they include:

- Reduction of labour costs by 15%;
- Reduction of the error frequency in product display and price tags by 90%;
- Reduction in write-offs due to PPP by 8%.
- Reduction of food waste by 15%.

Thereby, IT solutions related to monitoring indicators of energy consumption via sensors, electronic price tags, cameras, etc., can be used to the maximum of their potential only within an integrated approach. In this scenario, positive effects go far beyond operational control of equipment efficiency and provision of real-time information on downtime and energy load [18, 19]. The bigger impact is that they provide grounds for informed management decisions based on planograms and heatmaps via illustrating the movement of customer flows.Subsequent paragraphs, however, are indented.

5 Conclusion

In order to shape a comprehensive solution in energy management, it is necessary to structure the main elements of energy management optimization; identify connections between business-related, applied, and infrastructure services; and develop a functional model of the energy management system.

According to Table 2, optimization of processes incolved in energy management at retail outlets depends on the interconnection of a whole range of business, applied, and infrastructural services. A more comprehensive overview of these services in FMCG retail is presented by the authors in [20].

Connections between business, applied and infrastructure services are depicted below (Figs. 5 and 6).

The model presented in this research can be applied in shaping the information subsystem for energy managemnet at a retail outlet in order to ensure further development of a single digital platform in FMCG-retail as a whole.

Table 2 Features of energy management optimization in retail outlets (designed by the authors)

Optimization element feature	Digital planogram	Heatmap	Indicators of energy consumption monitoring	Remote control module
Implementation tool	Electronic price tags	CCTV cameras	Sensors, network cards	Network cards
Technology behind IT-solution	IoT	Computer vision	IoT	IoT
Essence of technology	Identification of objects (price tags, planograms) using radio signals	Automated identification of people and their movement in the recording	Automated collection and transmission of energy data between physical devices (current sensors) in a wireless mode	Control of equipment operational mode via signal transmission from the central server
Purpose of technology implementation	Price management; Planogram management; Assessment of customer traffic and interaction with products; Evaluation of merchandising efficiency; Placement of targeted ads	Assessment of customer traffic at a retail outlet; Tracking customer demand for goods; Detecting flaws in design and/or organization of merchandise in the store; Detecting key customer routes	Monitoring of enegy use by equipment; Automated data collection on operational mode of equipment	Automated control of engineering equipment at a retail outlet
Data generated as a result of implementation	Price register; Register of trade equipment (shelves, cameras, displays, etc.)	Data on intensity of customer traffic	Data on energy consumption in different retail areas, sections and/or electrical panels	–
Implementation outcome	Difital planogram of a retail outlet	Data on intensity of customer flow	Data on energy consumption in different retail areas, sections and/or electrical panels	Availability of remote control of equipment at a retail outlet, including zone control
Business-process	Management of energy consumption at a retail outlet			

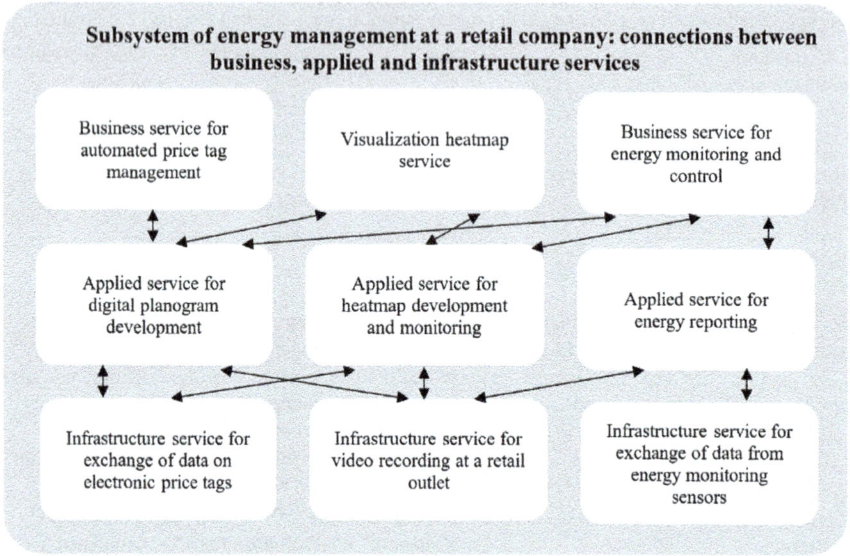

Fig. 5 Subsystem of energy management at a retail company: connections between business, applied and infrastructure services

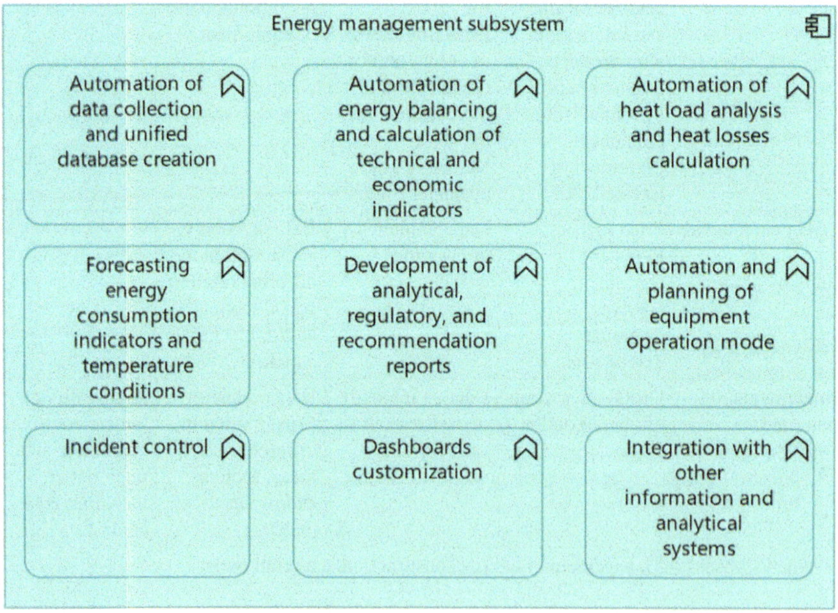

Fig. 6 Model of energy management subsystem functions of FMCG retail outlet (developed by the authors)

References

1. Hasan, M.: A new smart approach of an efficient energy consumption management by using a machine-learning technique. Indones. J. Electr. Eng. Comput. Sci. **25**(1), 68–78 (2022)
2. Sizova, E., Zhutaeva, E., Volokitina, O., Eremin, V.: Management of innovations in the field of energy-efficient technologies. Adv. Intell. Syst. Comput. **1258**, 521–531 (2021)
3. Strielkowski, W., Firsova, I., Lukashenko, I.: Effective management of energy consumption during the COVID-19 pandemic: the role of ICT solutions. Energies **14**(4), 47–59 (2021)
4. Voronova, O., Khareva, V.: Network retail FMCG-segment in the Russian federation: current state and problems of development. Int. Sci. J. **2**, 7–16 (2019)
5. Niyazbekova, S., Barykin, S., Sergeev, S.: Sustainable energy efficient human-centered digital solutions for ESG megacities development. Front. Energy Res. **10**, 87–96 (2022)
6. Madsen J.J.: Start saving energy: no reservations about energy management. Buildings **97**(3), 32 (2003)
7. Bakharev, V., Kapustina, I., Mityashin, G.: Green retailing: an analysis of strategies. Sib. J. Life Sci. Agric. **12**(5), 79–96 (2020)
8. Korablev, V.V., Ilin, I.V., Levina, A.I., Shemyakina, A.A.: Economic assessment of the life cycle of hydrogen production at hydroapplications. Int. J. Hydrogen Energy **68**, 696–704 (2024)
9. Ilin, I., Vasilenok, V., Marchenko, R.: Use of renewable energy and tax burden on CO_2 emissions in industrial enterprises. In: E3S Web of Conferences, vol. 110, pp. 02102 (2019)
10. Ilin, I., Levina, A., Iliashenko, O.: Enterprise architecture analysis for energy efficiency of Saint-Petersburg underground. Adv. Intell. Syst. Comput. **692**, 1214–1223 (2018)
11. Ilin, I., Laskin, M., Logacheva, I., Sarygulov, A., Tick, A.: Land plots evaluation for agriculture and green energy projects: how to overcome the conflict using mathematics. Mathematics **10**, 4376 (2022). https://doi.org/10.3390/math10224376
12. Politaeva, N.A., Illin, I.V., Oparina, A.M., Donetskova, A.S.: New energy approaches to the use of waste biosorbents of microalgae Chlorella kessleri. Povolzhskii Ekol. Zhurnal **2022**(3), 322–335 (2022)
13. Karakeçe, E.: The core of business: is it energy management or management energy? Contrib. Manag. Sci. **2**, 243–255 (2021)
14. Osheyor, G.J., Leonard, J.: Sustainability in energy maintenance: a global review of policies and technologies for sustainable energy infrastructure management. Econ. Growth Environ. Sustain. **2**(2), 117–121 (2023)
15. Krymov, S., Kapustina, I.: Review of modern tools and methods of merchandising of a trade enterprise. Practical marketing **12**(1), 76–83 (2016)
16. Liu, L.N.: Distributed optimal energy management for integrated energy systems. Trans. Ind. Inform. **18**(10), 6569–6580 (2022)
17. GOST R ISO 50001-2023. https://docs.cntd.ru/document/1200195836. Accessed 23 Aug 2024
18. GOST R ISO 14001-2016. https://docs.cntd.ru/document/1200134681. Accessed 23 Aug 2024
19. Ruiz, D.: Predictive energy management in internet of things: optimization of smart buildings for energy efficiency. J. Intell. Syst. Internet Things **10**(2), 8–17 (2023)
20. Voronova, O., Vasiliev, V.: Automation of retail outlets of network trading companies based on the development of an architectural model of IT services for offline sales management. Innov. Inf. Technol. Cond. Digit. Econ. 400–402 (2024)

Sovereign Bonds and Oil Shocks: An Empirical Analysis of Arctic Countries

Gulnara Romashkina, Yuliya Yukhtanova, Darya Agafonova, Djamilia Skripnuk, and Alexander Rybas

Abstract The relevance of this study is driven by the increasing uncertainty in the global economy. This article examines the dynamics of the level and volatility of sovereign bond yields in five Arctic countries during the period 2021–2024. The study is based on a dataset of average yields of 10-year sovereign bonds from Arctic countries, Brent oil prices, and yield spreads. Quantile regression models and GARCH were calculated. The scientific contribution of the article lies in the analysis of new data, which includes price crises, the period of sectoral sanctions against Russia. The results demonstrated that in the long term, the sovereign bond markets of Arctic countries are interconnected, depend on "past news," follow the U.S. bond market, and are influenced by oil price shocks. Oil price shocks disrupt the interconnectedness of sovereign bond markets. Based on the behavior of sovereign bond yields in response to oil price shocks, two types of countries can be distinguished—oil producers and oil consumers. Price volatility in the oil market tends to reduce bond yield volatility in Iceland and Sweden but increases it for the United States, Russia, and Norway. In the short term, oil prices are positively correlated with the volatility of Norwegian sovereign bond yields. Russian sovereign bonds experienced significant shocks due to the imposition of sanctions in the spring of 2022, followed by a return to the average trend. Russian sovereign bonds are strongly correlated with oil prices only during non-crisis periods.

Keywords Sovereign bonds · Crisis · Arctic countries · Oil prices · Influence · Shock · Volatility

G. Romashkina · Y. Yukhtanova · D. Agafonova
Tyumen State University, 6 Volodarskogo Street, 625003 Tyumen, Russia

D. Skripnuk (✉) · A. Rybas
Peter the Great St. Petersburg Polytechnic University, Polytechnicheskaya, 29, 195251 St. Petersburg, Russia
e-mail: skripnyuk.d@spbstu.ru

I. Ilin and M. Youzhong (eds.), *Digital Systems and Information Technologies in the Energy Sector*, Lecture Notes in Networks and Systems 1244,
https://doi.org/10.1007/978-3-031-80710-7_7

1 Introduction

The global economy is facing increasing uncertainty regarding access to financing opportunities. The rise in volatility is driven not only by financial factors but also by macroeconomic, political, behavioral, and other significant events [1, 2]. The oil market is one of the key factors significantly influencing global economic dynamics, including the yield of sovereign bonds and sovereign credit risks. Oil price volatility leads to changes in the macroeconomic indicators of countries dependent on oil exports and imports. Considering that these macroeconomic trends serve as predictors of bond yields, we aim to establish the role of oil shocks in determining the yields of sovereign bonds in Arctic countries. The dynamics of sovereign bonds from Iceland, Norway, Russia, Sweden, and the United States were analyzed using daily data for the period 2021–2024, covering both stable and turbulent times. The article investigates how the reactions of sovereign bonds from Arctic countries differ in response to oil price volatility.

The research hypothesis of our article is that the dynamics of the level and volatility of sovereign bond yields in Iceland, Russia, Norway, Sweden, and the United States are interrelated and respond to oil price shocks.

The scientific contribution of this article lies in the analysis of new data, which includes the period of the imposition of sectoral sanctions against Russia, the application of a novel approach, and the testing of original assumptions. Our approach is enhanced by the use of quantile regression to decompose periods into "crisis" and "non-crisis" phases, as well as the decomposition of dynamic volatility through a GARCH model. Contemporary research has confirmed the superiority of GARCH models over VAR models in cases of asynchronous and volatile financial markets [3, 4].

2 Literature Review and Research

The scientific literature actively discusses the issues concerning the relationship between the debt securities market and macroeconomic factors, such as changes in GDP, oil prices on global commodity markets, inflation, and the key interest rate. The example of forecasting GDP dynamics using bond yield curve spreads can be considered a classic [1]. It highlights two key aspects that have become the foundation for numerous subsequent studies of the bond market: the use of VAR models, variation in term spreads and curvature in the exogenous vector, and the use of lagged values of endogenous vectors. Later studies included the transmission of risks through financial markets [5]. The effectiveness of level, volatility, and curvature variables in predicting bond yields based on TP-VAR has been demonstrated [6, 7]. Thus, sovereign bond yield curves provide insights into broader economic processes, including GDP trends, inflation, external debt history, and overall economic development [8–11].

The study of the yield dynamics of 10-year government bonds across 40 major global economies enabled an assessment of effects such as changes in global debt volume, the debt situation of European economies during the COVID-19 period, the impact of interest rate inversion between China and the United States, and the degree of correlation among the world's largest economies amid economic globalization [12].

In later works, uncertainty in sovereign bond markets is modeled using VAR decomposition and the formation of dynamically changing indices [13]. One of the influential economic processes is the dynamics of the oil market [14]. For instance, a decline in oil prices reduces interest rates and inflation but increases the yield of the U.S. bond index, which in the medium term may lead to a recession [15]. Li et al. demonstrated how oil price uncertainty shocks statistically reduced economic activity levels in China [13].

Oil uncertainty predicts U.S. bond yields and volatility [16]. According to Nazlioglu et al., oil prices tend to predict bond prices in Canada, Mexico, Norway, Russia, Venezuela, China, and India [17]. The impact of oil price shocks can be studied, for example, using variance decomposition [18]. Research on the relationship between oil price shocks and the stock market during 1995–2013 indicates differences in impact across different periods [7].

High oil prices may strain government budgets through increased sovereign bond yields [19]. The impact on sovereign bonds has proven ambiguous (in terms of both magnitude and direction). Oil demand shocks positively affect stock market returns and increase the interconnectedness between stock and bond markets [20].

Statistical studies show that oil demand and risk shocks, along with components of the U.S. yield curve, are the main transmitters of shocks, while Japan, South Korea, and Brazil are the primary receivers of secondary shock effects [21]. Analysis of the dynamic effects of external economic shocks on African countries' bonds highlights the importance of factors such as crude oil and copper exports and wheat imports in Zambia, Nigeria, and Egypt [22].

The yield dynamics of 10-year government bonds across 40 major global economies provide information on such effects as changes in global debt volume, the debt situation of European economies during COVID-19, the impact of interest rate inversion between China and the U.S., and the degree of correlation among the world's largest economies in the context of economic globalization [12].

For countries like Saudi Arabia, Russia, and the United Arab Emirates, it has been shown that higher oil prices due to supply shocks can exacerbate inflationary pressures and widen current account deficits, increasing the risk of sovereign default. Demand shocks can disrupt revenue flows and hinder economic growth [23].

Bond yields, according to Maghyereh's research, demonstrate a short-term reaction to oil price fluctuations in developed countries. Furthermore, oil demand shocks exhibit cyclical effects. Monitoring the impact of oil shocks on bond yields across countries is essential to mitigate the potential consequences of oil price volatility and ensure financial stability in the markets [24].

In the Arctic countries, there are common institutional features of behavior in bond markets [25]. Based on a literature review, we hypothesized that the sovereign bond markets of Arctic countries are interconnected, and oil shocks affect bond yield volatility differently in oil-producing and oil-consuming countries during periods of relative stability and turbulence.

3 Data and Methodology

The paper analyzes the yields of 10-year government bonds from five Arctic countries.

Notation:

Y(n,t) Total return indices of 10-year sovereign bonds (YSB);

BR(t), the selling price of Brent crude oil in USD;

where n represents the country and t represents the current date. All prices used in the index calculations are as of the current date. The data source is the Cbonds information agency. Observation period: 28.08.2019–01.03.2024. If no date is indicated in the notations, general or partial averages are provided.

Data are standardized using the formula:

$$Z(x) = (x - m(x))/\varepsilon, \quad (1)$$

where m is the mathematical expectation, and ϵ is the standard deviation of x.

The selection of significant lags was performed based on the likelihood ratio (with a Chi-square significance level not exceeding 0.001), Akaike, and Schwarz criteria. The residuals in the model were standardized, and an iterative algorithm with quantile estimation and 90% confidence intervals was applied.

Panel data (1) represent the dynamics of sovereign bond yields for the U.S. (Y0), Iceland (Y1), Norway (Y2), Russia (Y3), Sweden (Y4), and Brent oil price (Y5). Standardized bond yields are denoted as ZY0, ZY1, ZY2, ZY3, ZY4, and the standardized oil price as ZY5 (Fig. 1).

Panel data (2) reflect the dynamics of spreads relative to the standardized U.S. sovereign bond yields for Iceland (SY1 = Z1–Z0), Norway (SY2 = Z2–Z0), Russia (SY3 = Z3–Z0), Sweden (SY4 = Z4–Z0), and oil price differences (DB) (Fig. 2).

The division of time periods into "non-crisis" and "crisis" phases was based on the upper quartile of oil price volatility, identifying 1035 and 91 observations, respectively. This allows us to examine how oil price shocks affect U.S. sovereign bond yields, which are considered risk-free, and their relationship with oil prices.

To apply OLS regressions, heteroscedasticity, and violations of the normality of variable distributions serve as limitations. Therefore, to assess the impact of oil price shocks on YSB, we employed quantile regression, which estimates the conditional median of the response variable.

Risks, measured through the volatility dynamics (variance) of the response variables, were assessed using the GARCH model (Generalized Autoregressive Conditional Heteroscedastic model). GARCH models volatility as the sum of a constant (A) base volatility (variance) and a linear function of the absolute values of several

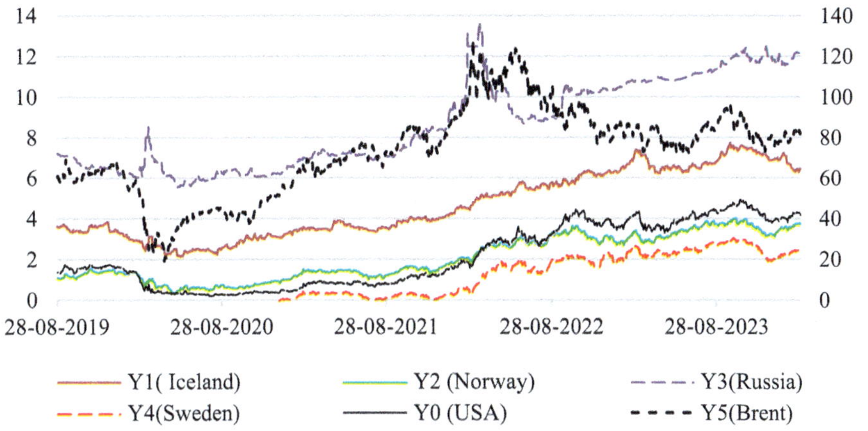

Fig. 1 Dynamics of sovereign bond yields for the USA, Iceland, Norway, Russia, Sweden (left axis), and Brent oil price (right axis), 28.08.2019–01.03.2024. *Source* Compiled by the authors based on Cbonds data (https://cbonds.ru/indexes/Cbonds-GBI-RU-YTM-eff/, https://cbonds.ru/indexes/9237/)

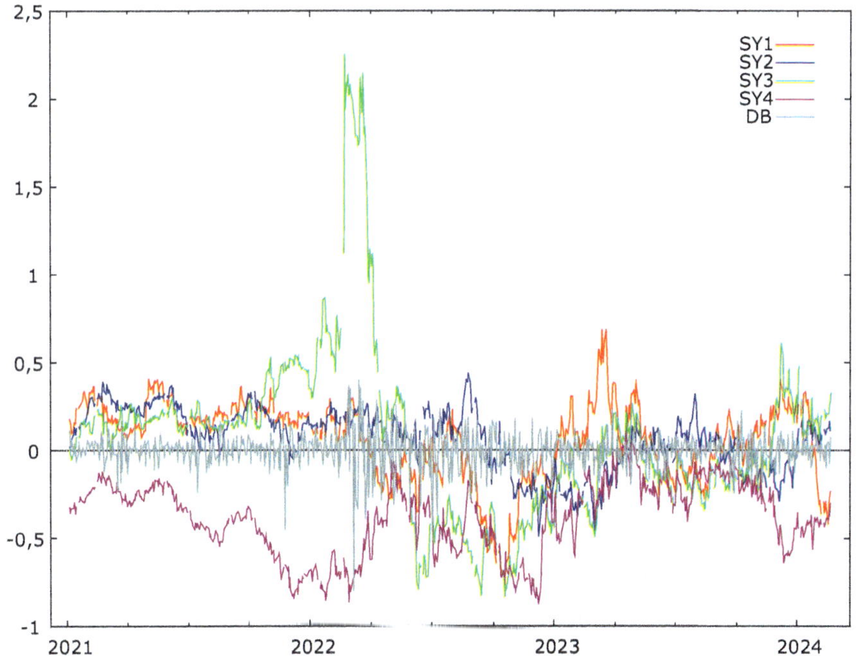

Fig. 2 Dynamics of spreads relative to the standardized yields of U.S. sovereign bonds for Iceland (SY1), Norway (SY2), Russia (SY3), Sweden (SY4), and oil price differences (DB), 03.01.2021–01.03.2024. *Source* Calculated by the authors based on Cbonds data (https://cbonds.ru/indexes/Cbonds-GBI-RU-YTM-eff/, https://cbonds.ru/indexes/9237/)

recent changes in the variance of the response variable and exogenous variables. According to the GARCH(p,q) model, the calculation of time-varying variance h_t is performed using the following formula:

$$h_t(\mathrm{Y(n,\,t)}) = \alpha_0 + \sum\nolimits_{i=1}^{q} \alpha_i \varepsilon_{t-j}^2 + \sum\nolimits_{i=1}^{p} \beta_i h_{t-j}, \tag{2}$$

where α_0 is the lag coefficient, or base volatility; q & p represents the number of prior evaluations of changes in the exogenous and endogenous variables {AR(p)} that influence the current value; β_i are the weight coefficients, reflecting the degree to which previous evaluations influence the current values, also referred to as "past events."

4 Research Results

Figure 1 presents the original data from August 28, 2019, to March 1, 2024. This period includes three major crises: a surge in oil prices and sovereign bond yields; the COVID-19 crisis; a subsequent drop in oil prices and a decline in bond yields across all countries in the sample; a period of rising yields and oil prices; and the crisis related to the imposition of sanctions on the Russian economy during 2022–2023.

In the first period, 2019–2020, associated with the corona crisis, oil prices fell sharply, and fluctuations in yields took place in a positive spectrum. Jumps in bond yields of Russia and Iceland are noticeable. Then there was a slight increase in the yields of Norwegian and US bonds. While the COVID-19 crisis has been extensively studied in the literature, there is very little reliable evidence on whether the overall trends persisted during the period of sectoral sanctions. For further calculations, we limited the period to January 3, 2021, to March 1, 2024.

The highest bond yields were observed for Russian sovereign bonds, which experienced a sharp increase starting in February 2022. However, within two quarters, the yields returned to their previous trend. From the spring of 2022, as sectoral sanctions were imposed on the Russian economy, U.S. sovereign bond yields (YSB USA) decreased, while the remaining countries in the sample continued to see rising yields on their sovereign bonds. The lowest yields, up until the spring of 2022, were observed for U.S. and Swedish sovereign bonds. Oil prices experienced a shock in February 2022, but by early 2023, the average trend had been restored.

The quantile (median) regression for the standardized yields of sovereign bonds during "non-crisis" and "crisis" periods is presented in Table 1. The response variables are the standardized sovereign bond yields for Iceland (ZY1), Norway (ZY2), Russia (ZY3), and Sweden (ZY4). The exogenous variables are the standardized oil price (ZY5) and U.S. sovereign bond yields (ZY0). The non-crisis period (N0) included 1035 observations, while the crisis period (N1) included 91 observations.

During non-crisis periods, all sovereign bond markets are strongly interconnected, as indicated by the high quantile regression coefficients with ZY_USA, ranging from

Table 1 Quantile regression in non-crisis (N0 = 1035) and crisis (N1 = 91) periods

	A		ZY5 (Brent)		ZY0 (USA)		Pseudo R^2	Band width	Sparsity
N1 (T = 91)	Coef	St. er	Coef	St. er	Coef	St.er			
ZY1	− 0.74***	0.09	0.46***	0.047	− 0.38*	0.17	0.62	0.21	0.5
ZY2	− 0.16*	0.07	0.17***	0.03	0.67***	0.13	0.73	0.15	0.34
ZY3	− 1.27***	0.2	1.1***	0.1	− 2.33***	0.37	0.49	0.44	1.1
ZY4	− 6.2***	0.08	− 0.07	0.04	0.99**	0.3	0.51	0.11	0.2
N0 (T = 1035)									
ZY1	0.06***	0.008	0.06***	0.01	0.89***	0.01	0.79	0.12	0.53
ZY2	0.06	0.006	0.16***	0.009	0.87***	0.01	0.85	0.08	0.37
ZY3	0.01	0.007	0.14***	0.01	0.82***	0.01	0.72	0.11	0.48
ZY4	− 0.31***	0.01	− 0.11***	0.01	1.06***	0.01	0.82	0.11	0.47

Observation period: 03.01.2021–01.03.2024
*$p < 0.05$, **$p < 0.01$, ***$p < 0.001$

0.82*** for Russia to 1.06*** for Sweden. Brent oil prices have the most significant positive impact on the sovereign bond yields of Norway and Russia. In crisis periods, the median yields across all analyzed markets decrease.

Sweden's YSB is closely tied to oil prices, but oil price shocks slightly reduce its correlation with U.S. YSB, moderately lowering the median bond yields.

Russian YSB is strongly correlated with oil prices only during non-crisis periods. During crisis periods, Russia's and Iceland's YSBs begin to move in opposition to the U.S. market, significantly increasing their volatility depending on oil prices.

Sweden's YSB reacts weakly to oil price shocks, with rising oil prices lowering the yields on its sovereign bonds.

The Dynamics of YSB Spreads for Iceland, Norway, and Russia are positioned above zero, meaning YSB yields exceed the average U.S. YSB yields (see Fig. 2). Russia's YSB spreads experienced significant shocks during the summer and fall of 2022, as Russian sovereign bond yields were quoted below U.S. YSB during the exodus of many foreign investors from the Russian market. However, by early 2023, Russian YSB had nearly returned to pre-crisis dynamics. Strong YSB volatility resumed at the end of 2023 and into 2024. Sweden's YSB, over the long term, remained below U.S. YSB levels.

The volatility of sovereign bond yields and yield spreads of Arctic countries relative to U.S. sovereign bonds, according to model (2), is presented in Table 2 for GARCH (1,2) and Table 3 for GARCH (1,5). The error variance for Russian bond yields and the yield spreads of Iceland and Russia is quite large. Overall, the evaluation criteria results are satisfactory for all countries except Russia. The GARCH (1,2) model shows significantly better performance than GARCH (1,5), though comparing the models allows for assessing the robustness of the results.

In the GARCH (1,2) model, the response variables are the standardized yields of sovereign bonds for Iceland, Norway, Russia, and Sweden, while the exogenous

Table 2 Components of GARCH (1,2) for the response of standardized bond yields

	Response variables			
Model parameters	ZY1(Ice)	ZY2(Nor)	ZY3(Rus)	ZY4(Swe)
M	0.45	0.45	0.46	0.005
St. dev	0.85	0.84	0.85	0.998
A	0.17**	0.15*	0.06***	− 0.29***
ZY5(1) (Brent)	− 0.07**	0.01*	0.05***	− 0.01***
ZY0(2) (USA)	0.95***	0.9***	0.85***	1.05***
Alpha(0)	0.002***	0.001***	0.0007***	0.003***
Alpha(1)	0.86***	0.62***	0.8***	0.65***
Beta(1)	0.1	0.36***	0.2***	0.28***
Log. plausibility	575	660	363	401
Akaike crit	− 1136	− 1306	− 712	− 789
Schwartz crit	− 1103	− 1273	− 679	− 756
Unconditional error variance	0.07	0.05	1E + 10	0.04

Observation period: 03.01.2021–01.03.2024
The lags of the exogenous variables of the model are indicated in parentheses
*$p < 0.05$, **$p < 0.01$, ***$p < 0.001$

Table 3 Components of GARCH (1,5) for yield spread responses

	Response variables			
Model parameters	SY1(Ice)	SY2(Nor)	SY3(Rus)	SY4(Swe)
M	0.07	0.07	0.08	− 0.37
St. dev	0.22	0.17	0.47	0.2
A	0.15***	0.12***	0.15***	− 0.31***
DB (1)	0.02	0.02	− 0.04	0.1**
DB (2)	0.03	0.08***	0.006	0.05
DB (3)	0.04	0.06***	− 0.01	0.05
DB (4)	0.06*	0.06***	0.044	0.01
DB (5)	0.04	0.08***	− 0.007	0.01
Alpha(0)	0.001***	0.001***	0.0004***	0.001***
Alpha(1)	0.66***	0.69***	0.6***	0.67***
Beta(1)	0.34***	0.28***	0.4***	0.31***
Log. plausibility	482	608	153	405
Akaike crit	− 945	− 1197	− 287	− 790
Schwartz crit	− 898	− 1150	− 240	− 743
Likelihood ratio	809***	622***	1363***	520***
Unconditional error Variance	1,0E + 09	0.044	1,4E + 11	0.12

Observation period: 03.01.2021–01.03.2024.
The lags of the exogenous variables of the model are indicated in parentheses.
*$p < 0.05$, **$p < 0.01$, ***$p < 0.001$.

variables are the oil price volatility (lags 1) and U.S. sovereign bond yield volatility (lags 2).

Including an endogenous variable in model (2) significantly improves the model quality for Iceland, Norway, and Sweden (GARCH (2,2)), lowering the conditional error variance below 1e-8, with an Akaike Criterion of < -2500). However, for Russia, such a model is not computable due to convergence issues. Given that our study aimed for a comparative analysis of yield levels and volatility, we opted to use GARCH (1,2) for analysis (see Table 2).

It is important to note that in the GARCH (2,2) model, the impact of sovereign bond yield dynamics approaches 1***, with alpha (0) < 6e-5***, alpha(1) = (0.05; 0.07; 0.05)***, beta(1) = (0.91; 0.92; 0.95). The coefficient for ZY5(1) is (0.03; 0.07; 0.07) for Iceland, Norway, and Sweden, respectively. This indicates that the influence of the bond yield dynamics on variance strengthens, thereby reducing the direct impact of U.S. sovereign bond yield dynamics.

We account for the time zone differences between the Western and Eastern hemispheres, which is why the impact of U.S. variables is included with a lag of at least two days, avoiding simultaneous precedence.

In the GARCH (1,5) model, the response variables are the yield spreads of sovereign bonds for Iceland, Norway, and Sweden relative to U.S. sovereign bond yields, while the exogenous variables are oil price volatility up to a 5-day lag (see Table 3). The situation in the U.S. bond market has a direct long-term effect on the volatility of the bond markets in the Arctic countries. Oil price volatility tends to reduce bond yield volatility in Iceland and Sweden but increases it for Russia and Norway.

The most strongly correlated with oil price volatility are the variance of the level and yield spreads of Norway's bonds. The volatility of Iceland's and Norway's sovereign bond yields responds to oil price volatility within a week and is influenced by U.S. bond yields, their own dynamics, and "past news."

In Russia, the volatility of bond yield spreads (but not the yields themselves) responds to oil price volatility only in the short term and is very unstable, depending on its own dynamics.

The volatility of Sweden's sovereign bond yields and their spreads responds to oil price volatility only in the short term and is significantly influenced by its own dynamics.

5 Conclusions

The two crises of 2019–2020 and 2022–2024 are fundamentally different in terms of the interactions of oil prices and sovereign bond yields. We do not reflect the causal relationship, because this will require the involvement of a significantly larger number of variables. In the period 2019–2020, there was a strong drop in oil prices, due to the general blockage of transport communications and industrial production. After the end of the crisis, the general trends quickly resumed. In the period after 2022, crisis trends had a shock effect only on Russian sovereign bonds. The spread of crisis processes was much slower but had significantly greater effects in absolute terms and in terms of volatility.

Next, we discuss the interaction of Arctic sovereign bond yields and oil price shocks only in the period 2021–2024.

In this study, we employed a step-by-step approach to analyze the impact on the level and volatility of sovereign bond yields (YSB) in Arctic countries, focusing on the U.S., Iceland, Norway, Russia, and Sweden. The U.S. serves as the largest and commonly regarded "risk-free" market. The models examined the influence of oil prices in conjunction with the effect of U.S. YSB, as we hypothesized that all markets to some degree absorb information from the U.S. market. Given the significant deviations from normality and stationarity, we utilized four different model types for the five countries: median regressions, GARCH for the level and spreads of YSB relative to U.S. YSB.

Our results demonstrated that the sovereign bond markets of Arctic countries are strongly interconnected, depend on "past news," and are influenced by the U.S. bond market and oil price shocks. Arctic countries exhibit varied reactions to oil price shocks.

The overall situation in the sovereign bond markets, which generally orients itself toward the U.S. market, plays the most significant role in affecting yield volatility dynamics. Based on the behavior of YSB in response to oil price shocks, we identified two types of countries—oil producers and oil consumers.

Dividing the study periods into "crisis" and "non-crisis" phases—where the former are periods of strong oil price shocks—showed that during non-crisis periods, all sovereign bond markets are closely interconnected.

In the long term, Brent oil prices have a significant positive impact on YSB in the U.S., Norway, and Russia. Russian sovereign bonds are strongly correlated with oil prices only during non-crisis periods. Since an increase in YSB reflects higher borrowing costs, in the long term, rising oil prices may have a restraining effect on economic development, despite providing a direct inflow of funds into the national budget.

During crisis periods, the median yields decrease across all observed markets, but the influence of external factors changes. Russian and Icelandic sovereign bonds begin to move out of sync with the U.S. market during crisis periods, with their volatility significantly increasing in response to oil price fluctuations.

Russian sovereign bonds experienced strong shocks due to the sanctions imposed in the spring of 2022, followed by a return to the average trend. However, strong volatility resumed after 2023. Sweden's bonds exhibit a weak reaction to oil price shocks, with rising oil prices decreasing sovereign bond yields.

Norwegian sovereign bond yields and the volatility of yield spreads are strongly linked to oil prices. In the long term, oil price shocks reduce the median YSB in Norway and disrupt the synchronicity with U.S. YSB dynamics. In the short term, oil prices are generally positively correlated with the volatility of Norwegian YSB.

Acknowledgements The study was carried out with the financial support of the Russian Science Foundation within the framework of project No. 23-28-01321 "Analysis of the behavior of debt market participants in crisis situations in related economies".

References

1. Ang, A., Piazzesi, M., Wei, M.: What does the yield curve tell us about GDP growth? J. Econ. **131**(1–2), 359–403 (2006). https://doi.org/10.1016/j.jeconom.2005.01.032
2. Polbin, A.: Econometric estimation of the impact of oil prices shock on the Russian economy in VECM model? Vopr. Ekon. **10**, 27–49 (2017). https://doi.org/10.32609/0042-8736-2017-10-27-49
3. Indryani, N., Usman, M., Nabila, S.U., Kurniasari, D.: Dynamic modeling data return by using BEKK-GARCH (Study: PT. Indofarma Tbk (INAF) and PT. Kimia Farma Tbk (KAEF) from June 2015 to July 2020). J. Phys. Conf. Ser. 1751(1), 012014 (2021). https://doi.org/10.1088/1742-6596/1751/1/012014
4. Abakah, E.J.A., Addo, E., Gil-Alana, L.A., Tiwari, A.K.: Re-examination of international bond market dependence: evidence from a pair copula approach. Int. Rev. Financ. Anal. **74**, 101678 (2021). https://doi.org/10.1016/j.irfa.2021.101678
5. Cevik, E.I., Terzioglu, H.C., Kilic, Y., Bugan, M.F., Dibooglu, S.: Interconnectedness and systemic risk: evidence from global stock markets. Res. Int. Bus. Financ. 102282 (2024). https://doi.org/10.1016/j.ribaf.2024.102282
6. Duncan, R.: Institutional quality, the cyclicality of monetary policy and macroeconomic volatility. J. Macroecon. **39**, 113–155 (2014). https://doi.org/10.1016/j.jmacro.2013.11.004
7. Antonakakis, N., Chatziantoniou, I., Filis, G.: Oil shocks and stock markets: dynamic connectedness under the prism of recent geopolitical and economic unrest. Int. Rev. Financ. Anal. **50**, 1–26 (2017). https://doi.org/10.1016/j.irfa.2017.01.004
8. Malliaropulos, D., Migiakis, P.: A global monetary policy factor in sovereign bond yields. J. Empir. Financ. **70**, 445–465 (2023). https://doi.org/10.1016/j.jempfin.2022.12.011
9. Huang, D., Jiang, F., Li, K., Tong, G., Zhou, G.: Are bond returns predictable with real-time macro data? J. Econ. **237**(3) (2023). https://doi.org/10.1016/j.jeconom.2022.09.008
10. Aydın, H.I., Özgür, Ö.: Term premium in Turkish lira interest rates: the role of foreign investors' share. Borsa Istanbul Rev. (2024). https://doi.org/10.1016/j.bir.2024.01.005
11. Shah, M.H., et al.: Exploring the interwoven relationship: property rights, financial freedom, government regulation, and stock market fluctuations in emerging economies—a novel system GMM perspective. Heliyon **10**(1) (2023). https://doi.org/10.1016/j.heliyon.2023.e23804
12. Zhuang, Y., Zhang, D., Tang, P., Peng, H.: Clustering effects and evolution of the global major 10-year government bond market structure: a network perspective. N. Am. J. Econ. Financ. **70**, 102064 (2024). https://doi.org/10.1016/j.najef.2023.102064
13. Li, Y., Chen, S., Sensoy, A., Wang, L.: Over-expected shocks and financial market security: evidence from China's markets. Res. Int. Bus. Financ. **68**, 102194 (2024). https://doi.org/10.1016/j.ribaf.2023.102194
14. Pesaran, M.H., Mohaddes, K.: Country-specific oil supply shocks and the global economy: a counterfactual analysis. Energy Econ. **59**, 382–399 (2016). https://doi.org/10.1016/j.eneco.2016.08.007
15. Mohaddes, K., Pesaran, M.H.: Oil prices and the global economy: is it different this time around? Energy Econ. **65**, 315–325 (2017). https://doi.org/10.1016/j.eneco.2017.05.011
16. Balcilar, M., Gupta, R., Wang, S., Wohar, M.E.: Oil price uncertainty and movements in the US government bond risk premia. N. Am. J. Econ. Financ. **52**, 101147 (2020). https://doi.org/10.1016/j.najef.2020.101147

17. Nazlioglu, S., Gupta, R., Bouri, E.: Movements in international bond markets: the role of oil prices. Int. Rev. Econ. Financ. **68**, 47–58 (2020). https://doi.org/10.1016/j.iref.2020.03.004
18. Cologni, A., Manera, M.: Oil prices, inflation and interest rates in a structural cointegrated VAR model for the G-7 countries. Energy Econ. **30**(3), 856–888 (2008). https://doi.org/10.1016/j.eneco.2006.11.001
19. Filippidis, M., Filis, G., Kizys, R.: Oil price shocks and EMU sovereign yield spreads. Energy Econ. **86**, 104656 (2020). https://doi.org/10.1016/j.eneco.2019.10465
20. Demirer, R., Ferrer, R., Shahzad, S.J.H.: Oil price shocks, global financial markets and their connectedness. Energy Econ. **88**, 104771 (2020). https://doi.org/10.1016/j.eneco.2020.104771
21. Umar, Z., Aharon, D.Y., Esparcia, C., Al Wahedi, W.: Spillovers between sovereign yield curve components and oil price shocks. Energy Econ. **109**, 105963 (2022). https://doi.org/10.1016/j.eneco.2022.10596
22. Xiao, H., Lin, J., Tang, X.: The spillover effects of external economic shocks on African international sovereign bonds. China Econ. Rev. 102238 (2024). https://doi.org/10.1016/j.chieco.2024.102238
23. Naifar, N.: Examining the nexus between oil shocks and sovereign credit risk: multidimensional insights from major oil exporters. N. Am. J. Econ. Financ. **74**, 102205 (2024). https://doi.org/10.1016/j.najef.2024.102205
24. Maghyereh, A., Ziadat, S.A., Al Rababa'a, A.R.A.: Exploring the dynamic connections between oil price shocks and bond yields in developed nations: a TVP-SVAR-SV approach. Energy **306**, 132475 (2024). https://doi.org/10.1016/j.energy.2024.132475
25. Romashkina, G.F., Skripnuk, D., Andrianov, K.: The attraction of bond financing by food producing companies in Arctic countries. Proc. Eng. **5**(4), 721–736 (2023). https://doi.org/10.24874/PES05.04.014

Optimization of Electro-Optical Parameters of Transparent Thin-Film Electrodes as a Method to Increase the Energy Efficiency of Transparent Electrical Devices

Nikita Parashchuk, Sergey Stakharny, Aleksandr Zaharov, Aleksandr Trubin, and Olga Chemeris

Abstract The synthesis and studies of multilayer conductive thin-film coatings of Oxide, Oxide/Metal, Metal/Oxide and Oxide/Metal/Oxide types have been carried out. The conditions for using the developed structures as electrodes for transparent electronic devices have been determined. Optimization of layer thicknesses has a positive effect on the general physical properties of transparent electrodes. The advantage of using Oxide/Metal/Oxide type structure on flexible substrates and the improvement of optical and electrical properties in comparison with the oxide type structure are shown. As a result, the quality index of transparent FoM electrodes was increased by 40% compared to monolayer ITO. As a result of this study, an electrode with a surface resistance of 12.1 Ohm/sq and an average transmittance of 76% in the optical range was synthesized.

Keywords Transparent electronics · ITO · Thin-film · Transparent electrode · Energy efficiency

---

N. Parashchuk (✉) · A. Zaharov · A. Trubin
Moscow University for Industry and Finance "Synergy", Moscow, Russia
e-mail: parashchuk.ns@mail.ru

S. Stakharny
State University of Education, Moscow, Russia

O. Chemeris
Peter the Great St. Petersburg Polytechnic University, St. Petersburg, Russia

I. Ilin and M. Youzhong (eds.), *Digital Systems and Information Technologies in the Energy Sector*, Lecture Notes in Networks and Systems 1244,
https://doi.org/10.1007/978-3-031-80710-7_8

1 Introduction

Currently, there is a trend of active growth in the transparent electrical devices' market (Fig. 1), the efficiency of which is determined by the combination of high electrical conductivity and high transparency in the optical wavelength range. This trend is driven not only by the increasing demand for transparent electrical devices, but also by the widening range of diversity in the use of these devices in a big variety of applications, which include not only traditional OLED devices [1–3] and solar panels [4–6], but also transparent heaters [7, 8], capacitors [9, 10], antennas [11–13], and wearable smart medical devices [14–16].

The energy efficiency of these devices can be increased by improving the properties of the electrodes required for their operation. These electrical contact systems most commonly take the form of thin-film electrodes made of semiconductor oxide ITO (Indium tin oxide) with characteristic order thicknesses of 100 nm. This material dominates the global market for transparent electronics due to its unique combination of relatively good electrical conductivity and optical transmittance, as well as relative ease of manufacture. The main disadvantages of ITO thin-film electrodes are: (1) the difficulty in manufacturing flexible transparent electrodes due to the low mechanical resistance of the ITO film to bending, (2) the high cost of the material due to the high demand for it [21].

These disadvantages of the ITO thin-film electrode can be solved by adding a thin intermediate metal layer to this film. The structure of this electrode then appears as Oxide/Metal/Oxide. The metal layer significantly increases the electrical conductivity of the transparent electrode and also increases its flexural strength [21]. The

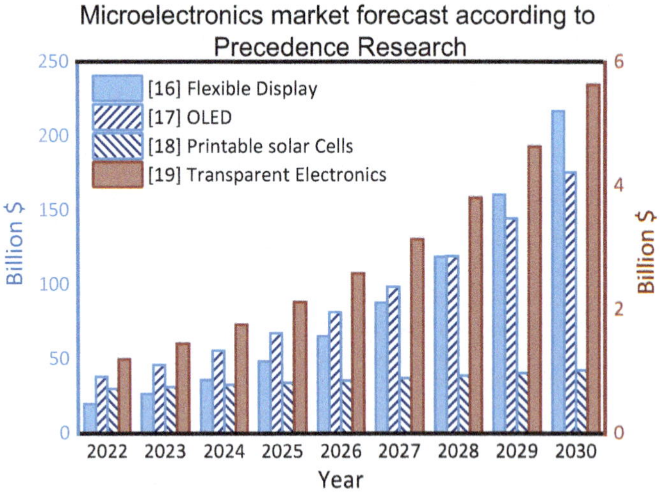

Fig. 1 Global microelectronic devices market growth forecast [17–20]

disadvantage of this solution is the deterioration of transparency in the optical wavelength range. Then, a calculated estimation parameter FoM is introduced for correct comparison of electrodes [22]:

$$FoM = \frac{\tau}{\sqrt[10]{R_{sheet}}}, \tag{1}$$

where τ is the average transmittance in the optical wavelength range, R_{sheet} is the surface resistance of the electrode. Obviously, FoM depends on the configuration of the Oxide/Metal/Oxide type structure. This configuration includes not only the conditions of structure formation, but also the final thicknesses of the oxide and metal layers. Optimization of these parameters will increase the FoM and thus the efficiency of the produced electrode.

Thus, the purpose of this study is to optimize the electro-optical parameters of transparent Oxide/Metal/Oxide type electrodes to improve the energy efficiency of transparent electrical devices.

2 Materials and Methods

One of the most popular methods of producing transparent electrodes is sputtering. The sputtering of Oxide/Metal/Oxide type structures is most commonly performed by a combination of thermal, electron beam and magnetron methods. In this study, the experimental setup shown in Fig. 2 was used for sputtering transparent electrodes.

The transparent electrodes were sputtered on 2.5 × 2.5 cm "Corning Eagle Xg" glasses and on 1.0 × 0.5 cm cut silicon wafers (Fig. 3). The fixation to the 20 cm diameter silicon substrate is achieved with a temperature resistant adhesive tape. The position of the adhesive tape relative to the glasses and silicon wafers is chosen so that an area without sputtered layers remains on each sample.

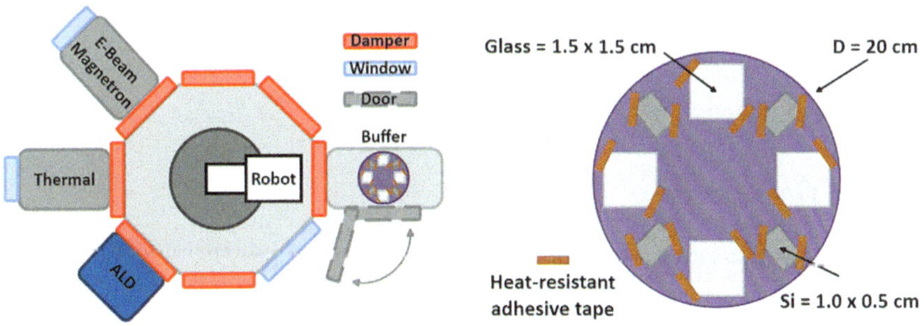

Fig. 2 On the left is a block diagram of the experimental setup, on the right is a diagram of substrates arrangement on the silicon wafer

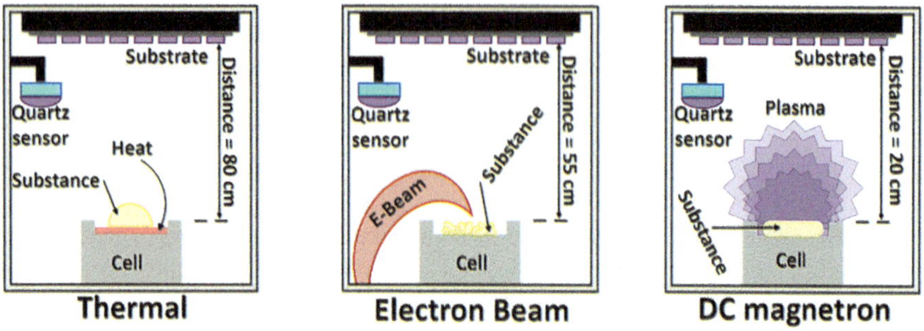

Fig. 3 Schematic representation of the sputtering methods

Throughout the entire process of sputtering transparent electrodes, the silicon wafer with glass and silicon substrates attached to it did not leave the vacuum conditions, which were stably set at 10^{-3} Torr for magnetron sputtering and 10^{-6} Torr for thermal and electron beam sputtering. Moving the electrodes between the sputtering chambers was accomplished with a robotic arm that carried the samples between the chambers within a buffer zone with a vacuum level of 10^{-6} Torr.

Quartz sensors (Fig. 2) located in each sputtering module are used to detect the sputtering rate and layer thickness in real time. The sensor is based on the piezoelectric effect, whose resonant frequency depends on the thickness of the piezo element. Detecting the change of its frequency during the sputtering process provides information about the time dependence of the sputtered layer thickness, i.e. also about the sputtering rate. The discreteness of these sensor values is 1A and it is assumed that the thicknesses of the sputtered layers on the sensor surfaces and on the substrate with glass and silicon wafers fixed on it are equal.

Varying the controlled parameters of each method allows to control the material sputtering rate and, consequently, the layer thickness with satisfactory accuracy [23–26]. Thus, it is possible to obtain multilayer thin-film coatings with different microstructural, optical and electrical properties.

3 Results

3.1 ITO/substrate

The study starts with the sputtering of ITO monolayer with the highest FoM. In all cases ITO sputtering was performed by the magnetron method. Since the microstructural properties of ITO are strongly correlated with the presence of oxygen vacancies in the film volume, we first studied the dependence of FoM on the O_2 feed rate to the sputtering chamber (Fig. 4). Ar with a fixed feed rate of 50 sccm was chosen as the working substance. It was found that the highest FoM is achieved at an O_2 feed rate of 1 sccm.

The formation of the ITO film microstructure is also affected by the sputtering rate, which, in the case of magnetron sputtering, is determined by the power P. In the case of varying the sputtering power, it was necessary to vary the sputtering time in order to maintain the condition of equal thickness of the final ITO films. It was found that the film sputtered at $P = 400$ W has the highest FoM (Fig. 5).

Thus, the best ITO monolayer configuration for magnetron sputtering was found, namely O_2 flow rate of 1 sccm, Ar flow rate of 50 sccm and magnetron power $P = 400$ W. It should be noted that thermal annealing method is usually used in sputtering of oxide compounds to increase transparency and conductivity. However, this method

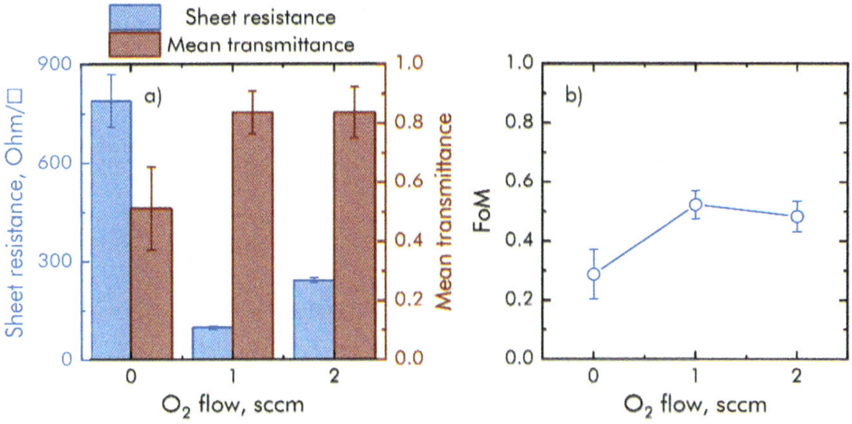

Fig. 4 Study of the effect of O_2 feed rate on the structural properties of ITO

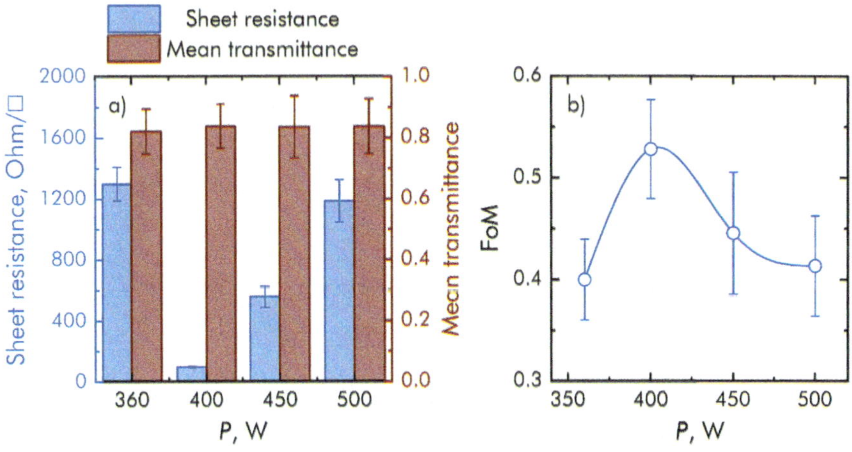

Fig. 5 Study of the effect of permanent magnetron power on the structural properties of ITO

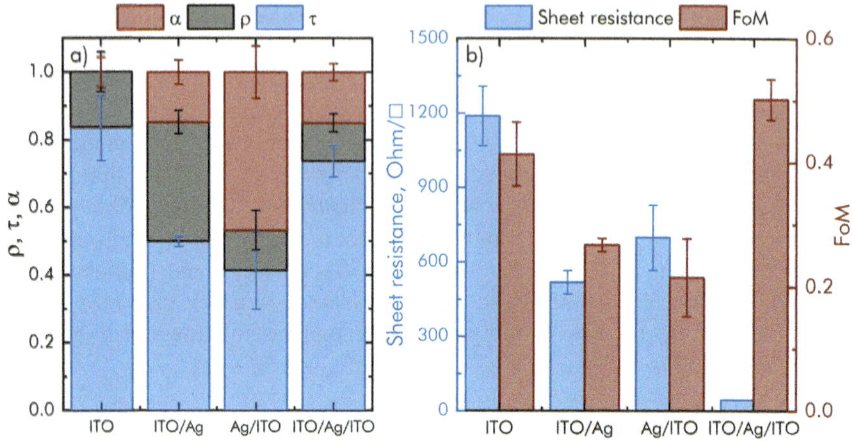

Fig. 6 FoM comparison for different configurations of ITO and Ag multilayer structures

is not applicable for devices whose manufacturing technology requires continuous presence of sputtered structures during the whole manufacturing process.

3.2 Oxide and Metal Containing Structures

The next step is the sputtering of multilayer structures containing thin metal layers. Figure 6 shows a study of the arrangement configuration of identical ITO layers ~ 40 nm and Ag ~ 6 nm sputtered by electron beam method. It was found that the best FoM is achieved with the Oxide/Metal/Oxide type structure.

The thickness of the metal layer largely determines the conductive properties of a transparent electrode with an Oxide/Metal/Oxide type structure, but properly matched thicknesses of oxide and metal layers are necessary to effectively reduce the reflection coefficient and increase the transmittance. Therefore, a series of sputtering experiments were performed keeping the thickness of the ITO layers ~ 40 nm constant and varying the thickness of the Ag interlayer (Fig. 7).

It was found that ITO/Ag/ITO type film with thicknesses of 40/10/40 nm has the best *FoM*.

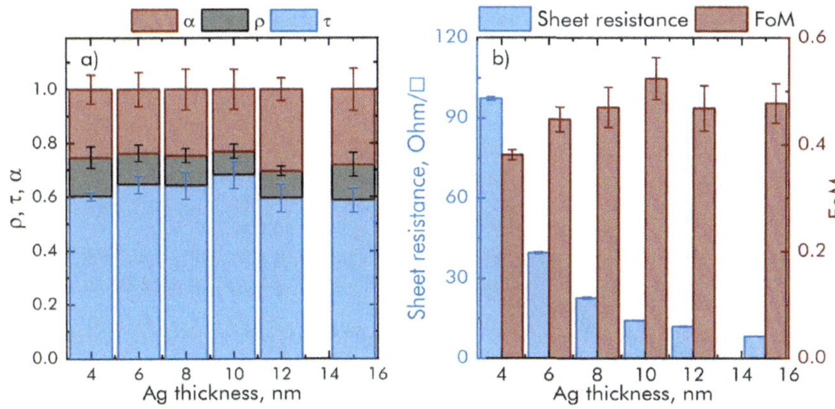

Fig. 7 Study of the effect of Ag thickness on the *FoM* Oxide/Metal/Oxide type structure

4 Conclusion

The synthesized Oxide, Oxide/Metal, Metal/Oxide and Oxide/Metal/Oxide type structures have been studied by magnetron and electron beam sputtering methods. The transformation patterns of these structures in open and oxygen-free atmospheres have been determined. The optimum layer thickness ratios depending on the conductivity and transmittance of optical range waves have been determined.

The use of the Oxide/Metal/Oxide type structure significantly increases the flexural strength of the transparent electrode, which positively affects the prospects of its use in flexible transparent electrical devices. In addition, it is shown that the establishment of optimal layer thicknesses allows a 20% increase in FoM compared to the control sample and a 40% increase compared to the monolayer ITO oxide.

The surface resistance of these electrodes (Fig. 6, "ITO/Ag/ITO") is 12.1 Ohm/sq with a light transmittance of 76%, which meets the requirements for modern electro-optical devices such as OLEDs, solar cells, and wearable smart sensors.

References

1. Salehi, A., et al.: Recent advances in OLED optical design. Adv. Func. Mater. **29**(15), 1808803 (2019)
2. Liu, Y.F., et al.: Recent developments in flexible organic light-emitting devices. Adv. Mater. Technol. **4**(1), 1800371 (2019)
3. Sun, N., et al.: Performance of OLED under mechanical strain: a review. J. Mater. Sci. Mater. Electron. **31**, 20688–20729 (2020)
4. Hu, Z., et al.: A critical review on semitransparent organic solar cells. Nano Energy **78**, 105376 (2020)
5. Li, X., et al.: Review and perspective of materials for flexible solar cells. Mater. Rep. Energy **1**(1), 100001 (2021)

6. Ilin, I., Laskin, M., Logacheva, I., Sarygulov, A., Tick, A.: Land plots evaluation for agriculture and green energy projects: how to overcome the conflict using mathematics. Mathematics **10**, 4376 (2022). https://doi.org/10.3390/math10224376
7. Papanastasiou, D.T., et al.: Transparent heaters: a review. Adv. Func. Mater. **30**(21), 1910225 (2020)
8. Lee, D., et al.: Highly flexible, transparent and conductive ultrathin silver film heaters for wearable electronics applications. Thin Solid Films **697**, 137835 (2020)
9. Nie, B., et al.: High-performance transparent and conductive films with fully enclosed metal mesh. ACS Appl. Mater. Interfaces **13**(34), 40806–40816 (2021)
10. Cheng, T., et al.: Flexible transparent bifunctional capacitive sensors with superior areal capacitance and sensing capability based on PEDOT: PSS/MXene/Ag grid hybrid electrodes. Adv. Func. Mater. **33**(5), 2210997 (2023)
11. Kim, J.W., et al.: Efficiency-improved UWB transparent antennas using ITO/Ag/ITO multilayer electrode films. IEEE Access **9**, 165385–165393 (2021)
12. So, K.K., Chen, B.J., Chan, C.H.: Microwave and millimeter-wave MIMO antenna using conductive ITO film. IEEE Access **8**, 207024–207033 (2020)
13. Serra, C.C., Medeiros, C.R., Costa, J.R., Fernandes, C.A.: Mirror-integrated transparent antenna for RFID application. IEEE Antennas Wirel. Propag. Lett. **10**, 776–779 (2011)
14. Bing, P., et al.: Photonic crystal fiber-based plasmonic sensor with ITO for medical applications. Eur. Phys. J. D **77**(4), 68 (2023)
15. Mitsubayashi, K., et al.: Optical-transparent and flexible glucose sensor with ITO electrode. Biosens. Bioelectron. **19**(1), 67–71 (2003)
16. Bu, C., et al.: Research progress and prospect of triboelectric nanogenerators as self-powered human body sensors. ACS Appl. Electron. Mater. **2**(4), 863–878 (2020)
17. Flexible Display Market Size, Share, and Trends 2024 to 2034. https://www.precedenceresearch.com/flexible-display-market. Accessed 21 Aug 2024
18. OLED Market Size, Share, and Trends 2024 to 2034. https://www.precedenceresearch.com/oled-market. Accessed 21 Aug 2024
19. Printable Solar Cells Market Size, Share, and Trends 2024 to 2034. https://www.precedenceresearch.com/printable-solar-cells-market. Accessed 21 Aug 2024
20. Transparent Electronics Market Size, Share, and Trends 2024 to 2034. https://www.precedenceresearch.com/transparent-electronics-market. Accessed 21 Aug 2024
21. Wang, T., et al.: Recent developments in flexible transparent electrode. Crystals **11**(5), 511 (2021)
22. Haacke, G.: New figure of merit for transparent conductors. J. Appl. Phys. **47**(9), 4086–4089 (1976)
23. Ilin, I., Vasilenok, V., Marchenko, R.: Use of renewable energy and tax burden on CO_2 emissions in industrial enterprises. In: E3S Web of Conferences, vol. 110, pp. 02102 (2019)
24. Kabaklı, Ö.Ş, Kox, J., Tutsch, L., Heydarian, M., Bett, A.J., Lange, S., Goldschmidt, J.C.: Minimizing electro-optical losses of ITO layers for monolithic perovskite silicon tandem solar cells. Sol. Energy Mater. Sol. Cells **254**, 112246 (2023)
25. Saeidi, M., Eshaghi, A., Aghaei, A.A.: Electro-optical properties of silver nanowire thin film. J. Mater. Sci. Mater. Electron. **34**(2), 110 (2023)
26. Sarma, B.K., Rajkumar, P.: Al-doped ZnO transparent conducting oxide with appealing electro-optical properties–realization of indium free transparent conductors from sputtering targets with varying dopant concentrations. Mater. Today Commun. **23**, 100870 (2020)

Digital and IT Solutions for Industrial, Functional and Regional Challenges

Economic Efficiency Evaluation of Digital Projects Management Improvement

Denis Bezruchko, Igor Ilin, Tatiana Kudryavtseva, and Dayana Gugutishvili

Abstract The purpose of this work is to develop proposals for improving the efficiency of digital project management for a company in the energy industry. In the context of increasing sanctions pressure on Russian companies and the withdrawal of foreign software suppliers from the market, many large companies are deciding to conduct digital projects based on Russian software. The number of digital projects of the company in simultaneous development is tens of units. At the same time, the total number of energy company employees involved in the development and implementation of projects covers hundreds and sometimes thousands of company personnel representatives. The companies studied are working to improve their digital project management processes, guided by the project management methodology ("body of knowledge") PMBOK and ISO 21500. The classical project management approach needs to modify processes in order to provide the necessary flexibility to manage the timing and quality of project products. The existing organizational structures for managing such projects did not include the position of a digital services design manager, which led to an increase in the time and cost of developing digital services. One possible solution is to divide the service manager's functionality between two positions: service design manager and service manager itself. At the same time, the introduction of a new position into the organizational structure of the project requires the development of existing functional process models and an assessment of the economic efficiency of their implementation. This solution has made it possible to significantly improve the quality of digital project management, which is reflected in the elimination of delays in the commissioning of digital projects, as well as the timely start of servicing digital projects.

Keywords Digital project management · Digital technologies · Digitalization in energy · Economic efficiency

D. Bezruchko (✉) · I. Ilin · T. Kudryavtseva · D. Gugutishvili
Peter the Great St.Petersburg Polytechnic University, Politekhnicheskaya St. 29, Saint-Petersburg 195251, Russia
e-mail: dbezru@mail.ru

I. Ilin and M. Youzhong (eds.), *Digital Systems and Information Technologies in the Energy Sector*, Lecture Notes in Networks and Systems 1244,
https://doi.org/10.1007/978-3-031-80710-7_9

1 Introduction

Digitalization of business processes of any type of economic activity has been a current trend of recent decades, and, as noted in many studies [1], there are no prerequisites for suspending or slowing down this process. Also, many researchers note the importance of digitalization processes and their positive impact on the country's economy [2, 3]. The introduction of sanctions against Russian companies in 2022 and the departure of foreign software suppliers from the market caught many companies in the energy industry in middle of the process of digital transformation. Under these conditions, companies tend to decide to independently develop the software they need, which leads to the need to pay special attention to the quality of digital project management.

In general, project quality management is described as a system of methods, tools and activities aimed at meeting the requirements and expectations of project clients for the quality of the project itself and its products. However, this definition is more consistent with quality from the point of view of clients of the project products. In the framework of this study, we are interested in the quality management from the point of view of the project sponsor, the main criteria for which is the ratio of the result and costs of the project. Therefore, the project quality management here and below refers to the following aspects of quality:

- Achieving project goals;
- Achieving the optimal number of project team members;
- Reasonableness of project management costs;
- Timely completion of projects;
- Timely placement of projects for maintenance.

The basis for this study was data from a large Russian corporate group in the energy sector implementing digital projects. At the beginning of 2024, the number of digital projects of the corporate group in simultaneous development was 150, and the total number of members of project teams reached 1030 people. A significant part of these projects is aimed at finding alternatives for imported software solutions, which is directly related to the information security of the corporate group. The corporate group has already achieved high productivity of project teams, but project deadlines are often shifted. Thus, the objectives of this study are:

- development of proposals to achieve timely completion of projects,
- reduction of costs for revision and maintenance of projects at the pilot operation stage;
- ensuring the economic results from the implementation of the projects are achieved in a timely manner;
- justification of the economic efficiency of the proposed measures.

The studied project is a "project management improvement project". To avoid misunderstanding, here and below, projects are understood as the actual digital projects of the company, and the "project management improvement project" is called an "event".

Statistical and analytical methods were used as research methods. To assess the effectiveness of the project for the implementation of the proposed measures, methods for assessing the economic efficiency based on the tree of key performance indicators (hereinafter—KPI), reinforced by methods of investment analysis and sensitivity analysis were used [4].

In solving the set task, the following contribution to scientific knowledge in the subject area was obtained:

- New functional positions in the project were defined, allowing for increased flexibility of project management in terms of deadlines and quality;
- A description of an adaptive mechanism for managing the quality and deadlines of the project was given (based on the introduction of feedback into the project management system);
- The understanding of the quality of project management from the point of view of the project sponsor was clarified.

2 Materials and Methods

2.1 Analysis of Existing Approaches to Digital Project Management

Almost all companies currently use some digital solutions for business process management. At the same time, there is a dependence between the scale of the company's business and the number of digitalization projects. If the number of digital projects in simultaneous development exceeds several dozen, then there is a need to analyze and systematize the work of project teams. At the same time, the quality of project management directly affects the pace of development of the company as a whole.

As is known, the formation of project teams is a task associated with the modeling of poorly formalized processes. Fuzzy cognitive models are often used to support personnel decision-making [5]. At the same time, the project teams must include the position of project manager (service architect), developers, tester and technical support specialist (service manager). Depending on the specifics of the project, the team may also include other positions.

At present, the structure of a typical project team in the corporate group under consideration generally corresponds to the widely accepted approaches of the PMBOK [6] and ISO 21500 and is presented in Fig. 1.

The process of working on a project in the corporate group is carried out sequentially in six stages:

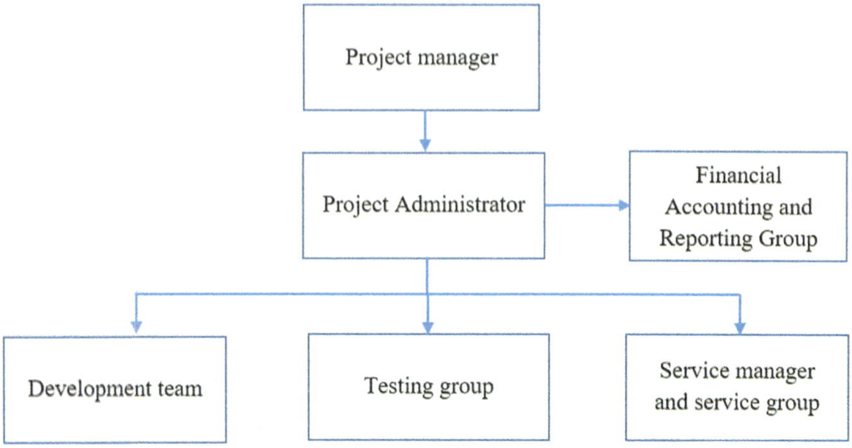

Fig. 1 The structure of the corporate group project team

1. Assessment,
2. Selection,
3. Development and implementation,
4. Pilot operation,
5. Completion,
6. Operation.

The "bottleneck" in this process is the pilot operation stage, where most errors or discrepancies with the terms of reference are identified, and time and money are spent on eliminating them.

The human resources for project teams are pulled from the structural divisions of the corporate group's IT department. The matrix form of team organization is used, and the matrix form that is used is rigid. In this case, the project manager bears full responsibility for the implementation of project tasks, and the heads of structural divisions are responsible for assigning personnel to project tasks.

It is important to note that setting tasks for project teams is a task to develop criteria for optimal prioritization of requirements in the context of continuous integration of software into the company's business processes [7]. In our case (and, in general, in similar divisions of other companies), a process approach is used as a personnel management system, which is a good applied tool for improving the efficiency of the IT department [8, 9]. Also, when managing costs, IT department management usually uses the well-proven tools of lean manufacturing, Six Sigma, Gantt charts, Ishikawa diagrams, etc. [10]. In other words, the work of project teams in the corporate group is already well structured, and most of the tools for improving efficiency have already been used.

At the same time, all existing approaches do not provide for the possibility of adjusting the terms and methods of achieving the project goals (considering them unchanged) depending on the deviation of the current status of the project from the original implementation schedule. Considering that the average implementation period of a digital project in the studied corporate group is three years (data from internal analytics), and during this time the project goals and their relevance may change, there is a need to develop an adaptive mechanism for managing the timing and costs of digital projects.

3 Results

3.1 *Developing Proposals to Improve Digital Project Management*

The statistical sample of this study consisted of 150 digital projects. The total number of overdue projects was 28. It was found that the reason for the delay is the partial non-compliance of the functionality of the digital project result (hereinafter referred to as the service) with the requirements for its service maintenance. The average delay time was 150 days, which led to an increase in the duration of the investment stage of the projects.

In addition, it was found that the formal project delivery does not always lead to the implementation of the service into industrial operation. 13.8% of projects remain at the trial operation stage, exceeding the planned period by an average of 150 days. These costs are included in the investment costs. The reason for this usually is the unpreparedness of the project service function, expressed in the lack of trained specialists in the field of software service.

In changing the digital project management methodology, the previously known mechanism for adapting the digital innovation potential of an enterprise to the strategic goals and objectives of its innovative development was used, taking into account the factors of the innovation climate and internal environment [11, 12]. Thus, the project management audit we conducted in the interests of the corporate group revealed the need to add functional feedback to the project cost and deadline management system. Considering the planned project delivery dates to be unchanged, the project team can apply certain technical solutions that can manage the speed and costs of developing a digital project [13, 14].

According to the PMBOK project management standard [6] used in the corporate group, the project team has a sufficient number of tools at its disposal to manage the quality of the project (Fig. 2).

At the same time, project teams could not use the above tools to manage project costs and deadlines due to the lack of authority and competence within project teams.

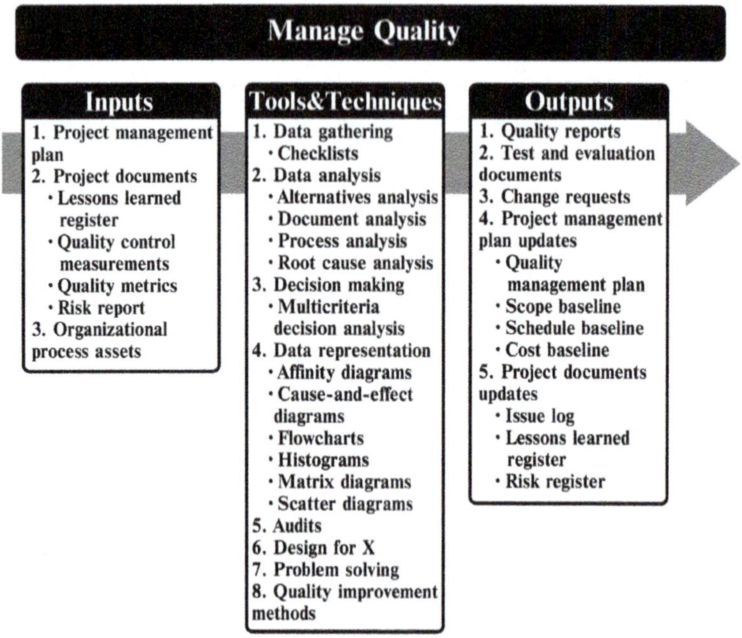

Fig. 2 Manage quality: inputs, tools and techniques, and outputs

This can be mitigated by dividing the service manager's functions between the positions of "Service Design Manager" and "Service Manager" for service management (Fig. 3).

According to the new structure, the service design manager operates at the stages of evaluation, selection, development and implementation, as well as the pilot operation. The service manager is included in the project at the pilot operation stage,

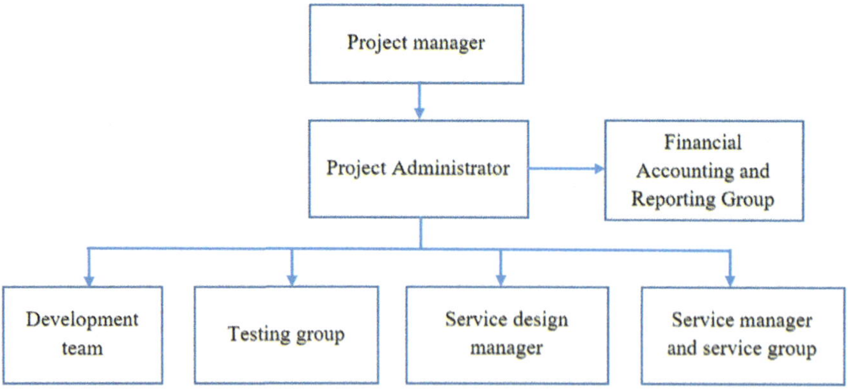

Fig. 3 The new structure of the corporate group project team

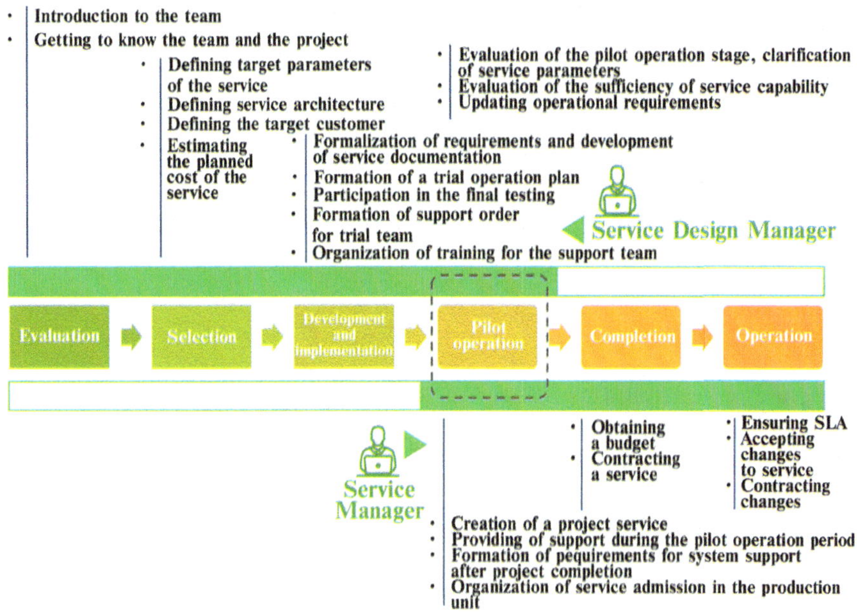

Fig. 4 Distribution of functionality of new positions at different stages of the project

ensures the completion of the project and continues to work at the operational stage (Fig. 4).

The new structure also provided a solution to the project "bottleneck" problem at the pilot operation stage by strengthening the project teams with the simultaneous and coordinated work of two different specialists: service design manager and service manager.

How is the breakdown of functions between the two positions implemented in practice? Previously, a service manager was required to be included in the project team and supported an average of 5 projects. According to the recommendations we developed, service managers are transferred to technology centers for project support, and the costs of their maintenance are attributed to the operation of projects. The deadlines are defined in Table 1. A new position of service design manager is introduced for each project, which involves hiring 30 specialists (at the rate of 5 projects per person), see Table 2.

It should be noted that the improvement of the quality of digital project management is achieved by introducing a feedback mechanism into the project management structure, allowing for prompt quality and cost management depending on the magnitude of the deviation of the expected project delivery dates and costs from the planned ones. In earlier studies on digital project management [5, 7, 11], such opportunities were not noted.

The costs of creating jobs for service design managers are included in investments in each project (Table 3). The main tasks that are set for managers in this position are:

Table 1 Diagram of project implementation stages (developed by author)

	Stage	Notes	2022	2023	2024	2025
1	Development of a service concept	Completed	X			
2	Justification of feasibility, obtaining a budget	Q1 2023		X		
3	Pilot implementation	2023		X		
4	Full-featured solution (general implementation)	2024			X	X

Table 2 Timetable for the implementation of the service design manager (recruitment plan) (developed by author)

	Total	2023	2024	2025
Schedule of implementation of positions	30	10	20	0
Number of implemented positions	–	10	30	30
Effect of implementation of positions	–	5	20	30
Volume of upcoming implementation	–	20	0	0
Number of implemented positions	30	10	30	30

designing and developing a service in accordance with the requirements for an IT service, optimal resources and using efficiency and quality management practices.

According to [15], the following key tasks (responsibilities) were proposed, as well as types of activities of the service design manager:

- Conduct an examination of design documents,
- Analyze the impact of the results of the projects on existing services,
- Identify service level needs from customers, manage service development and a set of service documentation,
- Determine existing operational requirements and form a service structure,
- Ensure the formation and coordination of service architecture with expertise centers,
- Ensure the implementation of operational requirements within the project,
- Ensure the launch of project management and testing processes for new services,
- Form a plan for transfer to maintenance,
- Determine the schedule for involving service resources/service processes,
- Form service model options,
- Attract, prepare and organize a service team for the start of trial operation,
- Estimate the costs of various options for organizing operation,
- Optimize the level of operating costs,
- Ensure the acceptance of new information systems and solutions for service,

Table 3 Investment in the project (developed by author)

	Total	2023	2024	2025
Creation of new jobs, thousand rubles per year	4800	1600	3200	0

- Ensure the development service competencies,
- Form strategies and programs for the development of information systems and solutions services.

The decisions and issues that fall into the service design manager's area of expertise are defined:

- Coordination of the transfer of project results and new information systems and solutions to the service,
- Determination of the required service quality indicators and setting up their monitoring,
- Estimation of costs for design and operation of the service,
- Participation in the development, examination and selection of key solutions in projects,
- Conducting negotiations with service providers/developers on the cost and structure of services required for operation.

The provided value of the service design manager is as follows:

- Support for users of supervised services, information systems and solutions,
- Management of the volume and quality of provided services,
- Control over the efficiency of the provision of services.

As a result of the implementation of the new position, it was possible to reduce the amount of delays in the delivery of projects, as well as the placement of delivery for service maintenance to zero.

3.2 Evaluation of the Economic Efficiency of Implementing Changes

To assess the economic efficiency of implementing the proposed changes, a project ("event") was formed to implement the service design manager position in the corporate group's project management structure, that is, a "project to change project management" was created. The event in question has the characteristics of an investment project, since it includes one-time initial (investment) costs. However, all the considered digital projects of the energy corporate group have an internal focus and are not intended for sale on the market, i.e. the product of the project is cost savings [16]. In this case, it becomes necessary to choose the right methodology for assessing the economic efficiency of the project.

The methodology for assessing the economic efficiency of internal projects is based on function cost analysis (FCA) [10]. These methods have begun to be widely used in IT departments, since they have serious potential for conducting technical and economic analysis. Given our experience, the most correct results of assessing the economic efficiency of projects are provided by investment analysis methods taking into account the specifics of digital projects, which is also noted in [17]. In this case, the logic of constructing financial models is based on the theory of interdisciplinary analysis in economics, described in [18, 19].

In our case, the economic efficiency of the event was assessed using the investment analysis method of the internal project based on the construction of a KPI tree. At the same time, as recommended in [20], all economic effects are calculated for the end user of the product (beneficiaries of the corporate group) and take into account both the costs of implementing the product and the savings achieved from its use.

To assess the economic efficiency of the event, it was necessary to make an important assumption. Given that all digital projects of the company have different scales and, accordingly, different costs, all calculations for future projects and for future periods were made for a set of conditional average projects, the parameters of which correspond to the average values of actual projects from 2018 to 2022. The reason for this assumption is that old projects in the corporate group are gradually ending, and the parameters of new ones have not yet been determined. At the same time, there is an understanding that the volume of work for the IT department of the corporate group will not decrease in the next 10 years. Therefore, our assumption about maintaining the average values of the effects of the projects is acceptable.

In the process of assessing the economic efficiency of our proposals, the following were selected as the basic influencing factors:

- Reduction of costs for project refinement at the investment stage,
- Reduction of costs for project servicing at the pilot operation stage,
- Timeliness of achieving the planned economic effect from the project.

The influencing factors are described in more detail in Table 4, and the numerical values of the effects are given in Table 5.

Table 4 Project implementation effects (developed by author)

Direct effects	Notes
Reducing the costs of project development at the investment stage	The delay in the project delivery leads to an increase in the costs of its completion. The effect is expressed in the need to pay the salaries of 5 FTE project team, who complete the project within 150 days
Reducing project service costs at the pilot operation stage	As a result of the unpreparedness of the service function, the project implementation is delayed (the pilot operation stage is extended). The average number of FTE is 2.5 people, the average delay time is 150 days
Indirect effects	Notes
Timeliness of achieving the planned economic effect from the project	The project did not bring the expected economic effect to the company until it was put into operation

Table 5 Calculation of the effects of the project implementation and their cost (developed by author)

Effect	Unit of measurement	Length of delay prior	Length of delay afterwards	Cost of effect, thousand rubles	% of projects	The magnitude of the effect per service design manager, thousand rubles
Direct effects						
Reducing the costs of project development at the investment stage	Year	0.411 (150 days)	0	16834.5	18.7	6457
Reducing project service costs at the pilot operation stage	Year	0.411 (150 days)	0	8417.3	14	2378
Indirect effects						
Timeliness of achieving the planned economic effect from the project	Year	0.411 (150 days)	0	65000.0	18.7	24,932
Total						33,767

Based on the monetary assessment of the achieved effects and the cost of the initial investments in the event (Table 6), its key performance indicators (NPV for direct effects and NPV for direct and indirect effects) were calculated (Table 7).

4 Discussion

In order to assess the effectiveness of the event on the implementation of the service design manager position, we calculated the results of its economic efficiency using investment analysis methods. The result of the calculation of the economic effect was the KPI of the event and intermediate indicators. Considering that the implementation of one service design manager position gives an effect of 33.8 million rubles, as well as a large number of projects in the corporate group (~150), very significant effects were obtained. Thus, the total annual effect, expressed in savings of the corporate group budget, will be 912 million rubles, and the net present value of the event is 3991 million rubles. The payback period of the project occurred already in the first year of the project and was not calculated, since the calculation step in the financial model was one year.

Table 6 Calculation of effects in monetary terms by years (developed by author)

Effect	Total for 8 years	2023	2024	2025 and beyond
Reducing the costs of project development at the investment stage				
Current value (as is)	1,549,696	193,712	193,712	193,712
Expected value (to be)	225,997	161,427	64,571	0
Monetary assessment of the effect	– 1,323,699	– 32,285	– 129,141	– 193,712
Reducing project service costs at the pilot operation stage				
Current value (as is)	570,759	71,345	71,345	71,345
Expected value (to be)	83,236	59,454	23,782	0
Monetary assessment of the effect	– 487,523	– 11,891	– 47,563	– 71,345
Timeliness of achieving the planned economic effect from the project				
Current value (as is)	5,983,562	747,945	747,945	747,945
Expected value (to be)	872,603	623,288	249,315	0
Monetary assessment of the effect	– 5,110,959	– 124,658	– 498,630	– 747,945
Total	– 6,922,181	– 168,834	– 675,335	– 1,013,002

Table 7 Key performance indicators of the project (developed by author)

KPI	Direct effects	Direct and indirect effects
NPV	675,848	3,991,342

In addition, the following difficult-to-measure effects were identified:

- Reduction in the overall risk of damage due to project delays,
- Reduction in the risk of damage due to the lack of a full-fledged service,
- Reduction in the risk of business downtime,
- Improvement in the quality of project development and user satisfaction,
- Higher quality of project investment assessment (previously, the volume of investment was underestimated due to incomplete data),
- Increase in the overall level of competence of project teams,
- Possible reduction in the costs of operating project products as a result of using more resource-saving solutions.

5 Conclusion

The study shows that the improvement of the quality of digital project management is achieved by introducing a feedback mechanism into the project management structure, which allows for prompt quality and cost management depending on the magnitude of the deviation of the expected project delivery dates and costs from the planned ones. In earlier studies on digital project management [5, 7, 11], such opportunities were not noted. We proposed the implementation of this mechanism in a corporate group company by introducing a new position of service design manager.

The process of the introduction of the service design manager position is characterized by the achievement of the following quantitative effects:

- Reduction of costs for project revision at the investment stage;
- Reduction of costs for project servicing at the pilot operation stage;
- Timeliness of achieving the planned economic effect of the project.

Of particular interest is the developed methodology for assessing the economic efficiency of internal projects, which include most digital projects of large corporate groups. We used the construction of a KPI tree (consisting of 4 levels) with subsequent assessment of investment KPIs. Unlike classical methods of investment analysis, the source of economic efficiency was not the revenue from product sales, but the effect of project implementation, expressed as a change in cash flow. Based on the obtained value of the change in cash flow, economic efficiency indicators were calculated and a sensitivity analysis was carried out.

The estimated value of the economic effect from the implementation of one service design manager position is characterized by a payback period of less than one year, and in monetary terms it amounted to 33,767 thousand rubles per year. The total economic effect of the event for 8 years is 6,176,663 thousand rubles.

Also, as a result of the implementation of measures to improve project management, noticeable effects have been achieved that do not have a numerical expression. These effects are expressed in the reduction of technological and management risks of the corporate group company, as well as the general increase in the competence of project teams.

Acknowledgements The research was supported by Russian Science Foundation grant No. 24-28-01796. https://rscf.ru/project/24-28-01796/.

References

1. Nikolaevskii, V.: Digital economy: the essence, definition and ontology of the economic system in the context of information and communication technologies. Financ. Credit. **28**(11), 2593–2628 (2022). https://doi.org/10.24891/fc.28.11.2593
2. Gasanov, O.: The impact of digitalization on the Russian economy. Financ. Credit. **28**(7), 1534–1553 (2022). https://doi.org/10.24891/fc.28.7.1534
3. Ponomareva, E.: Digitalization as a driver of economic growth: does only infrastructure matters? J. New Econ. Assoc. **3**, 51–68 (2021). https://doi.org/10.31737/2221-2264-2021-51-3-3
4. Ilin, I.V., Levina, A.I., Dubgorn, A.S., Abran, A.: Investment models for enterprise architecture (EA) and IT architecture projects within the open innovation concept. J. Open Innov. Technol. Mark. Complex. **7**(1), 1–18 (2021)

5. Azhmukhamedov, I., Azhmukhamedov, A.: Methodology for forming a team for the implementation of IT projects based on a fuzzy cognitive model of competency assessment. J. Appl. Inform. **4**(34), 70–76 (2011)
6. A guide to the project management body of knowledge (PMBOK guide). Project Management Institute (2017)
7. Nepomnyashchikh, A., Nepomnyashchikh, E., Lavrov, D.: Prioritizing software requirements in a continuous integration environment. J. Appl. Inform. **1**(37), 20–27 (2012)
8. Matrosova, E., Tikhomirova, A.: Improving the work of the IT department through the implementation of a process approach and a system for assessing the performance of employees. Audit. Financ. Anal. 1 (2021). https://doi.org/10.38097/AFA.2021.93.34.019
9. Bril, A., Kalinina, O., Ilin, I., Dubgorn, A., Iliashenko, O.: Forecasting the turnover growth in the risk management system as management decisions support. In: Proceedings of 2017 20th IEEE International Conference on Soft Computing and Measurements, vol. 7970692, pp. 692–693 (2017)
10. Knyazkova, V.: Development of value analysis and features of ITS application in software products design. π-Economy **15**(5), 111–121 (2022). https://doi.org/10.18721/JE.15507
11. Shmeleva, A., Suloeva, S.: Development of a mechanism for adapting digital innovation potential of an organisation with allowance for peculiarities of digital innovation projects. Sustain. Dev. Eng. Econ. **2**, 63–80 (2022). https://doi.org/10.48554/SDEE.2022.2.5
12. Ilin, I., Levina, A., Iliashenko, O.: Enterprise architecture analysis for energy efficiency of Saint-Petersburg underground. Adv. Intell. Syst. Comput. **692**, 1214–1223 (2018)
13. Kukharev A.: Digital Project Technology: Key Principles and Practical Tips for Effective Implementation (2017)
14. Stroykov, G., Babyr, N., Ilin, I., Marchenko, R.: System of comprehensive assessment of project risks in energy industry. Int. J. Eng. Trans. A **34**(7), 1778–1784 (2021)
15. National standard of the Russian Federation GOST R ISO 21500-2014 "Project management guidelines" (approved by order of the Federal Agency for Technical Regulation and Metrology dated November 26, 2014 N 1873-st)
16. Poljanskihh, A., Levina, A., Dubgorn, A.: Investment in renewable energy: practical case in Estonia. In: MATEC Web of Conferences, vol. 193, pp. 05065 (2018)
17. Plaskova, N.: The development of the economic analysis methodology in the digital economy. Account. Anal. Audit. **5**(2), 36–43 (2018). https://doi.org/10.26794/2408-9303-2018-5-2-36-43
18. Subtselnaya T., Tishchenko S., Negrii V.: On the applicability of interdisciplinary methods in economic research. Financ. Credit. **28**(4), 788–805 (2022). https://doi.org/10.24891/fc.28.4.788
19. Guzhev, D.: Methodology for calculating the net present value of investment project, taking into account the variability of determining the cash flow of capital investments. Financ. Credit. **28**(9), 2016–2031 (2022). https://doi.org/10.24891/fc.28.9.2016
20. Semenova, Y.: Theoretical aspects of investment attractiveness: content of the concept, factors and methods of evaluation. Techno Econ. **3**(2), 65–75 (2022). https://doi.org/10.57809/2022.3.3.6

Building an Operational Resource Management System in the Equipment Reliability Management Loop Within the Framework of IT-Transformation

Konstantin Frolov, Daniil Ivanov, and Alexandr Nasonov

Abstract The paper proposes a new approach to resource management to improve the efficiency of production equipment management in the digital transformation of the IT landscape. Important attention is paid to the use of formal methods that can achieve more significant results compared to traditional approaches. Digital transformation is considered not only in the context of corporate information system development, but also within the framework of the DIKIW pyramid model, which is widely used in transformation processes. Theoretical and methodological aspects of planning and operational resource management for maintaining target performance indicators (KPIs) in the equipment reliability management loop are described. It is shown that existing methods based on combinatorial models are close to the limit of their effectiveness in enterprise information systems (EIS). To expand the capabilities of EIS and increase their efficiency, the use of methods based on the digital layer of the DIKIW model is proposed. The paper develops a conceptual logical-algebraic model of operational planning and adaptive resource management, and justifies its implementation in EIS components, indicating its differences from traditional approaches. The proposed methods demonstrate high efficiency in solving problems of production systems management, which confirms their theoretical and practical significance in the digital era.

Keywords Asset reliability management · Domain ontologies · Multi-agent systems · Adaptive management · Operational resource planning · Digitalization · Digital economy · Personnel bypass system · Digital transformation · Automation · Business processes · Predictive analytics

---

K. Frolov (✉) · D. Ivanov · A. Nasonov
Peter the Great St. Petersburg Polytechnic University, Saint Petersburg, Russia
e-mail: frolov_kv@spbstu.ru

I. Ilin and M. Youzhong (eds.), *Digital Systems and Information Technologies in the Energy Sector*, Lecture Notes in Networks and Systems 1244,
https://doi.org/10.1007/978-3-031-80710-7_10

1 Introduction

One of the key areas of economic sector development in the context of the global economic infrastructure is the digital transformation of the main economic sectors (light and heavy industry, services, banking sector, etc.). The modern world exists in the paradigm of mass integration of information technologies into various public and private business structures, which is a consequence of the transition of the global world into the Fourth Industrial Revolution or the so-called Industry 4.0 and further into Industry 5.0 and Society 5.0.

This transition entails a number of peculiarities:

- Increasing the productivity of the main industries through the development and integration of information technologies that provide full or almost full automation of the main business processes of enterprises;
- Orientation of production to meet individual customer needs, which in a competitive environment implies the possibility of its precise and quick changeover to promptly meet production requirements and constraints;
- Development and implementation of methods of adaptive resource management, which in the real sector of the economy is expressed in the transformation of a wide range of enterprises into virtual structures, each element of which is a real company embedded in a consortium of independent business players working as a single organisation;
- Using innovative methods to solve selection problems, digital technologies to implement them to reduce losses resulting from overproduction, excess output to compensate for rejects, unproductive movements in supply chains, unavailability of inputs due to breakdowns or unplanned maintenance;
- Development of technologies that improve personnel safety and reduce environmental risks;
- Considering human beings as a key driver of innovation in the real economy;
- Increasing the range of digital solutions and the market for their supply as a fundamental factor of innovative development;
- Aggregation of computing resources in order to technologically and organisationally transform them into cloud data centres and hyperscalers.

These features are the result of the emergence of a sustainable innovation wave, which will be the basis of a new technological cycle that is taking shape, the beginning of which has actually been formed [1].

Within the framework of this article, we will consider only one of the noted aspects of the emerging new technological reality—the reduction of losses in the real sector caused by the inaccessibility (limited availability) of productive resources. We will propose methods to reduce these losses, emphasising the innovativeness of these methods, which are clearly digital in nature.

2 Reliability Management of Production Assets—Current State of the Art

Under production assets we will understand any technical means of the enterprise that are involved in value-adding, supporting and auxiliary operations, as well as means that are in the management of the enterprise, can be used but are not actively used (for example, reserve capacities).

In this sense, production assets include the means involved in the performance of logistic operations in the supply and sales loop, prototyping of products under development, and production itself (machine tools, elements of the energy complex, quality control tools, computing devices, etc.) [2].

Considering the sustainability of its functioning in a competitive environment as a fundamental task of a modern enterprise, we assume that the characteristic of sustainability will be a generalised efficiency indicator, which depends on the result of realising the value of goods/materials/services produced and the costs of their creation in the market. We will unite goods/materials/services by the category 'product'. Product creation costs are the costs arising in supply chains, in the distribution and sale of the product, in the performance of supporting activities: financial accounting, recruitment and development of personnel, implementation of environmental measures, etc. Costs in supply chains are well enough described in product life cycle management models, and the information layer of the subsystem is developed on the basis of recommendations developed within CALS models. Among all functional areas covered by CALS, we will limit ourselves within the framework of this article to the functional area corresponding to the tasks of production assets reliability management (see Fig. 1). It should be noted that the conclusions to be drawn within this functional area are applicable to the service management area under certain assumptions [3].

The set of activities carried out in the functional area of reliability management is developed and fixed in international standards, in particular in ISO 55000, ISO 55001 and, in this sense, does not require revolutionary changes, as it has proved its applicability. The SCOR model of processes in this area is presented in Fig. 2 [4, 5].

An analysis of the process architecture of this functional area, which thanks to standards has been called EAM (Enterprise Asset Management), shows that within the Deming cycle, all phases of this cycle except 'P' (Planning) are well supported by information systems both methodologically, algorithmically and programmatically [6]. The main difficulties arise when solving planning tasks are:

- Long-term resource planning: what resources, with what characteristics, in what locations, should be available to be able to utilise them for operational asset maintenance tasks? The solution to this problem is complex and should be oriented towards fixed maintenance strategies according to the strategy options on the 'Gartner Strategy Ladder' (see Fig. 3). The equipment planning horizon is long-term to medium-term. Technical assets (diagnostic equipment, spare parts, consumables, repair equipment, personnel involved in maintenance) are considered as resources;

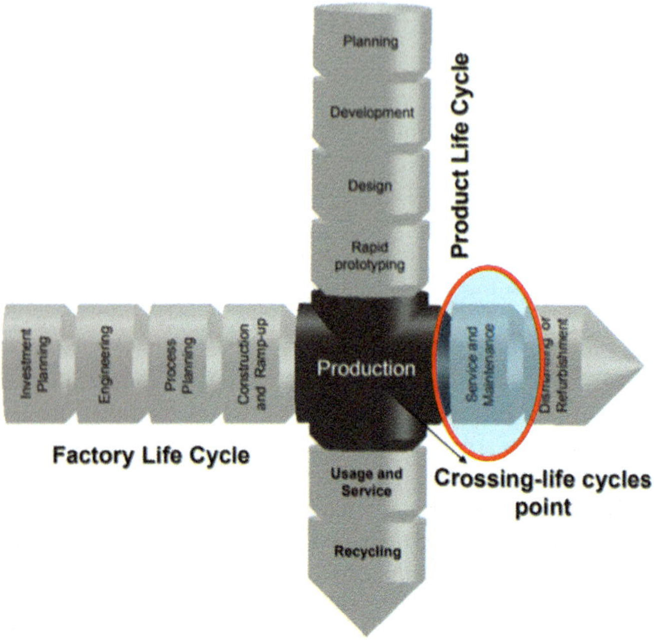

Fig. 1 Life cycle management model and functional area of reliability and service management

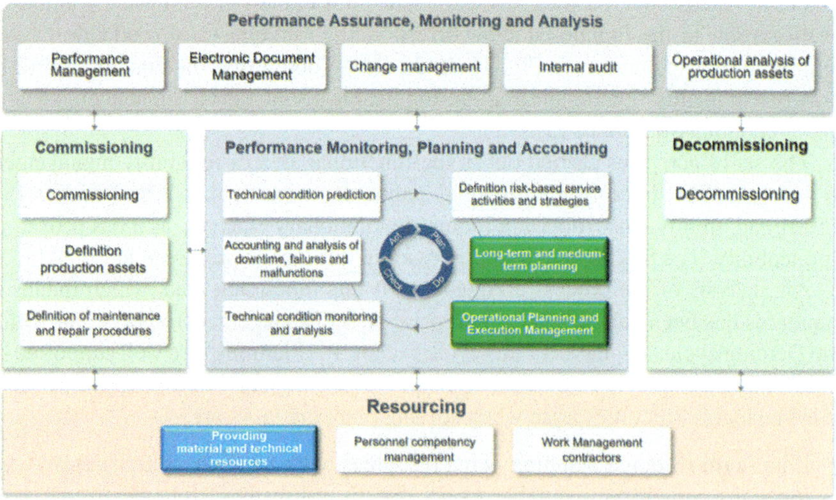

Fig. 2 SCOR model of processes in the functional area of reliability management

Fig. 3 Ladder of service strategies

- Long- and medium-term planning to determine which strategies for which equipment, in which time periods, with which resources, to achieve which targets in the medium-term horizon are applicable;
- Operational planning of maintenance operations and adaptive resource management for unplanned events in the control loop: calculation of planned costs, planning of operations - who performs which operations, planning for replenishment of materials required for maintenance, what preparatory activities (e.g. training, briefings) should be performed in the operational horizon [7].

These problems can be solved by traditional methods based on the combinatorial formulation of the problem and the use of combinatorial or heuristic methods. However, even for problems of small dimensionality the automated search for a variant of resource allocation to perform maintenance tasks becomes an intractable task, in this connection we have to resort to Gantt tools for constraint checking, and the scheduler manually searches for allocation variants independently [8].

Above we talked about the task of resource allocation (personnel, consumables, spare parts, diagnostic equipment) and noted its complexity in terms of operational resource allocation. At the same time, we proceeded from the fact that we excluded some real requirements from consideration during planning, for example:

- Taking into account the speed of arrival of the crew to the technical facility in planning;
- Selection of the most important from the point of view of minimisation of the loss function of the object to be serviced when parallel servicing is impossible due to the lack of resources;
- Consideration of alternatives to an integrated resource: for example, several specialists with one set of equipment or one specialist with a different set of equipment for diagnosis and repair [9, 10].

We have noted that the problem of operational resource planning for a set of operations, being classified as a combinatorial problem of the theory of schedules, encounters almost insurmountable difficulties when trying to solve it in an automated way. However, the problem of adaptive resource management, which arises when unplanned events occur in the resource management loop, is even more difficult. To date, DSS (Decision Support System) class systems are used to automate decision making, oriented on forming a base of typical situations with elaborated variants of their resolution. Taking into account the fact that the combinatorial number of situations requires the development of not less than a number of options to get out of them, this way is not the best. Let us analyse which options for solving the problem of adaptive resource management have the best prospect of use.

3 Operational Planning and Adaptive Management of Asset Servicing Resources

Combinatorial methods for solving the planning problem are implemented by NP complete algorithms, which limits their applicability to the search for optimal resource allocation for asset maintenance. An unplanned event in the resource management loop is, in terms of the Deming cycle, the need for new planning that takes into account the state of the system. For example, if planning of resources for asset maintenance in the strategies 'by workload' and 'by time' is carried out and after the plan calculation an unplanned breakdown occurs, the simplest variant of resource allocation for maintenance in the new situation is to start planning, in the initial conditions of which the new need is fixed.

Such an algorithm of actions is extremely costly in terms of IT resources and, in case of an operational need for response, in most cases unacceptable in terms of impossibility to find the allocation of resources within the acceptable time. For example, if the permissible time of response to the breakdown of a pump in the heating circuit of the workshop is 15 min, and preparation of data for planning, execution of the planning procedure and delivery of the planning result to the executor is 2 h, then the value of such a system is zero [11].

When analysing unplanned events, we will distinguish the following classes of events:

- Events related to the serviceability of a productive asset;
- Events related to the status of service resources (serviceable employee, serviceable spare part, serviceable delivery vehicle for service crews).

An analysis of the Gartner Maintenance Strategy Ladder shows that unplanned events include not only reactive events such as 'The machine broke down', but also events corresponding to higher strategies. For example, if a state or future state analysis reveals an approaching fault, then, depending on the timing of its occurrence and the nature of the future fault, such events can be classified as unplanned events.

With a large number of assets and a powerful maintenance system (many personnel, many crew locations, large assortment of spare parts), the batch mode of scheduling the allocation of resources to tasks is inefficient due to the impossibility of scheduling in an automated mode [12].

An alternative to combinatorial methods of solving the planning problem is the use of a dynamic network of programme agents of needs and capabilities (SP-network), in which service orders (needs) and resources (capabilities), such as repair personnel and equipment for diagnostics and repair, can be represented. Agents of orders and resources can enter into virtual contracts (of course, having nothing to do with a legal document) and communicate with each other or initiate a process of mutual revision and renegotiation of plans as expected or unforeseen events occur. The SP-network introduces agents of orders (see Table 1), technological tasks and processes, products, resources including machines and workers, materials and components and any others that are required to solve different types of business tasks [13].

For each class of agents, their own satisfaction and bonus-fee functions are introduced to regulate their elasticity of concessions to achieve a balance of interests. For example, the order goal is the cheapest or fastest execution, the resource goal is the maximum utilisation, and the product goal is the highest turnover, but depending on the situation one has to compromise and trade off cost for time, etc.

At the heart of the proposed technology, which supports agent competition and co-operation, is the possibility for agents to continuously, asynchronously and in parallel resolve arising conflict situations, when several orders or tasks claim to use the same resource or product (or vice versa), by offering to pay compensation for freeing up a time slot. Receiving compensation allows the yielding party to the conflict to find a new slot without loss of satisfaction or with the least possible loss of satisfaction while receiving an additional amount of money into an account to pursue options for improving their condition in the future [14].

The system created on such principles claims to realise 'Emergent Intelligence (EI)', which manifests itself in the collective interaction of agents to develop and make coordinated decisions. From the software point of view it is necessary to speak not about a product, but about a solution of a new class, which is positioned as the next stage of realisation of corporate decisions. Such EI can be defined as a property

Table 1 Table captions should be placed above the tables

Type	Objectives and preferences	Restrictions
Ordering agent	To be accomplished with minimum delay *(c)* and cost *(p)*	Timing, scope, cost cap
Agent of the task within the process	To be completed on a suitable resource within the specified timeframe in a minimum amount of time	Characteristics of resources and products required, start and finish dates, linkages with other tasks
System agent (enterprise as a whole)	System agents activity management, interaction with external systems	Scheduling time, depth of chains of scheduling permutations

of open self-organising systems, in which the solution to any complex problem is formed in the course of competition and cooperation of system agents until the 'competitive equilibrium' (consensus) occurs, when none of the agents can no longer improve the solution [15].

As a basis for decision-making in PV networks, a 'virtual market' of the system is used, where order agents can buy resource time and resolve conflict situations, when several orders claim to use the same resource, by means of compensation in case of loss of satisfaction for releasing a time slot. The virtual market model is formulated as follows: there is a set of agents A = {Ai}, a set of agents' goal functions C = {Ci}, i = 1, ..., n, where n is the number of agents, and a set of tasks T. For any task set T, the function Ci(T) is defined as the cost for agent Ai to perform all tasks of T. Initially, each agent chooses some arbitrary task set for which the sum of all agents' costs is not minimal, resulting in a suboptimal allocation. The agents then enter into negotiations in order to improve the allocation, as a consequence of which sooner or later a «competitive equilibrium» is established in the virtual market, which is considered to be the solution to the minimum cost problem, since no other solution will lead to an improved outcome. Negotiation can be viewed as an iterative process of give-and-take, at each step of which agents enter into a 'contract' to exchange tasks and virtual currency [16].

The goals of each agent are defined through the satisfaction function Yi (plani), which is a weighted sum of M components corresponding to different criteria—KPI indicators and calculated on the basis of the current plan of work plani associated with the agent of the system:

$$Y_i = \sum\nolimits_{j=1}^{M} w_{ij} y_{ij}, \tag{1}$$

where y_{ij}—component of the satisfaction function by criterion, $j = 1..M$, w_{ij}—weighting factor, $0 \leq w_{ij} \leq 1$ and $\sum\limits_{j=1}^{M} w_{ij} = 1\forall_i$.

Depending on the achieved satisfaction, the agent is given a bonus (penalty), the amount of which is determined through the bonus and penalty function defined for him: *Bi(Yi)*. The expected bonus and the current budget can be spent by the agent to compensate agents who agreed to concessions and whose state was changed during the schedule rearrangement.

Each order (need, task, etc.) j has several individual criteria x_i and assumed ideal values $x_{ij}{}^{id}$. At each ordering agent j, a bonus/penalty normalised function on component i ('virtual value') is computed.

The problem of each ordering agent can be transformed to the problem of maximising the satisfaction function $y_j{}^{task}$ of order j:

$$y_j^{task} = \sum_i \alpha_{ij}^{task} \cdot f_{ij}^{task}\left(x_i - x_{ij}^{id}\right), \tag{2}$$

Thus, in the SP-network of agents of orders and resources the problem of optimisation by several criteria—Pareto optimization—is solved. In other words, in the course of self-organisation of the solution (from bottom to top), the sum of minima of deviations from the ideal is first sought both for each single agent and for the whole community of agents [17].

Bonus and penalty functions determine the agent's ability to rearrange the schedule according to its interests.

The type of these functions is chosen in such a way that bringing the agent's state closer to its ideal of indicators increases the agent's satisfaction and bonus size.

Given these functions, each agent can earn virtual money, which it can spend on improving its state and transforming the schedule in the direction it wants.

At the same time, if one of the agents sees by its satisfaction function that it can significantly improve its performance in the course of conflict resolution, then it will be able to estimate also the gain in virtual currency that it will get from the expected placement, and from this 'expected' money (profit-win) to compensate the conflicting agent for the search for a new position and even for the deterioration of its position, so that in general the latter has not lost anything [18].

The resulting schedule is the result of collective decision-making and negotiation of order and resource agents (repair service agents), who make coordinated decisions and find trade-offs in the virtual market of the multi-agent system.

The essence of the compensation-based adaptive scheduling method is as follows:

- For each order and resource a different software agent is created, acting on its behalf and on behalf of it;
- The order agent starts planning by searching for the resources it needs in a scene that describes the current situation in the external world;
- If suitable resources (performers) are occupied, a conflict is recorded and negotiations are initiated to resolve it by shifting and releasing time slots;
- In the course of negotiations, variants are possible: a new task will go to a less suitable service technician, a previous task will leave or shift, performance will change or the availability time of the service technician (diagnostic equipment) will be extended;
- Changes produced by agents to the plan under mutual concessions are accepted if the gain in the value of the new solution exceeds the loss from the deterioration of other agents' satisfaction;
- Even after solving their problem and building a plan, each agent does not stop and continues to try to improve their situation [19].

When conflicts of interest arise, the depth and elasticity of movements or transitions of agents are regulated 'according to the situation' by virtual bonuses and penalties that play the role of virtual energy. As a result, a small order, temporarily pushed back by other large orders, for which it receives compensation, can accumulate such 'energy' that will allow it to 'beat off' a much better resource from a large order at a convenient time, according to the context of the situation.

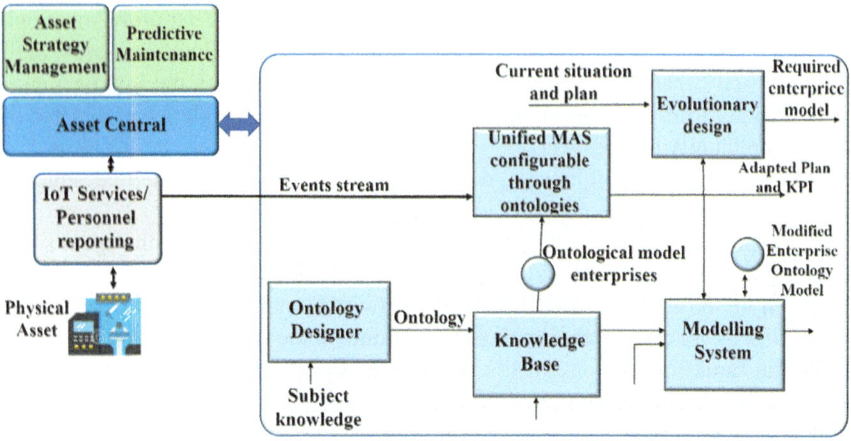

Fig. 4 Component architecture of the solution for planning and adaptive resource management in the EAM loop

Thus, the ultimate goal of finding a solution is to establish maximum satisfaction of the system as a general balance of interests of orders and resources taking into account their individual requirements [20].

The following component architecture of the solution (only when applied to the EAM domain) can be proposed for implementation in a corporate information system (see Fig. 4):

From the point of view of the correspondence of service strategies to the technological implementation of their implementation in the corporate information system, a model based on the 'strategy ladder', extended by the DIKIW pyramid, can be proposed, which allows linking the IT solution with the informatisation layer (business processes) and with digital services (see Fig. 5) [21].

The proposed model illustrates a simple principle: the more technological asset management strategy is supposed to be used in the enterprise, the more complex management model is required to support the composition of the selected strategies, the wider the coverage of the digital layer of the corporate IT landscape. We make one further observation.

Although our argumentation is based on the business case of operational planning of asset maintenance activities, the same multi-agent model can be applied to the planning tasks noted in Fig. 2. By extending the solution architecture, the ontology of the subject area, it is possible, in particular, to solve the problem of finding an effective assignment of maintenance strategies to units of equipment and medium-term resource planning (taking into account qualifications, mobility, equipment) in such a way that a balance between the composition and utilisation of mobile teams and remote technical diagnostics and repair facilities is found within the available capabilities (technical, budgetary).

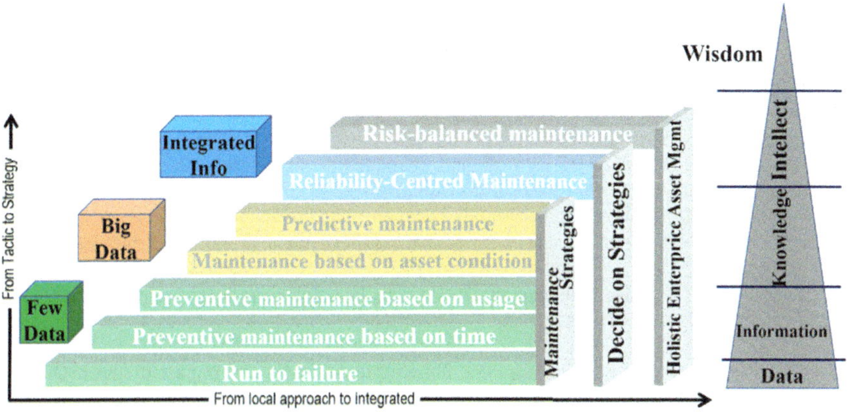

Fig. 5 Correspondence of data model levels to service strategies in the EAM

4 Conclusion

The article shows that effective resource planning in the EAM loop is an important tool to ensure the sustainability of the enterprise. With effective planning, a balance between the reliability of production assets and their industrial utilisation is achieved. Classical planning models can be applied to solve the problems formulated in the article, but only for significant constraints. Classical planning tools are not able to provide adaptive resource management in the EAM loop. The proposed planning and adaptive management methods do not have the main disadvantages of combinatorial models and provide the achievement of better resource utilisation effect. Multi-agent approaches provide the solution of planning tasks not only at the operational level of management, but also at the tactical and strategic levels. The evolution of adaptive management methods corresponds to the development of the layers of the DIKIW pyramid: if classical systems are oriented on the use of layers from data to the knowledge level, the proposed tools claim to use the capabilities of the intelligence level as well.

The successful use of multi-agent solutions in production and service (there are relevant references) forms a conviction in the justified expectations of success of their application in the field of EAM—both in the operational beast and in the long term.

Acknowledgements The research was supported by the Ministry of Science and Higher Education of the Russian Federation (Agreement No. 075-15-2022-1136 dated 01.07.2022).

References

1. White, B.: Managing Complexity by George Rzevski and Petr Skobelev, vol. 18, pp. 49–50. WIT Press, Southampton, Boston (2015). ISBN: 978-1-84564-936-4. https://doi.org/10.1002/inst.12050
2. Starukhin, V.: Digitalization as a method of reducing transaction costs of a manufacturing enterprise. Kant **41**, 95–99 (2021). https://doi.org/10.24923/2222-243X.2021-41.18
3. Mezina, T.: The impact of "Industry 4.0" on economy and production. In: European Proceedings of Social and Behavioural Sciences, pp. 670–677 (2020). https://doi.org/10.15405/epsbs.2020.12.04.77
4. ISO 55000:2014. Asset Management—Overview, Principles and Terminology. IDT, 2014
5. ISO 55002:2018. Asset management—Management Systems—Guidelines for the Application of ISO 55001. IDT, 2018
6. Alsuf, I., Alsuwaidi, M., Hamdan, S., Shamsuzzaman, M.: The impact of the ISO 55000 standard on the effectiveness of the organization: certificates of certified firms of the UAE. Total Qual. Manag. Bus. Excell. **32**(6), 1–19 (2018)
7. Ivanov, D., Frolov, K.: Optimizing the process of defining areas of responsibility in the context of small aircraft leasing using EAM systems. Technoeconomics 2, **3**(6), 4–15 (2023). https://doi.org/10.57809/2023.2.3.6.120:09
8. Iorsh, V.I., Kryukov, I.E.: Approach to building a process model of asset management. Lin-Technol. Lean Manuf. **3**, 22–27 (2016)
9. Ivanov, D., Pelipenko, E., Dubgorn, A. Improving the efficiency of drug supply management of a medical organization through the use of a business intelligence system. Technoeconomics, **2**(3), 79–89 (2023). https://doi.org/10.57809/2023.2.3.6.7
10. Fadeev, A., Kalyazina, S., Levina, A., Dubgorn, A.: Requirements for transport support of offshore production in the Arctic zone. Transp. Res. Procedia **54**, 883–889 (2021)
11. Tejada, A., Frolov, K., Overes, E.: On cloud solution to improve business performance for product deliveries. In: Jahn, C., Ungvári, L., Ilin, I. (eds.) Algorithms and Solutions Based on Computer Technology, pp. 79–89. Springer (2022). https://doi.org/10.1007/978-3-030-93872-7_7
12. Ivanov, D., Frolov, K., Frolov, A.: Implementation of ILS methodology on the basis of ISO 55000 series standards. In: Devezas, T.C., Berawi, M.A., Barykin, S.E., Kudryavtseva, T. (eds.) Understanding the Digital Transformation of Socio-Economic-Technological Systems, pp. 79–91. Springer (2024). https://doi.org/10.1007/978-3-031-56677-6_6
13. Li, A., Xie, Y., Li, S., Tsung, F., Ding, B., Li, Y.: Agent-oriented planning in multi-agent systems (2024). https://doi.org/10.48550/arXiv.2410.02189
14. Baddam, V., Gumussoy, S., Boker, A., Eldardiry, H.: Learning emergence of interaction patterns across independent RL agents in multi-agent environments (2024). https://doi.org/10.48550/arXiv.2410.02516
15. Youvan, D.: Emergent intelligence: a comparative study of Physarum polycephalum and AI models in distributed decision-making and global behavior (2024). https://doi.org/10.13140/RG.2.2.32942.73286
16. Brussel, H., Wyns, J., Valckenaers, P., Bongaerts, L., Peeters, P.: Reference architecture for holonic manufacturing systems: PROSA. Comput. Ind. **37**, 255–274 (1998). https://doi.org/10.1016/S0166-3615(98)00102-X
17. Skobelev, P.: Multi-agent systems for real time resource allocation, scheduling, optimization and controlling: industrial application. In: 10-th International Conference on Industrial Applications of Holonic and Multi-Agent Systems (HoloMAS), pp. 5–14. Springer (2011)
18. Shirzadeh Chaleshtari, A., Shadrokh, S.: A branch and bound algorithm for resource constrained project scheduling problem subject to cumulative resources. World Acad. Sci. Eng. Technol. **6**, 23–28 (2012)
19. Yanishevskaya, A., Pesterev, P.: The architecture of the multi-agent search system of the company. J. Phys. Conf. Ser. **1210**, 012161 (2019). https://doi.org/10.1088/1742-6596/1210/1/012161

20. Rowley, J.: The wisdom hierarchy: representations of the DIKW hierarchy. J. Inf. Sci. **33**(2), 163–180 (2007). https://doi.org/10.1177/0165551506070706
21. Arunarani, A.R., Manjula, D., Sugumaran, V.: Task scheduling techniques in cloud computing: a literature survey. Futur. Gener. Comput. Syst. **91**, 407–415 (2019)

Modelling the Effect of Additive Technologies: Logistic Aspect

Liubov Nefedova

Abstract As is known, the delivery of components to the production site requires time and money. In addition, it is necessary to carefully work out and optimize logistics processes to improve production efficiency. Today, the above tasks can be eliminated through the use of additive technologies. An innovative approach helps to significantly simplify the production process. The article discusses the efficiency of using additive technologies in production processes. Articles on additive technologies, their application are analyzed and conclusions are made. Examples of successful implementation of additive technologies in various industries, such as automotive, aerospace, electronics and medicine are considered. Also, "as is" and "as should be" models are built to practically evaluate the efficiency of using additive technologies. It is emphasized that additive technologies help reduce logistics costs, increase production flexibility and the ability to create complex geometric shapes in order to individualize production. In conclusion, conclusions are also made on the prospects and trends in the development of additive technologies in the future.

Keywords Additive technologies · Additive manufacturing · 3D printing · Transportation costs · Supply chain

1 Introduction

The delivery of spare parts and components to remote sites is a complex and multi-stage process that requires careful coordination and planning. In the absence of additive technologies, effective supply chain management becomes extremely important, since all spare parts and components must be physically delivered to the point of use. This imposes certain restrictions on time and financial resources that must be taken into account when planning deliveries. More often than not, traditional delivery

L. Nefedova (✉)
Obukhov Plant, Saint Petersburg, Russia
e-mail: Avantage2001@yandex.ru

I. Ilin and M. Youzhong (eds.), *Digital Systems and Information Technologies in the Energy Sector*, Lecture Notes in Networks and Systems 1244,
https://doi.org/10.1007/978-3-031-80710-7_11

methods impose significant costs on enterprises related to transportation, storage, and distribution, which in turn complicates overall logistics.

Using additive technologies such as 3D printing to manufacture components and spare parts directly at the point of use can significantly reduce transportation costs. These technologies, based on the layer-by-layer addition of material, open up new horizons in the production process, allowing a company to quickly respond to changes in demand and produce parts as needed. Switching from traditional production and delivery methods to additive manufacturing can significantly reduce transportation costs and shorten delivery times, which is important in industries that require a high degree of flexibility and rapid response to market conditions.

In addition to reducing costs and time, additive technologies offer the possibility of customized production, which allows for the creation of unique parts that meet the specific requirements of the end user. This approach involves not only the physical reduction of distances, but also the optimization of the production process itself, which in turn contributes to the increase in the efficiency of the enterprise. In the context of globalization and the rapid development of technology, the need to adapt to modern conditions becomes especially relevant.

In order to substantiate the importance and feasibility of using additive technologies, it is necessary to conduct a comprehensive analysis at the theoretical and practical levels, using scientific articles, real cases, as well as architectural and mathematical models. Such an analysis will identify the key advantages of additive technologies compared to traditional methods, as well as determine strategies for their implementation in existing production chains. Research on this topic seems important not only for understanding modern trends in the field of production, but also for forming the basis for future innovative solutions that contribute to the effective management of production processes.

2 Methodology

In this study, the economic effect of using additive technologies from the point of view of the transport aspect is assessed using a comprehensive approach. The project involves theoretical and practical proof of the effectiveness of additive technologies for the production of spare parts and components. The comprehensive approach includes an analysis of existing publications on the topic of the project, a study of real cases of using similar technologies to reduce transport costs and a comparative analysis of costs in both cases and identifying conditions for achieving savings.

The research methods include literature and scientific article analysis, real case studies, process modeling and comparative analysis. The search for scientific articles is carried out using scientific article databases, such as Google Scholar or ResearchGate, searching for publications by key queries "additive technologies and economic effect", "additive technologies and transport aspect", "additive technologies and logistics costs", "additive technologies and transport costs", Russian and foreign sources and publications are considered. Then, the obtained information is systematized to write a coherent text with links to primary sources and conclusions.

Process modeling consists of reviewing the current ("as is") and optimized ("to be") process diagrams. The former involves creating an architectural model of the process of delivering spare parts to remote sites without using additive technologies. The model includes all key elements: production processes, vehicles, delivery routes, etc. The "to be" model includes process diagrams using additive technologies, where such elements as 3D printers on site, raw materials for printing spare parts and components, and other changes in the infrastructure are added to the processes. The models are visualized using business process modeling software (for example, Archi).

The expected results consist of theoretical and practical parts. Theoretical results should include systematization of knowledge about the impact of additive technologies on logistics and transportation of components and spare parts to a remote site. Practical results should include construction of architectural models and derivation of mathematical formulas confirming the identified impact of using additive technologies and development of recommendations for optimization of logistics and transport processes.

3 Literature Review

3.1 Article Analysis

According to the article [1], additive technologies are considered part of the 6th technological paradigm and will be significantly integrated into various areas of production, including heavy and light industries and the transport sector.

The integration of additive technologies into manufacturing can lead to a number of positive effects [1] emphasizes that the method of processing the material and turning it into a part affects the final strength of the product, and additive technologies are mentioned as a serious competitor to traditional methods. For some types of products, the use of additive technologies is mandatory to achieve the desired strength and wear resistance characteristics.

In [2] additive manufacturing is described as a process of manufacturing components from 3D computer models by selectively depositing, curing, or compacting materials layer by layer. Each layer is the cross-sectional geometry of the component at a set height.

One of the most important effects of using additive technologies in manufacturing is the reduction of logistics costs due to the production of parts and components directly at the point of use. This eliminates the need to transport goods from the manufacturer to the distributor, to the consumer, which is especially important for large and heavy products. A literature review [3] notes that additive technologies contribute to a significant reduction in logistics costs by reducing the need to transport finished products. The study [4] also emphasizes that 3D printing allows goods to be produced closer to the consumer, thereby reducing transportation costs. The article [5] also notes savings on fuel costs, while reducing delivery distances due to the use of additive technologies. The article [6] lists the following advantages and justifications for the revolutionary nature of these technologies: instantaneous transfer of part models over the network to anywhere in the world, a reduction in the number of component parts for parts, and individual production of products of any shape. These factors help to reduce transportation and warehouse costs, there is a saving in raw materials, and production waste is minimized [6].

The article [7] describes the stages of additive manufacturing. They are as follows: the emergence of an idea for a necessary part, development and approval of the concept, creation of a computer 3D model that will be used in production, printing of a prototype (sometimes a full-fledged product), refinement and identification of deficiencies by engineers, application of the model in production: serial launch or creation of a single individual final product [7].

The author of the article [5] identifies 2 types of additive manufacturing: centralized and decentralized. Centralized manufacturing is similar to classical manufacturing and is the implementation of alternative technology into classical manufacturing. Decentralized manufacturing involves the distribution of production capacities across various locations physically close to consumers. It is noted that it is the decentralized type that significantly reduces logistics costs. The study [8] notes that logistics costs typically account for more than 5% of the final cost of the product. According to the work, this amount can be significantly reduced by integrating additive manufacturing. The author notes that delivery costs can be reduced by AT due to a reduction in: warehouse space; transportation costs, delivery distance, and the minimum effective batch size of the product, leading to a reduction in the load on equipment used in cargo transportation [8]. It is also emphasized that additive manufacturing is less dependent on the location area than traditional manufacturing and can be deployed virtually anywhere in the world, thus reducing the cost of not only domestic logistics, but also international logistics. An additional effect of production localization is a significant increase in delivery speed.

The article [9] describes the main methods of additive manufacturing, which include the extrusion printing method (FDM/FFF), powder method (SLS, EBM, SLM), inkjet printing (3DP) and lamination (LOM), which allow the creation of complex parts directly at the place of their use, which significantly reduces transportation costs.

In addition to the ability to manufacture products at the point of consumption, some researchers note the possibility of reducing delivery times by creating local production using additive technologies. The article [3] notes that additive technologies allow spare parts to be manufactured right near the point of use, which creates conditions for "Just in Time" deliveries and significantly reduces equipment downtime and associated costs. The materials [10] also note that additive manufacturing allows for a reduction in the time required to develop and manufacture new products, which speeds up their delivery to the consumer.

Studies [11] and [12] show that additive technologies can reduce the need for finished product warehousing. By reducing production volumes and shortening supply chains, which in turn reduces storage and logistics costs. The works note that digitalization and the introduction of additive technologies allow for more efficient inventory management and reduced storage costs [11]. Also, the use of 3D printing allows for the production of the exact number of required parts, which reduces storage costs due to a reduction in permanent warehouse stocks [12]. The work [13] points to the economic aspects of additive manufacturing development, emphasizing its ability for flexible production and reduced logistics and warehousing costs due to localization of production.

Reducing the length of supply chains and localising production through the introduction of additive technologies help to reduce carbon footprints. This is achieved through precise dispensing of materials and minimising waste in 3D printing. A study [3] emphasises that additive technologies can significantly reduce the environmental burden by reducing transportation and material usage.

Due to its technological features, additive technologies can adapt to changes in demand and customize production to the needs of a specific customer, which reduces the need for large batches of products and long-term storage and corresponds to the trend of mass customization. Articles [4] and [10] agree that 3D printing allows for the production of products as needed, which reduces storage costs and improves production flexibility. The work [5] also notes the high speed of production reconfiguration to maintain a balance between supply and demand in both quantitative and qualitative terms.

The work [1] notes that the integration of additive technologies into production processes can significantly reduce production lines, minimize the manufacturing time of complex technological parts and improve the performance characteristics of the final product. This, in turn, helps reduce production and logistics costs.

Among the limitations of additive manufacturing [8] is the volume limitation. The specifics of the technology limit the speed, while increasing the possible level of customization of similar products.

Thus, for the production of large batches of identical products, the technology is less efficient than traditional production methods, while being more efficient for small and medium batches of similar, but not absolutely identical, goods.

Additive technologies can significantly improve manufacturing efficiency, shorten supply chains and localize production, thereby reducing the logistics costs of finished products and/or processing resources using traditional methods, reducing the carbon footprint, increasing production efficiency and flexibility, and improving the quality of finished products.

3.2 Analysis of Real Cases/Examples of Savings on Transportation Costs Through the Use of Additive Technologies

The authors of the article [1] emphasize that the use of additive technologies is a new global trend in production in various industries, namely in the automotive industry, construction, aerospace, medicine, etc. For example, in the construction industry, one of the promising areas in the field of construction of all kinds of structures is construction 3D printing [14]. Also, 3D printed implants are already used in medicine, which are ideal for each patient [15]. In addition, 3D printing makes it possible to precisely control the dose of drugs and produce dosage forms with complex drug release profiles and prolonged action [16].

In the aerospace industry, additive technologies are already widely used for 3D printing of parts. Airbus uses additive technologies in the production of components for its aircraft [17]. This allows the company to produce parts closer to their final assembly locations, reducing logistics costs and delivery times. The use of 3D printing also allows Airbus to quickly respond to spare parts needs, reducing the need for large inventories in warehouses and cutting transportation costs. One of the developments—partition panels for the aircraft cabin—not only reduces the storage and transportation time of spare parts, but also, due to its lighter design, helps reduce the fuel costs consumed by aircraft during flights.

There are many examples of additive manufacturing applications in the automotive industry. Companies such as Nissan, Opel, BMW and others use AT for prototyping, tool making, coating, etc. [18]. BMW uses 3D printing to produce prototypes and components for its cars: from individual orthoses for employees to large robotic grippers [19]. This has reduced logistics costs by 20% and significantly accelerated the development and testing of new models. 3D printing has allowed BMW to reduce the need for international transportation of components and increase the speed of response to changes in the design of cars.

In the electronics sector, Siemens is actively implementing additive technologies in production processes, which allows the company to produce spare parts and components at the point of use. This significantly reduces equipment downtime and reduces transportation costs by up to 25%. For example, in the Siemens Mobility division, the use of 3D printing has reduced the production time of customized repair parts by 95%. In particular, it has "reduced our dependency on outsourcing tools via suppliers" [20]. This, in turn, has optimized the supply chain and reduced logistics costs.

These cases demonstrate how the use of additive technologies allows companies to significantly reduce transportation costs, improve logistics and increase the overall efficiency of production processes.

4 Results

4.1 Description of the Model «as is»

The general process model without additive technology in industry is presented below in terms of transportation and decomposition of transportation and production processes, they are shown in Figs. 1, 2 and 3.

The figure shows the stages of supply of components, parts to a remote object. This model includes the following points:

1. Planning and preparation
   a. Requirements and limitations
   b. Identification of the list of required materials, components and parts
2. Logistics organization
   c. Sourcing and contracting suppliers
   d. Choice of delivery method (air, road, sea, rail)
   e. Calculation of transportation costs
   f. Organization of delivery to the final destination
3. Transportation of cargo

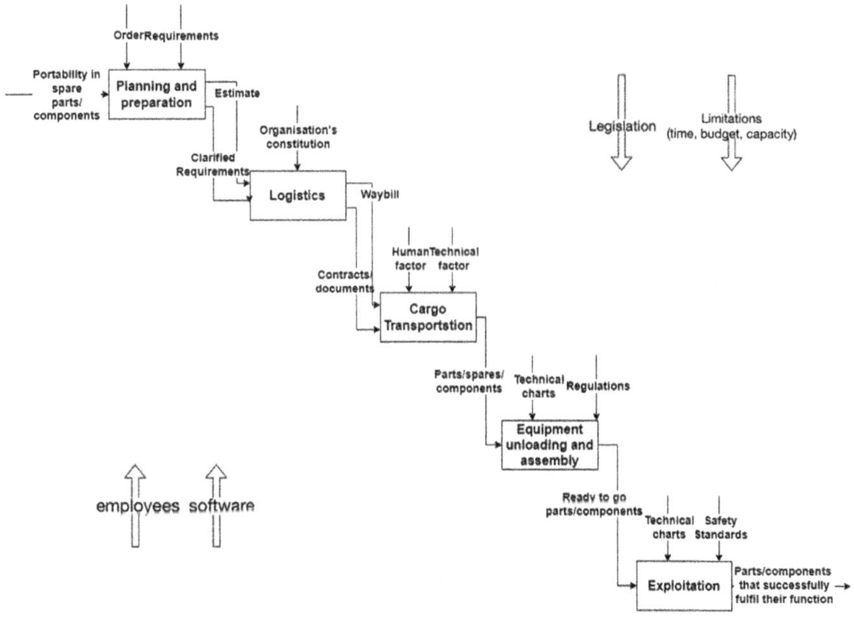

Fig. 1 Model «as is»

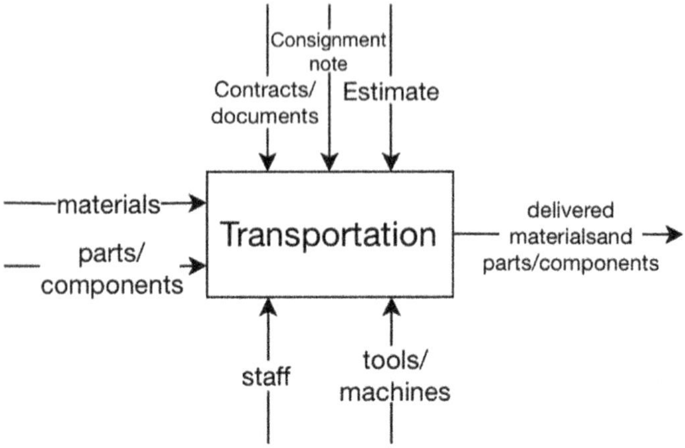

Fig. 2 Transportation «as is»

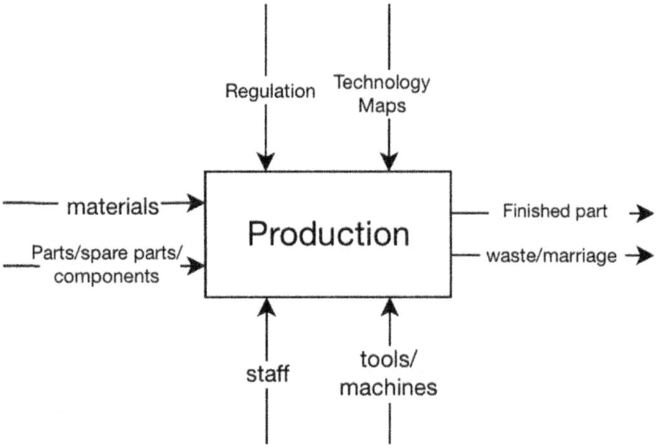

Fig. 3 Production «as is»

  g. Delivery of the cargo to a remote location

4. Unloading and assembling equipment

  h. Verification of conformity of the goods with the documents
  i. Inventory and distribution on site
  j. Installation and adjustment of parts and components on site
  k. Carrying out the necessary checks

5. Exploitation

  l. Completion of tests and commissioning of equipment

Based on the illustration in Fig. 2, it can be said that without additive technologies there is a need to transport large quantities of spare parts and components, including for vehicles involved in delivery. Because they cannot be created directly on the object. This results in the possibility of additional time and money costs, due to the likelihood of marriage among the cargo. Also, more transportation is needed due to the large number of parts and supplies supplied.

Figure 3 shows the decomposition of the processes of unloading and assembling parts and components into the ready equipment, and partially reflects the process of its operation, since at the output we have a finished product. The processes are combined for ease of analysis. The model reflects that production without additive technologies entails marriage, which leads to the costs of its recycling, that is, it is not waste-free production.

4.2 Description of the Model «to Be»

In this article, the diagrams are similar to those of the previous one, they are shown in Figs. 4, 5 and 6, but for production using additive technologies. The only difference is that there is a process to create parts directly on site.

The creation of parts is included in this model, because we do not have to transport already finished parts—it will be easier to produce them on site. This process reduces the cost of transportation due to reduced volume of delivered cargo; the main material is printing powder, printing equipment also need less. At the same time, additive

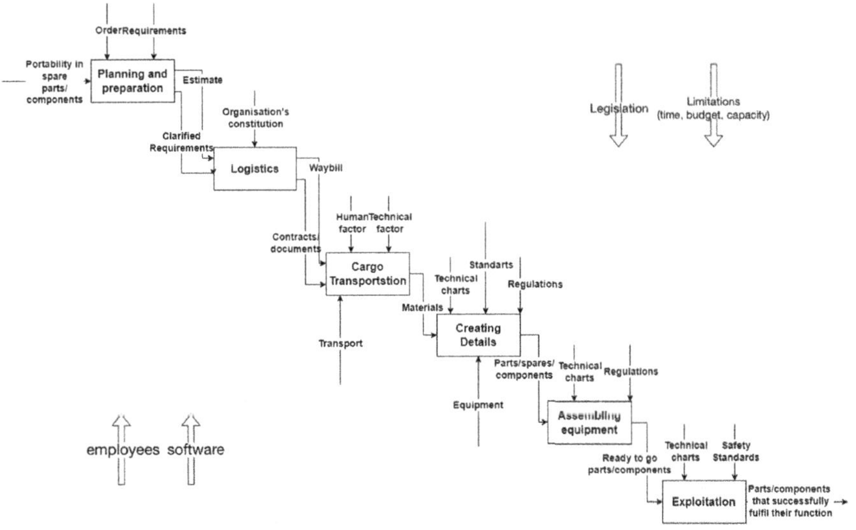

Fig. 4 Model «to be»

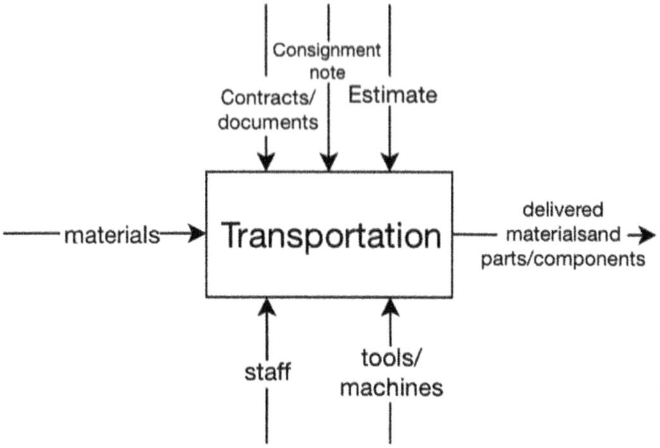

Fig. 5 Transportation «to be»

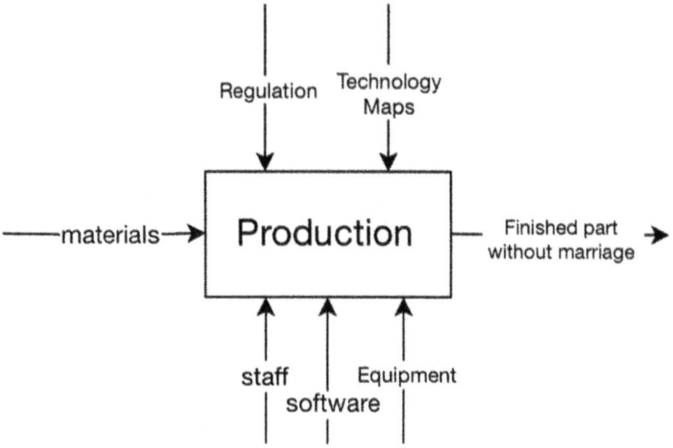

Fig. 6 Production «to be»

technologies reduce the cost of processing scrap as this production is waste-free—defective parts are converted into original material and reused.

In Fig. 5 we see a large-scale sub-process of transporting cargo to a remote site using additive technologies. There is no strong difference in this paragraph, the main points regarding costs have been outlined above.

This figure combines the processes of part creation, assembly and operation to facilitate their analysis. In general, these steps are faster and require less cost due to the availability of additive technologies than the same parts production but without them. How to eliminate the likelihood of marriage: when a marriage is detected, it is much faster to rework the defective make and create a new one from these materials immediately on site than to order and wait for a new one.

5 Model Comparison

Above we have considered two models: «as is» and «as should be», and also in more detail the most important points for this work and aspects of these models. Now we will compare them.

1. Planning

The «as is» model emphasizes the capabilities and conditions of suppliers, in the «as should be» model first define the capabilities and production capacity of 3D-technologies.

2. Production

Without additive technologies, the production of parts and components is only possible in-plant according to reference forms and drawings, the introduction of additive technologies makes the design and development process more flexible through its digitalization. It also opens up the possibility of designing unique, almost impossible to execute without 3D printing parts, removes seams and overall improves production quality. Additive technologies greatly reduce time costs, eliminate the likelihood of marriage, make production waste-free.

3. Transportation

In the model «as is» a large number of spare parts and components must be delivered, to further assemble from them equipment on site, the model «as should be» allows only to deliver to the object raw materials for 3D printing, which significantly reduces costs.

4. Storing

Due to the lack of need for storage of large volumes of materials and parts, additive technologies allow you to eliminate large storage areas and leave a minimum for storage of raw materials for printing.

5. Unloading

The components and parts when applying additive technologies are created directly on site, which excludes the financial costs of disposal of scrap and time required to wait for a new delivery to replace defective parts. In the «as is» model, the commissioning of equipment depends on checking the load already at the site and any defective parts have a strong impact on the process of assembly and commissioning of equipment.

6 Mathematical Formulas

The overall cost formula for supplying parts and components to a remote site is as follows:

$$C = C_{mat} + C_{transp} + C_{storage} + C_{insurance} + C_{else}$$

C_{mat}—cost of materials and parts.
C_{transp}—cost of transportation.
$C_{storage}$—cost of storage.
$C_{insurance}$—cargo insurance costs.
C_{else}—unforeseen expenses, taxes and fees etc.

When using additive technologies this formula is transformed:

$$C = C_{mat} + C_{transp} + C_{else} + C_{insurance} + C_{else} + C_{print} + C_{equip}$$

C_{mat}—cost of materials and parts.
C_{transp}—cost of transportation.
C_{elec}—cost of electricity to run 3D printer.
C_{equip}—maintenance and repair costs of equipment.
C_{print}—printing costs, including consumables.
$C_{insurance}$—cargo insurance costs.
C_{else}—unforeseen expenses, taxes and fees etc.

Savings are achieved by reducing transportation costs, the model «as is» includes high transportation costs, especially for remote sites, since most parts are produced directly at the site, additive technologies significantly reduce these costs; Reduction of storage space and stocks—the costs for storage and maintenance of warehouses are reduced, as spare parts and components are produced when necessary, respectively, and the need for large storage spaces is eliminated; Reduction of logistics and administrative costs—the cost of logistics and insurance is reduced.

So, we analyzed articles and real cases. Research shows that additive technologies reduce transport costs by reducing the need to move finished products and localize production closer to the consumer. This reduces fuel costs and allows "Just in Time" deliveries to be organized, reducing downtime of equipment. In addition, the need for product storage is reduced and inventory management is improved.

In the study and comparison of two models of transportation of parts and components to remote sites—«as is» and «as should be» were drawn up process diagrams, models and differences between traditional production and production with additive technologies are considered. The work was detailed in the most important aspects of these models and the influence of 3D-technologies on them. The cost structure of both models and its variation depending on the availability or absence of additive technologies is also considered.

Thus, additive technologies are an effective way to reduce logistics costs, increase production flexibility and improve product quality, which confirms the feasibility of their introduction into modern production.

7 Discussion

The results of this study show that additive technologies can significantly reduce transportation and related costs in the production and delivery of spare parts and components to remote sites. Despite the positive findings, there are still some aspects that require further research and analysis.

This section is devoted to a discussion of possible directions for further research and other potential applications of the results of this work.

One of the most important factors influencing the profitability of using additive technologies is the scale of production. In large-scale production, the costs of traditional methods of transportation and manufacturing can be lower than those of 3D printing. It is necessary to conduct research that will determine the boundaries of the production sizes at which additive technologies are economically feasible. This may include the construction of models of the dependence of savings on the number of products produced.

In addition to economic advantages, additive technologies can also bring environmental benefits. By reducing carbon dioxide emissions and energy consumption during transportation. Given the relevance of research on reducing the negative impact of production on the environment, assessing the environmental impact of additive technologies compared to traditional production and transportation methods is also an important area for further research.

One of the most important advantages of using additive technologies is the ability to produce parts and components in extreme conditions. Where logistics is difficult or impossible. Such conditions include, for example, polar expeditions, remote objects in deserts and space missions. In such cases, additive technologies may be the only way to produce the necessary parts.

For small and medium-sized businesses, additive technologies can open up opportunities for creating local production capacities. Developing business models based on the use of these technologies will help entrepreneurs optimize transportation costs and reduce the time to market for their products.

This project has shown that additive manufacturing has a high potential to reduce transport and logistics costs, but there are many areas for further research. Further work on this topic will help to better understand the limitations, conditions and advantages of using additive manufacturing, and will also show new areas of application.

8 Conclusion

Additive technologies, as part of the 6th technological paradigm, have significant potential for integration into various industries. Their implementation can lead to increased production efficiency, improved quality of manufactured products, as well as create conditions for mass customization of products and increase the adaptability of production to rapidly changing conditions of the external and internal business environment. In addition, one of the main advantages is a significant reduction in logistics and storage costs due to the ability to produce parts and components directly at the place of use.

Additive technologies make it possible to adapt to changes in demand and customize production, which also reduces storage costs and increases production flexibility.

In conclusion, additive manufacturing is revolutionizing logistics by offering more flexible and cost-effective solutions. Reducing inventory, lowering transportation costs, and increasing production flexibility make additive manufacturing an attractive choice for a variety of industries. The introduction of 3D printing allows companies to respond more quickly to changes in demand and better cope with logistics challenges, especially in remote and hard-to-reach areas.

References

1. Armashova-Telnik, G.S., Sokolova, P.N., Degterev, D.V.: Additive technologies: an innovative effect in industry. Bull. Vguit **82**(4), 347–353 (2020)
2. Sokolnikov, V., Andryukhina, D., Zinovkina, D.: Analysis of additive technologies. Integral **3**, 610–622 (2023)
3. Woldesilassiea, T.L., Lemu, H.G., Gutema, E.M.: Impacts of adopting additive manufacturing process on supply chain: systematic literature review. Logistics **8**(1), 3 (2024)
4. Hofmann, E., Sternberg, H.: Supply chain management and industry 4.0: conducting research in the digital age. Int. J. Phys. Distrib. Logist. Manag. **49**(10) (2019)
5. Calignano, F., Mercurio, V.: An overview of the impact of additive manufacturing on supply chain, reshoring, and sustainability. Cleaner Logistics and Supply Chain **7**, 100103 (2023)
6. Rostec: Additive manufacturing in action. https://rostec.ru/news/additivnye-tekhnologii-v-deystvii/?ysclid=lzwrz6hcp8915248731. Accessed 20 July 2024
7. Smirnova, A.: Additive Manufacturing: Stages, Technologies, Advantages and Applications. https://top3dshop.ru/blog/additivnoe-proizvodstvo.html?ysclid=lzws04n2bg100799443. Accessed 10 July 2024
8. Attaran, M.: Additive manufacturing: the most promising technology to alter the supply chain and logistics. J. Serv. Sci. Manag. **10**(03), 189 (2017)
9. Kiryushkin, D.A., Anashkin, V.V.: Basic methods of additive manufacturing. Actual Probl. Aviat. Cosmonautics **1**, 191–193 (2022)
10. Bauernhansl, T., Emmrich, V., Paulus-Rohmer, D., Döbele, M., Schatz, A., Weskamp, M.: Geschäftsmodell-Innovation durch Industrie 4.0 (2015)
11. Hermann, M., Pentek, T., Otto, B.: Design principles for industrie 4.0 scenarios. In: 2016 49th Hawaii International Conference on System Sciences (HICSS), pp. 3928–3937. IEEE (2016, January)

12. Haggarty, D.R., Shurin, J.B., Yamanaka, K.L.: Assessing population recovery inside British Columbia's Rockfish Conservation Areas with a remotely operated vehicle. Fish. Res. **183**, 165–179 (2016)
13. Boychenko, O.V.: Economic aspects of development of innovative digital production based on additive technologies. Eur. Res. **2**(25), 42–44 (2017)
14. Samandasyuk, G., Slesarev, I., Kozhen, M.: Additive technologies in construction. Globus 57–58 (2018)
15. Atmommedia: Additive Technologies. https://atommedia.online/reference/additivnye-tehnologii/. Accessed 15 July 2024
16. Yulusov, I., Papko, S., Nikulin, I., Klimenko, I.: Application of additive technologies in modern medicine. StudNet **2**, 860–869 (2022)
17. Kudryashov, N.: Airbus with 3D Printing on Board: Bionic Design, Weight Reduction, Efficiency. https://blog.iqb.ru/Materialise-Airbus-Case/. Accessed 18 June 2024
18. Teshaboev, U.M.: New trends in the development of the automotive industry. Universum: Tech. Sci. **10**, 33–36 (2021)
19. BMW Group. https://www.Press.Bmwgroup.Com/Global/Article/Detail/T0442351en/Bmw-Group-Expands-Use-Of-3d-Printed-Customised-Robot-Grippers'. Accessed 20 June 2024
20. Pulling Its Weight: Siemens Mobility Boosts Custom Production Capability with FDM Additive Manufacturing. https://www.stratasys.co.jp/Contentassets/7b2dc97a9d74463e874db0c920bacba0/Cs_Fdm_Au_Siemensmobility_A4_0219a.Pdf?V=48f938. Accessed 1 July 2024

Characteristics of Digital Infrastructure Development in the Arctic

Anastasia Levina, Sofia Kalyazina, Alissa Dubgorn, and Anastasia Gorbacheva

Abstract The Arctic zone of Russia is a region characterised by high economic potential, importance for the development of the Russian economy and national security, but also by a number of problems, environmental fragility and underdevelopment of logistics, digital and other infrastructure. Russia has adopted the Strategy for the Development of the Arctic Zone, the State Programme "Socio-Economic Development of the Arctic Zone of the Russian Federation" and the Programme "Digital Platform Arctic 2035", which aim to comprehensively address the existing problems of the region and unlock its potential. At the same time, measures for the development of digital infrastructure are a very important solution. This paper analyses the available cases of digital technology application in the Arctic zone and the key digital technologies used. On this basis, aspects of consideration of the digital transformation of the Arctic zone of the Russian Federation are highlighted and factors of sustainable development of the Arctic in relation to the possibilities of digital technologies are identified. Digital infrastructure can and should be used to transform the logistics infrastructure. The paper considers the main components and ways of integrating digital infrastructure using the example of the key Arctic transport hub of Murmansk. In the future, the integration of digital and transport infrastructure for other ports and other transport hubs in the Russian Arctic will be considered.

Keywords Arctic · Digital infrastructure · Digital technology · Digital transformation · Murmansk

A. Levina · S. Kalyazina (✉) · A. Dubgorn · A. Gorbacheva
Peter the Great St.Petersburg Polytechnic University, Saint-Petersburg, Russia
e-mail: kalyazina_se@spbstu.ru

I. Ilin and M. Youzhong (eds.), *Digital Systems and Information Technologies in the Energy Sector*, Lecture Notes in Networks and Systems 1244,
https://doi.org/10.1007/978-3-031-80710-7_12

1 Introduction

The Arctic zone of Russia is an important region for several reasons. Among the economic reasons are rich natural resources, the development of which can bring Russia significant economic benefits and ensure energy independence; promising sea routes (climate change and melting ice open up new opportunities for using the Northern Sea Route, which reduces the time and cost of shipping goods between Europe and Asia), fisheries; tourism. The Arctic is a strategically important region for Russia in terms of national security, ensuring control over the country's northern borders [1]. At the same time, it is necessary to take into account environmental factors related to climate change and the importance of preserving biodiversity. Thus, the Arctic zone of Russia has enormous economic, political, environmental and social significance and requires special attention and effective policies aimed at its sustainable development and preserving its unique nature.

In the era of digital transformation, there is a unique opportunity to use technology as a tool to achieve these goals.

This paper explores the concept of digital support for sustainable development in the Arctic zone, highlighting the various ways in which digital solutions can contribute to preserving the Arctic ecosystem, empowering local communities and promoting responsible economic activity [2, 3]. To this end, the potential of digital infrastructure to transform Arctic logistics is explored. Digital infrastructure, including technologies such as artificial intelligence (AI), the Internet of Things (IoT), Big Data, blockchain and 3D printing, offers a path to achieving greater efficiency and sustainability in the Arctic logistics sector.

The integration of digital technologies is examined using the case of Murmansk, a key Arctic transport hub.

2 Materials and Methods

This study delves deeply into the complex interrelations between digital technologies, digital transformation and sustainable development in the Arctic zone of Russia. The research methodology is based on the analysis and synthesis of existing literature. A review of scientific papers for the period from 2011 to 2023 was conducted to cover the latest research in this area. For this purpose, scientific publication databases such as Web of Science and Google Scholar were used. The literature analysis revealed key aspects of digital transformation in the Arctic zone of the Russian Federation. Based on these aspects, factors of sustainable development of the Arctic related to the potential of digital technologies were identified.

3 Literature Review

The article by [4] discusses the potential role of technology in promoting sustainable development in the Arctic and the importance of addressing the unique challenges facing the region. Investments in R&D related to innovative technologies can have ripple effects on socioeconomic development.

Many scientific articles are devoted to the topic of digital twins. PolArctic, an Alaskan startup, is developing a digital twin of the Arctic Ocean to combat climate change and better manage Arctic fish stocks [5], predicting the impact of climate change on Arctic fisheries. Using artificial intelligence and machine learning, PolArctic plans to integrate different types of data, including indigenous knowledge and scientific research. The initiative is expected to help the fishing industry by optimising catch volumes, setting accurate catch limits, and helping vessels navigate through dynamic protected areas that take into account species biomass and climate change. The digital model will mimic real ocean conditions such as acidification and temperature changes. The potential benefits of the digital twin model extend beyond the fishing industry to the shipping, climate research and defence sectors. The increased accuracy of the model could help the shipping industry by predicting navigation problems such as sea ice, potentially leading to safer routes and lower insurance costs. In the defence sector, the model can help identify trends in illegal fishing and support coastguard operations. In addition, the model will support local Arctic communities by supporting subsistence activities and food security, as well as providing valuable data for climate research.

The paper by [6] explores the development of a digital clone to support the development of Arctic regions.

The article by [7] analyses the integration of digital technologies in the oil and gas and mining industries in the Russian Arctic, taking into account the current state and projected future trends in the industry. Several trends and perspectives are identified, highlighting the importance of digital technologies to improve business process management and address challenges such as harsh climatic conditions, cybersecurity threats and sensitivity to foreign technological dependence. The adoption of digital technologies is seen as a response to the need for improved operational optimisation, particularly in the harsh and remote environment of the Russian Arctic. The paper notes that while companies such as Gazprom Neft are leaders in digital innovation, the sector's overall digital transformation is hampered by a lack of skilled personnel, inadequate facilities and growing cybersecurity threats. Key concerns include the potential impact of sanctions, which could either spur domestic software innovation or hinder the development of the Arctic project due to increased operational costs. In addition, the COVID-19 pandemic has catalysed the adoption of more automated digital methods in production and business processes.

Digital infrastructure needs a reliable, adaptive power supply due to power outages in the Arctic. The paper by [8] is devoted to this problem.

The article by [9] examines digital transformation in the supply chain for offshore oil and gas projects in the harsh Arctic environment. The study shows that Gazprom

Neft is currently leading the digital transformation efforts. However, even Gazprom Neft lags behind global leaders in some areas, such as well operations. High dependence on imported equipment, which is further constrained by sanctions, limits the progress of digital transformation. The authors conclude that digital transformation alone is not enough to make Arctic offshore projects economically viable in the medium term. They recommend focusing on disruptive technologies that significantly reduce production costs, and suggest further research on how to implement such technologies in the face of sanctions pressure. The paper postulates that while renewable energy sources (RES) will play a greater role in Arctic regions, the oil and gas sector, supported by digital technologies, will dominate the energy market for the foreseeable future. Despite progress, digital transformation in the sector remains challenging, with virtual and augmented reality, the Internet of Things and artificial intelligence being integral but underutilised technologies due to operational integration issues.

The document that sets the context based on two key strategies—the State Policy of the Russian Federation in the Arctic until 2035 and the Strategy for the Development of Artificial Intelligence in the Russian Federation until 2030 [10]. This strategic framework emphasises the Russian government's intention to harness the potential of artificial intelligence and robotics to further exploit the rich mineral resources of the Arctic, while mitigating environmental impacts and overcoming extreme local conditions. The role of AI in the Arctic is diverse, spanning mining, security, healthcare, transportation, construction, energy, historic preservation, environmental monitoring and even tourism. Known applications include intelligent automated process control systems to solve complex gas extraction problems and the development of robotic systems for search and rescue operations. The paper also discusses improving health care through telemedicine, which is critical for remote Arctic communities.

With regard to the development of artificial intelligence and robotics in the Arctic, several initiatives stand out:

- Oceanos marine robots for environmental monitoring [11].
- Robotic underwater and surface vehicles with adjustable hull geometry for versatility in research and logistics applications [12].
- Chinese development of the Tan'so 4500 autonomous deep-sea robot for Arctic exploration [13].
- IceNet, a new artificial intelligence sea ice forecasting tool from the British Antarctic Survey and the Alan Turing Institute [14].

The article by [15] discusses the need for a strategic and sustainable approach to the development of Russia's Arctic territories, with a special focus on the importance of sea port infrastructure within the framework of blue economy principles. It emphasises the need to use the Arctic's water potential for sustainable development, while maintaining ecological balance and minimising anthropological impact. Central to the discussion is the concept of smart ports, which combine advanced information technology with modern management techniques through digitalisation. This approach aims to increase environmental friendliness and efficiency while reducing negative impacts on the ecosystem. Environmental control becomes the main focus of risk management, enabling the monitoring and mitigation of negative environmental impacts. The second vector of control in risk management relates to information security. The article highlights the importance of intelligent decision support systems based on ontological models, which facilitate the creation of databases and knowledge bases necessary for effective risk management.

Examples from Anchorage (USA) and Bude (Norway) [16] show how Arctic cities are gradually adopting smart technologies to improve their transport systems. In Russia, the authors note the importance of maritime transport, especially the Northern Sea Route, and the importance of port cities such as Murmansk and Sabetta. Innovative projects such as Yamal LNG and the Integrated Development of the Murmansk Transport Hub underline the Russian government's commitment to Arctic infrastructure development. The authors propose introducing smart city elements through digital platforms in major Arctic cities as a starting point for improving logistics services, with the ultimate aim of expanding these platforms for full-scale smart city development. These proposals are in line with the Russian 'Strategy for the Development of the Arctic Zone of the Russian Federation and National Security until 2035'.

The article also highlights the importance of implementing information and communication infrastructure as an important step in the development of the Arctic, pointing to ongoing projects such as the Arctic Connect link and the Arctic Labs digital platform as key initiatives to achieve this goal. The authors conclude that the development of Arctic transport infrastructure in line with smart city concepts is important for creating a comfortable and safe living environment in the digital economy.

The paper also suggests using digital economy concepts to transform Arctic cities into smart cities that improve efficiency and quality of life. Special attention is given to the development of digital platforms that ensure the coordination of different modes of transport in the Arctic. The development of an integrated communication infrastructure is an integral part of promoting smart city principles in the Arctic region. Current initiatives and projects reflect significant efforts to introduce smart technologies into Arctic development in line with Russia's Arctic Development Strategy until 2035.

The paper [17] also discusses the transformative potential of the Internet of Things (IoT) and machine learning in the context of smart cities. It emphasises the shift towards a data-centric approach, harnessing the vast amounts of data generated by sensors installed throughout the urban environment. The focus is on embedding intelligent data into the fabric of smart cities and using this data to improve various aspects of urban life.

The following article by [18] examines the current state and challenges of digital connectivity in the Arctic region. It highlights the efforts of the Arctic Council and the Arctic Economic Council (AEC) since 2016 to improve connectivity in this remote area, and underscores the urgency given the region's diverse and often inadequate infrastructure. The connectivity landscape in the Arctic is a patchwork of different technologies, tailored to each region's geographic and economic circumstances. For example, Alaska relies on a combination of satellites, microwave towers and undersea fibre-optic cables, but still has the slowest connection speeds in the United States. Within Canada, the Northern Territories show stark differences, with the Yukon having better terrestrial cable connectivity than Nunavut, which relies solely on satellites and experiences frequent and lengthy internet outages. In Greenland, a combination of satellite, digital radio and undersea cable connects almost every village, while Iceland's extensive fibre-optic networks and high-speed broadband make it an attractive location for data centres and bitcoin mining companies. Nordic countries such as Norway, Sweden and Finland also demonstrate a variety of connectivity options, with Norway leading the way in fibre and 4G mobile internet coverage in its Arctic regions. Russia, meanwhile, has connected its Arctic cities with land-based cables, although many areas of the east coast remain unconnected, except for strategic locations such as military bases and oil drilling sites. The digital divide in the Arctic has significant social implications, increasing feelings of isolation among local communities. In recent years, however, the telecommunications sector has made active efforts to bridge the gap.

The paper [19] examines the potential and challenges of developing digital service platforms for the Northern Sea Route (NSR). It describes the advantages of the NSR as a faster maritime transport corridor connecting China with Europe, reducing geopolitical risks and providing an alternative to the Suez Canal. However, the NSR faces significant challenges, such as high costs due to the need for an icebreaker fleet for winter navigation, expensive vessels designed for Arctic conditions and underdeveloped multimodal transport infrastructure in the Arctic. To address these challenges, the study proposes the creation of digital service platforms inspired by the successful e-platforms in the maritime container shipping industry. These platforms are designed to improve supply chain transparency and efficiency, support multimodal transport and provide integrated services that meet the high standards expected by carriers. The paper uses an Enterprise Architecture (EA) approach, in particular TOGAF, and a Capability-Driven Approach (CDA) to analyze requirements and propose suitable digital tools. Digital business ecosystems are identified as critical for modern economic cooperation, with e-platforms playing a key role in their development. These platforms facilitate various business processes, from vessel and cargo management to regulatory compliance, by enabling efficient information flow between stakeholders. The study categorizes existing maritime e-platforms into those created by commercial ports, government authorities, and major industry players, each with specific roles in the business ecosystem. The authors suggest that the implementation of such e-platforms for the NSR will enhance its competitiveness by offering services that meet global standards. This includes the creation of cloud-based maritime community platforms and single-window systems for regulatory compliance, which are already in use in other parts of the world. In addition, the integration of advanced technologies such as cloud computing, the Internet of Things, Big Data, and blockchain will be essential for the effective functioning of these digital platforms. The article highlights that the development of e-platforms for the NSR is not only about the implementation of new technologies, but also about creating a holistic digital business ecosystem. This ecosystem would streamline operations, reduce costs and improve service quality, making the NSR a viable and competitive international transport corridor.

The paper by [20] examines the use of digital tools for decision support in large-scale environmental logistics projects in the context of Industry 4.0, climate change and the recent global economic crisis. The authors emphasise the importance of digital technologies for effective planning and implementation of these projects to meet today's challenges.

4 Results

Developing digital infrastructure is key to sustainable development in the Arctic.

The unique conditions of the Arctic region define specific challenges and solutions:

- Infrastructure development, including high-speed Internet and satellite communications. This will allow remote communities to participate in the digital economy. In the Arctic, this requires significant investment and innovation, such as the use of satellite Internet services and the construction of resilient data centers that can withstand extreme temperatures. This will require the construction of fiber-optic networks, improved satellite coverage, and perhaps even the use of low-orbit satellite constellations to expand Internet access.
- Increased data management requirements: the ability to process vast amounts of data generated from various sources. With the proliferation of IoT devices, sensors, satellites and remote monitoring systems in the Arctic, ever-increasing amounts of data are being generated every day.
- Scalability and flexibility of digital infrastructure, which may include the use of modular components, cloud services and open source code, agile development methodologies.
- Cybersecurity measures to protect critical infrastructure, sensitive data and digital assets from cyber threats and attacks, including strong encryption protocols, multi-factor authentication mechanisms, intrusion detection systems and regular security audits; promoting a culture of awareness and training in cybersecurity.
- Remote monitoring and management, control of infrastructure assets, logistics operations and environmental parameters to optimize resource use and minimize environmental impact, taking into account the remote and hostile Arctic environment. By implementing IoT sensors, satellite imagery and telemetry systems, stakeholders can remotely monitor the performance, operability of critical infrastructure assets such as energy facilities, transport networks and environmental monitoring stations. This enables predictive maintenance, real-time decision-making and rapid response to emerging situations, thereby improving operational efficiency, safety and environmental sustainability in the Arctic.

- Digitalization of services such as healthcare, education and government services to improve accessibility and quality of life for Arctic residents while reducing the need for physical infrastructure. Using telemedicine platforms, online educational resources and e-government portals, stakeholders can overcome geographical barriers, increase access to essential services and provide local communities with greater autonomy and self-reliance. Moreover, digitalization enables the provision of personalized and culturally sensitive services tailored to the unique needs and preferences of Arctic populations, thereby promoting social inclusion, cohesion and resilience in the region.

Key technologies that can realize the formulated requirements are artificial intelligence, machine learning, the Internet of Things, cloud computing and Big Data.

1. Artificial Intelligence (AI) and Machine Learning (ML)

AI-powered solutions enable predictive maintenance, route optimization and resource allocation, making Arctic logistics and digital infrastructure operations more efficient, resilient and environmentally friendly. Examples: machine learning algorithms, predictive analytics, autonomous systems. The harsh Arctic environment poses a constant threat to equipment functionality. Traditional maintenance schedules are often changed, leading to unexpected breakdowns and costly downtime. Predictive maintenance powered by AI changes the scenario. By analyzing sensor data from equipment in real time, AI systems can detect anomalies and predict potential failures before they happen. This proactive approach enables predictive maintenance, minimizes downtime, extends equipment life, and improves overall operational efficiency. Arctic logistics involve navigating complex terrain and unpredictable weather conditions. AI-powered route optimization takes these factors into account by analyzing weather, road, and traffic data in real time. This allows for the most efficient routes to be created, minimizing travel time, fuel consumption, and environmental impact. AI can also take safety considerations into account, avoiding routes during extreme weather events and keeping personnel safe. Autonomous drones can be used to manage inventory, conduct automated inspections of remote sites, or deliver critical materials to hard-to-reach locations. This frees up personnel to perform more complex tasks, improves safety in potentially hazardous environments, and improves overall operational efficiency. Specific training datasets should be developed for the Arctic.

2. Internet of Things (IoT)

IoT can be used to collect data on the health of the environment, infrastructure, and equipment. It is emerging as a transformative technology for collecting real-time data on the health of the environment, infrastructure, and equipment in the Arctic region. By implementing sensors, connected devices, and telemetry systems, stakeholders can collect vast amounts of data on environmental parameters, asset performance, and operating conditions. Temperature, humidity, wind speed, and visibility are all critical to understanding the impact of the harsh Arctic climate on infrastructure and personnel. Sensors embedded in roads, bridges, and warehouses monitor for potential cracks, wear, or other structural issues, enabling predictive maintenance and safe operations. Sensors on vehicles, machinery, and other equipment monitor performance metrics such as engine temperature, fuel consumption, and vibration levels. This data enables early detection of potential issues and the development of predictive maintenance strategies. This data supports insight-based decision making, enabling predictive maintenance, asset optimization, and adaptive management strategies to improve efficiency, reliability, and sustainability across sectors including energy, transportation, and the environment. The data collected would be of no value without a robust communications network. Connected devices and telemetry systems ensure seamless data transfer from sensors to a central platform. This real-time data flow provides stakeholders with a holistic view of their operations, allowing them to make informed decisions based on the latest information.

3. Cloud Computing and Big Data

Big data is becoming an asset for innovation and optimisation in the Arctic, offering opportunities to develop new products, services and business models while improving the efficiency of different sectors of the economy. Using the vast amounts of data generated by Internet of Things devices, satellite imagery and digital platforms, stakeholders can gain deeper insights into market trends, consumer behaviour and environmental dynamics. Big data analytics will enable personalised services, supply chain optimisation and targeted interventions to drive economic growth, competitiveness and sustainability in the Arctic region.

4. Blockchain can be used to provide transparency, traceability and security in Arctic supply chains, reducing risks associated with fraud, counterfeit goods and unauthorised access. Unlike traditional databases controlled by a single entity, blockchain distributes information across a computer network. By creating decentralised and immutable ledgers, blockchain allows stakeholders to track the movement of goods, authenticate products and enforce contractual agreements with unprecedented integrity and trust. Blockchain-based solutions increase supply chain resilience, reduce transaction costs, and promote accountability and trust among participants, thereby contributing to ethical sourcing, fair trade, and sustainable development in the Arctic.
5. 3D printing is emerging as a technology for the local production of spare parts and other products, offering cost-effective and flexible solutions to the logistical challenges of the Arctic region. Using portable 3D printers and digital design files, stakeholders can produce customised parts, tools and equipment on demand, reducing reliance on centralised supply chains and improving lead times. 3D printing enables just-in-time production, inventory management and repair services, increasing operational flexibility, sustainability and cost-effectiveness in remote areas of the Arctic.
6. Cloud computing can be used to provide a scalable and flexible platform for service delivery. By using cloud infrastructure and services, stakeholders can dynamically allocate resources, scale operations and adapt to changing demand

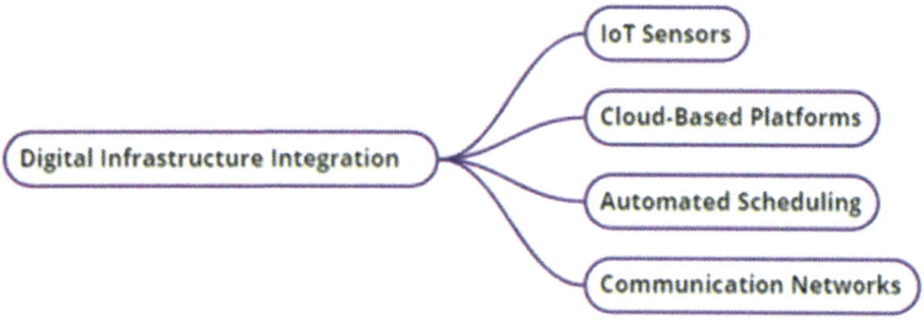

Fig. 1 Key components for digital infrastructure integration

patterns without significant up-front investment in hardware or infrastructure. Cloud computing enables rapid deployment, flexibility and cost-efficiency in service delivery, giving stakeholders the ability to innovate and iterate quickly in response to changing business needs and market dynamics.

7. APIs (Application Programming Interfaces) serve as a bridge that enables seamless integration and communication between services, applications and systems in the Arctic ecosystem. By providing standardised interfaces and endpoints, APIs facilitate communication, data exchange and collaboration between different systems. APIs allow stakeholders to integrate services with each other and with existing systems, enabling seamless data exchange, workflow automation and service organisation in the Arctic region.

8. Digital twins are virtual models of objects that exist in the real world. Using IoT sensors, real parameters of the object are transmitted to the server and processed by specialised software, and potentially possible situations are calculated using IT infrastructure. Digital twins are used to simulate different situations, allowing decisions to be made based on the resulting predictions, thus optimising operations, performing predictive maintenance, etc.

Using the example of Murmansk, a major transport hub in the Arctic, it is possible to formulate the main components for digital infrastructure integration (Fig. 1):

- IoT sensors on vehicles and cargo for real-time monitoring of location, condition and usage.
- Cloud platforms for real-time data access and analysis.
- Automated planning: Artificial intelligence-driven systems for dynamic planning and routing based on real-time data.
- Communications networks: reliable satellite and 5G networks for always-on connectivity.

As an example, we will detail the component "satellite networks and 5G networks" with a display of the components of the digital infrastructure, implemented information systems and services, based on the corresponding digital technologies (Fig. 2):

The implementation of the concept of digital support for sustainable development in the Arctic is based on the implementation of a digital platform.

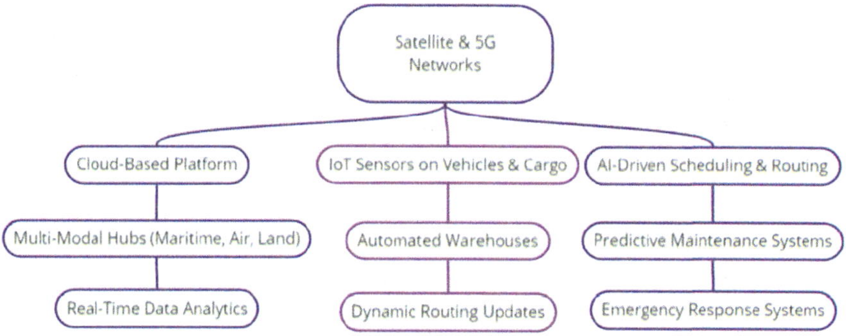

Fig. 2 Detailing of the digital architecture component "satellite networks and 5G networks"

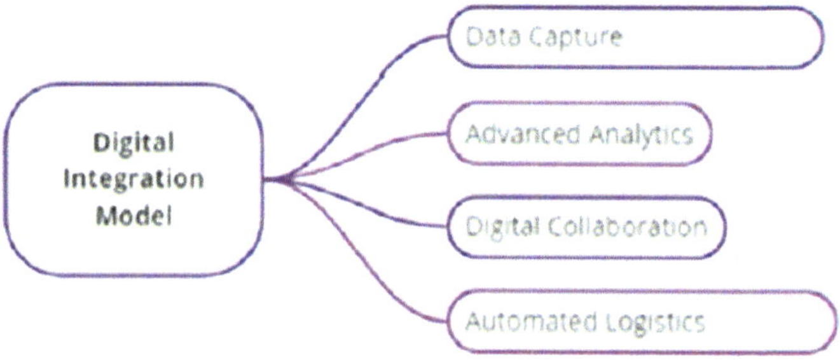

Fig. 3 The main functions of the digital platform

The digital platform provides (Fig. 3):

1 Data collection: on personnel, equipment and resources in real time.
2 Advanced analytics using artificial intelligence to forecast supply needs, optimise transport schedules and identify potential equipment maintenance issues, with integration of weather and environmental data for risk assessment and proactive planning.
3 A cloud-based platform for communication and document sharing, remote consultations, including medical consultations, and staff training.
4 Automated logistics, autonomous vehicles for transporting supplies within the site, remote monitoring and control systems for equipment operation.

5 Conclusion

Innovative technologies, especially artificial intelligence and digital twins, play a key role in adapting to the changing Arctic environment.

Digital platforms provide a roadmap for revolutionising governance and engagement, accelerating social change and contributing to a more inclusive and efficient socio-economic ecosystem in the Arctic. The Arctic 2035 Digital Platform was developed by the Arctic Development Project Office to gather ideas for the strategic sustainable development of Russia's Arctic zone. A careful balance between innovation and regulation is considered necessary to protect personal data and ensure the overall success of digital platforms as tools for sustainable progress.

The successful implementation of digital technologies depends on a careful balance between innovation and the preservation of the fragile Arctic ecosystems.

Further research will focus on the integration of digital and transport infrastructure, using the example of major ports in the Russian Arctic.

Acknowledgements The research was supported by Russian Science Foundation grant No. 23-78-10190, https://rscf.ru/project/23-78-10190/.

References

1. Sergeev, V., Ilin, I., Fadeev, A.: Transport and logistics infrastructure of the Arctic zone of Russia. Transp. Res. Procedia **54**, 936–944 (2021)
2. Fadeev, A., Kalyazina, S., Levina, A., Dubgorn, A.: Requirements for transport support of offshore production in the Arctic zone. Transp. Res. Procedia **54**, 883–889 (2021)
3. Ilin, I., Kersten, W., Jahn, C., Weigell, J., Levina, A., Kalyazina, S.: State of research in arctic maritime logistics. In: Data Science in Maritime and City Logistics: Data-Driven Solutions for Logistics and Sustainability. Proceedings of the Hamburg International Conference of Logistics (HICL) 30, pp. 383–407. epubli GmbH, Berlin (2020)
4. Kim, J.D., Lee, S., Kim, M., Kim, J.: Technological innovations for a sustainable Arctic. Global Asia **15**(4), 40–43 (2020)
5. Stepanova, I., Vorotnikov, A., Doronin, N.: The potential of digital platforms for sustainable development using the example of the arctic digital platform 2035. In: IOP Conference Series: Earth and Environmental Science, pp. 012004. IOP Publishing (2020)
6. Skripnuk, D., Kikkas, K.: The concept of creating a digital clone of the Arctic territories. In: IOP Conference Series: Materials Science and Engineering, pp. 012108. IOP Publishing (2020)
7. Samylovskaya, E., Makhovikov, A., Lutonin, A., Medvedev, D., Kudryavtseva, R.-E.: Digital technologies in arctic oil and gas resources extraction: global trends and russian experience. Resources **11**(3), 29 (2022)
8. Levkina, A., Gladun, E.: New approaches to the development of the Russian Arctic: sustainable innovative development of collaborative local economies. Innovations **11**, 45–52 (2018)
9. Gafurov, A., Skotarenko, O., Nikitin, Y., Plotnikov, V.: Digital transformation prospects for the offshore project supply chain in the Russian Arctic. In: IOP Conference Series: Earth and Environmental Science, pp. 012009. IOP Publishing (2020)
10. Maksakova, M.A.: Revisiting the Arctic strategy of Russia up to 2035. In: The Handbook of the Arctic: A Broad and Comprehensive Overview, pp. 53–73. Springer (2022)
11. Yuh, J., Marani, G., Blidberg, D.R.: Applications of marine robotic vehicles. Intel. Serv. Robot. **4**, 221–231 (2011)
12. Wang, W.H., Engelaar, R.C., Chen, X.Q., Chase, J.G.: The state-of-art of underwater vehicles—theories and applications. In: Chen X.Q. (ed.) Mobile Robots—State of the Art in Land, Sea, Air, and Collaborative Missions. ISBN: 978-953-307-001-8, InTech (2009) Available from: http://www.intechopen.com/books/mobile-robots-state-of-the-art-in-land-sea-air-and-collaborativemissions/the-state-of-art-of-underwater-vehicles-theories-and-applications

13. Zhang, T., Tang, J., Qin, S., Wang, X.: Review of navigation and positioning of deep-sea manned submersibles. J. Navig. **72**(4), 1021–1034 (2019)
14. McDonald, A. et al.: Pushing the Limits of Subseasonal-to-Seasonal Sea Ice Forecasting with Deep Generative Modelling. Copernicus Meetings (2024)
15. Makarova, I., Buyvol, P., Mukhametdinov, E., Boyko, A.: The construction of seaports in the Arctic: prospects and environmental consequences. J. Mar. Sci. Eng. **11**(10), 1902 (2023)
16. Ablyazov, T., Asaul, V.: Development of the Arctic transport infrastructure in the digital economy. Transp. Res. Procedia **57**, 1–8 (2021). https://doi.org/10.1016/j.trpro.2021.09.018
17. Ullah, A., et al.: Smart cities: the role of internet of things and machine learning in realizing a data-centric smart environment. Complex Intell. Syst. **10**(1), 1607–1637 (2024)
18. Delaunay, M., Landriault, M.: Connectivity and Infrastructure—The Arctic Digital Divide. Handb. Geopolit. Secur. Arct. High North Coop. Confront. 231–248 (2020)
19. Ilin, I., Maydanova, S., Levina, A., Lepekhin, A.A.: Northern sea route e-platforms: tools for competitive development. In: Data Science in Maritime and City Logistics: Data-driven Solutions for Logistics and Sustainability. Proceedings of the Hamburg International Conference of Logistics (HICL) 30, pp. 441–455. epubli GmbH, Berlin (2020)
20. Lukyanov, S., Abramov, V., Popov, N., Petrov, Y., Gogoberidze, G.: Decision support digitalization for large environmental logistic projects. In: XIV International Scientific Conference "INTERAGROMASH 2021" Precision Agriculture and Agricultural Machinery Industry 1, pp. 674–681. Springer (2021)

Indicators of the Development of Infrastructure for the Big Data Economy in the Arctic Zone

Anna Smirnova, Elizabeth Moshurova, and Sofia Kalyazina

Abstract The article considers indicators of the development of the big data economy in the Arctic zone of the Russian Federation and tests the hypothesis that there are prerequisites for the transition to this stage in the selected subjects. The big data economy is considered as a stage that follows the digital economy and is based on the infrastructure created within it. Three main indicators of readiness for the transition to the Big Data Economy are identified: Internet coverage and the degree of its use by economic agents, the presence of data processing centres and the development of virtual infrastructure, the intensity of registration of databases and computer programs by commercial organisations. Based on the analysis of statistical data, the consideration of practical examples of the use of infrastructure for data analysis, and the data on the registration of intellectual property objects by the residents of the Russian Arctic, it is concluded that the Arctic region has the prerequisites for the transition to the Big Data economy.

Keywords Digitalisation · Big data · Big data economy · Arctic regions · Data center · Intellectual property objects

1 Introduction

The last decade has been marked by active human activity in the digital realm. Petabytes of data are generated every day, recording the digital footprint of every internet user. Huge amounts of data are generated in the processes of industrial, political, social and economic life and are becoming an important source of information for research in various fields, leading to the development of increasingly complex

A. Smirnova · E. Moshurova
ITMO University, Kronverksky Pr. 49, St. Petersburg 197101, Russia

S. Kalyazina (✉)
Peter the Great St. Petersburg Polytechnic University, St. Petersburg, Russia
e-mail: kalyazina_se@spbstu.ru

I. Ilin and M. Youzhong (eds.), *Digital Systems and Information Technologies in the Energy Sector*, Lecture Notes in Networks and Systems 1244,
https://doi.org/10.1007/978-3-031-80710-7_13

and efficient methods and tools for collecting, organising, visualising and processing data [1].

This process is becoming a natural consequence of the digitalisation process, driven by technological advances and access to significant computing power on which machine learning, artificial internet and other high performance computing processes can operate [2].

By 'big data' the authors mean data sets characterised by high volume, velocity and/or variety of information. This term describes data that is difficult to process and analyse using traditional processing methods, requiring an infrastructure to work with it [3].

It is noted that the collection, storage and intelligent analysis of big data for information can create significant value for the global economy, increasing the productivity and competitiveness of companies and the public sector, and creating a significant economic surplus for consumers [4]. This fact makes it possible to summarise current processes in the concept of the 'Big Data Economy'.

The authors understand the big data economy as a stage of development that follows the digital economy and represents a deeper implementation of work with big data in the activities of organisations and the state to make management decisions based on it.

The increasing complexity of socio-economic systems and the growing amount of information in the twenty-first century make it necessary to work with modern tools in order to avoid economic crises, social changes and political instability [5]. The transition to the Big Data economy is conditioned by the impact of data on all spheres of the economy: public and private sectors, business and life activities of individuals, as well as on the reconfiguration of social science methods and cognitive practices [6].

The transition to the big data economy in the Arctic zone of the Russian Federation (hereinafter—AZRF) seems to be one of the stages of economic transformation within the concept of the cyclic economy working in symbiosis with climatic conditions, the search for new growth points in the preservation of natural capital, the implementation of ESG principles for sustainable development of territories. The view on the possibility of developing the Big Data economy as a tool for these changes forms the need to test the hypothesis that there are prerequisites for the development of the Big Data economy in the AZRF, through the evaluation of indicators reflecting the main trends in this field. The subject of the study are the regions grouped at the state level for spatial and climatic reasons into the Arctic zone, which is recognised as a priority geostrategic area of Russia.

The subject of the study is the digitisation process and related indicators, as well as the digital infrastructure of the Arctic.

2 Materials and Methods

A mixed approach, namely quantitative and qualitative methods, was used to investigate and analyse the development of the infrastructure of the big data economy in the Arctic zone of the Russian Federation.

Quantitative methods include statistical analysis and modelling, which have been used to synthesise data on technological development in the Arctic. Qualitative methods, such as document analysis, study of official information to interpret and explain the specific context of the object under study.

The authors rely on an extensive review of scientific literature and the results of the research work of master's and PhD students of ITMO University 623,111—'Study of the technological profile of the region within the framework of big data economy technologies'.

Data for statistical analysis were obtained from the official websites of the Federal State Statistics Service (Rosstat), the Unified Interdepartmental Information and Statistical System, and the Federal Service for Intellectual Property (Rospatent). To collect original information on registered databases and computer programs in the Arctic zone, data parsing was used at the place of registration of the legal entity for which the patent is registered.

The information collected is presented in the form of graphs and tables to enable comparison, analysis and generalisation of the data obtained.

The source of regulatory and legal information and official documents is the websites of public authorities. The data is supplemented by reports and comments from organisations involved in the creation and operation of infrastructure related to the Big Data economy, as well as analysis of public sources of information on the activities of regions in terms of their main components.

This methodology is in line with the aims of the article as a research question, since the study of big data issues is related to the problem of access to actual data. This leads to the use of available data related to the monitoring of the development of digital technologies and digital infrastructure, measuring patent activity as statistical material.

Data collection and analysis methods have not been without limitations, including the complex and controversial process of digitising the Arctic, the limited data available, the incomplete coverage of regional territories in the Arctic zone, and the availability of statistics for the entire territorial unit without administrative breakdown. All this requires careful attention to data and results.

3 Literature Review

The use of big data for the purpose of economic growth and sustainable development of individual regions and countries, and as a factor of production to gain competitive advantage in business and develop technological innovation, is a topic that is becoming increasingly popular among researchers.

Big data is a cross-cutting technology of the digital economy. The United Nations Organisation for Trade and Development (UNCTAD) sees the digital economy as a system that has driven the development of data as a tool for improving competitiveness and achieving the Sustainable Development Goals. The key is that both raw data and the ability to process it digitally are needed to create value from data [7]. This is part of digitalisation, which is defined as the process of using technology and data to improve and transform business processes [8].

The authors of the 2019 UNCTAD report rightly point out that with the development of digital technologies and their penetration into economic processes, the role and value of data and information is growing exponentially [7]. This category includes the infrastructure that enables the creation and storage of digital information as such (electricity, internet, etc.), as well as the technologies used to process it in order to extract value from the use of this data [9].

The trend of transitioning to a big data economy in the global economy has been noticeable since 2015. To promote the development of the digital economy, governments around the world are implementing policies related to big data, such as the US Federal Government's Big Data Research and Development Initiative, Australia's Public Service Information and Communication Technology Strategy for 2012–2015, China's Big Data Promotion Action Plan of 2015 [4]. In Russia, it is the Strategy for the Development of the Information Technology Industry in the Russian Federation for 2014–2020 and in the perspective until 2025.

Based on this, we can conclude that big data, as a cross-cutting technology for the development of the digital economy, dictates the need to create the necessary infrastructure, technologies and investments in human capital, which in turn increases its importance and leads to the transition to a new stage—the big data economy. This confirms the need to assess indicators of the development of its infrastructure, especially for regions with digitalisation difficulties, where this factor may become a key factor in the development of territories.

This is also emphasised in the study by [10]. The authors talk about the formation and development of a digital system of regional management through the creation of a "digital image" or "digital twin" of the entire Arctic socio-technical and natural system in the global world as a key principle of Arctic development, which is not limited to the collection of data obtained during the development and manufacture of a product, but continues to collect and analyse information throughout the life cycle of a real object using the Internet of Things.

Countries use different calculation methods to assess digital transformation. The purpose of introducing such indicators is to improve the quality of management decisions through digital technologies, including the introduction of a culture of data-driven management decision-making.

In Australia, for example, the methodology is based on: analysing investments in ICT development in state and local government; assessing the activities of private individuals involved in developing the state's digital macro-environment; and assessing the accessibility of the Internet to citizens. In Singapore, the methodology is related to assessing the availability of government electronic systems for various transactions; the ability of digital government systems to conduct digital transactions; considering digital transformation projects; assessing data flow and data processing systems; and assessing the overall integration of all digital systems into a single system [11, 12].

In Russia, data analytics has emerged as a tool for effective decision-making in public administration, public service delivery and business process management, as part of the national Digital Economy of the Russian Federation programme, which runs from 2017 to 2024, one of the objectives of which is to create a competitive infrastructure for data processing and storage.

With the transition to a new stage of economic development in Russia, illustrated by the development of a new national project 'Data Economy', the digital transformation of the state, the assessment of the penetration of end-to-end digital technologies, including big data, will become one of the key indicators of territorial development. This is linked to the goal of achieving 'digital maturity' in public administration and key sectors of the economy by 2030 through the automation of the majority of transactions on unified digital platforms using big data processing technologies, machine learning and artificial intelligence [13].

However, many of the indicators are still not available in open statistical databases for the Arctic regions. At the state level, the digital maturity of regions can be assessed using various ratings, such as the Rating of Russian Regions, Digital Maturity Rating, which are not calculated on an annual basis and therefore cannot be analysed retrospectively [14]. In this regard, it is necessary to assess the indicators of the development of the infrastructure of the big data economy in the Arctic Zone of the Russian Federation, taking into account the available data and ratings.

Currently, the digitalisation indicators of the AZRF regions are linked to the growth of indicators assessing the implementation of the national project "Digital Economy", such as: the use of the Internet by the population, the use of broadband access to the Internet by organisations and households, the use of information exchange of data in the electronic document management system by organisations in the total number of respondents, and the costs of implementing and using digital technologies. These indicators are also the basis for the development of the infrastructure of the Big Data economy.

The authors of the article [15] conducted an assessment of the level of digitalisation of regions using the method of integral assessment of the composite index. For this purpose, they used six indicators combined into three indices (digitalisation potential, digitalisation index of organisations and population) to assess the digitalisation potential of regions. The authors point out the dependence of the digitalisation potential on these factors. According to the main statistical indicators, the Yamalo-Nenets Autonomous Okrug, the Murmansk Oblast, and the Republic of Karelia are leading in the field of information and communication technologies.

The article by [14] proposes a ranking of the Arctic regions of Russia by the level of digitalisation based on the construction of an integral index with 4 sub-indices: digitalisation of households, digitalisation of the population, digitalisation of the activities of public authorities and digitalisation of organisations. According to the author's calculations, the Yamalo-Nenets Autonomous Okrug is the leader in almost all sub-indices, while the Chukotka Autonomous Okrug is among the outsiders.

The main problems of digitisation in the northern regions include the lack of digital infrastructure allowing the use of the Internet at certain latitudes, climatic conditions that can adversely affect the operation of communications equipment, the inability to guarantee uninterrupted communications and the lack of qualified personnel [16].

Thus, the development of regions, both on the part of public authorities and the business sector, depends to a large extent on the level of digitalisation, which is a requirement of the times. The proliferation of technology is generating large amounts of data on a daily basis, and working with it is becoming a necessity for effective decision-making. The intensity of the transition to a big data economy is closely linked to the availability of infrastructure, both at the national level and in individual regions. The Arctic region has its own peculiarities and limitations for the digitalisation process, in which certain results have been achieved, as can be seen from the relevant indicators.

On this basis, the authors of the article hypothesize that there are conditions for the development of the Big Data economy in the Arctic zone of the Russian Federation, the impact of which may affect not only the region itself, but also the country as a whole.

4 Results

4.1 On Big Data Infrastructure

The development of information and communication technologies and their widespread introduction into people's lives is leading to a 'big data explosion', coming from multiple sources and providing us with a continuous flow of fresh and digitised data on how people, businesses and other organisations interact.

The components of the big data economy can be divided into three blocks (Fig. 1).

The first block is the data itself, which is a heterogeneous collection of information that requires high technical capacity to process. Big data is characterised by the 5Vs: velocity, volume, value, variety and veracity. The second block highlights technologies for working with data, such as distributed file systems, frameworks, machine learning, etc. The third and most important block is the infrastructure of the Big Data

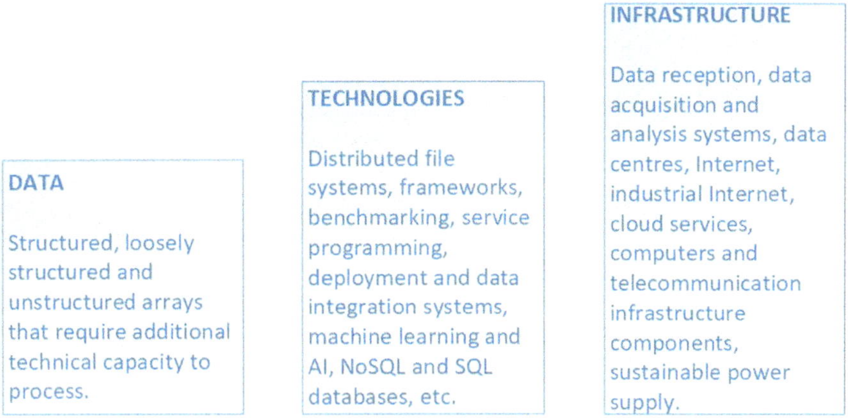

Fig. 1 The components of big data economics

economy, including: computing and telecommunications infrastructure components, the Internet, data reception, collection and analysis systems, the Internet of Things, the Industrial Internet, sustainable energy supply, data centres, cloud services, etc. The third and most important block is the infrastructure of the big data economy.

Internet coverage drives economic progress and innovation by enabling data to be accelerated, analysed and commercialised. Broadband improves the ability to collect data, makes it much easier to integrate data from different sources, and provides the impetus for sustained innovation. The internet enables businesses to work efficiently with cloud technologies, allowing them to scale resources and quickly analyse large volumes of data. Internet connectivity has a particular impact on the development of artificial intelligence and machine learning, where the amount and quality of data used is directly related to the accuracy and reliability of the results obtained. The better the connectivity, the more efficient the process of learning and implementing intelligent systems, for example in the implementation of smart city projects.

To work with big data, complex systems are used in which several components can be distinguished. Basically, there are four levels: collection and processing, storage, big data analysis and visualisation. Each level has its own set of technologies, tools and processes [16].

Data collection is enabled by the proliferation of the Internet and related technologies. The proliferation of social media, the Internet of Things, the Industrial Internet and other technologies generates data that vary in volume, speed of arrival, frequency of transmission, diversity and reliability.

Sources can also be roughly divided into three types: social (human), machine (sensory), transactional. Social includes human actions on the Internet, i.e. digital footprint (actions in social networks, search queries, country and city statistics, etc.). Machine—Internet of Things (IoT) and Industrial Internet (IIoT), i.e. data obtained from devices in use, connected to the Internet and exchanging data, including data used in production (IIoT). Transactional data comes from purchases, money transfers, deliveries of goods and ATM transactions.

Based on the impact of the Internet on the generation of big data, the indicator of Internet coverage and its use by economic entities is highlighted as one of the indicators of the development of the big data economy.

For the processing, storage and analysis of big data, a data processing centre is implemented as a cloud/physical computing infrastructure.

A physical data centre is a facility that houses computing resources that are networked together using a communications infrastructure to store data and host applications [17]. Data centres are critical components of modern information and communication technology (ICT) infrastructure around the world, providing a wide range of services for data storage and management, data protection, transaction validation, user-to-user communication and data security [18].

As a company grows, so does the amount of data it processes in the course of its activities, both internally and externally, requiring a high-performance IT infrastructure that takes into account load and data protection requirements. This factor is one of the drivers of growth in the cloud services segment, where the business need to reduce costs is leading to the search for convenient turnkey solutions—such as a virtual data centre [19]. Virtual data centres are physically located in traditional data centres and are leased to businesses under various models, which traditionally include IaaS (Infrastructure as a Service), SaaS (Software as a Service) and PaaS (Platform as a Service). Cloud solutions are a fast-growing market worldwide, particularly due to advances in artificial intelligence and the Internet of Things.

Data centres are classified according to their status as corporate, commercial and government data centres. Corporate data centres include the in-house data centres of organisations whose activities involve information technology, such as enterprise-level information systems, automated process control systems and business digitalisation solutions. Commercial data centres have all the necessary information, telecommunications and engineering infrastructure to provide customers with a range of services, from computing power and data storage to networking equipment and a dedicated communications channel. Government data centres are specifically designed to meet the needs of government agencies and to host government information systems.

Thus, the proliferation of cloud-based information system hosting services and the growing demand for them by enterprises require the creation of physical infrastructure (data centres) and its commercial development according to modern business models related to the provision of IT service as a service. This fact can be attributed to the indicators of infrastructure development in the Big Data economy, as the reason for their development is the growing flow of information and the need to work with it.

Organisations work with data through technology, in specialised databases and through computer software.

Databases are organised collections of structured information or data, usually stored electronically in a computer system. To meet the information needs of an organisation, it is essential that this data is logically linked and accessible at all times. It is the unique arrangement, arrangement and positioning of data in relation to each other in a database that is the subject of copyright.

A computer programme is a set of data and instructions, presented in an objective form, which is intended for the functioning of a computer in order to achieve a specific result. In other words, in the context of the Big Data economy, it is a unique combination of instructions that allow certain information to be obtained from the data in databases, which, when analysed, allow conclusions to be drawn on the basis of which management decisions can be made.

Thus, databases and computer programs are an important indicator of companies' activity with big data and orientation towards data-driven approach in management decision-making. Their registration as objects of intellectual property and use is an indicator of the transition to the big data economy as a sign of the accumulation and registration of valuable data arrays and the creation of unique functionality to achieve certain results in data processing (data value creation).

4.2 Indicators of the Development of the Big Data Economy in the Russian Arctic Regions

4.2.1 Internet Penetration

The spread of the Internet as part of the infrastructure of the big data economy in the Arctic zone of the Russian Federation faces a number of serious problems and challenges due to the presence of hard-to-reach and remote areas (Fig. 2).

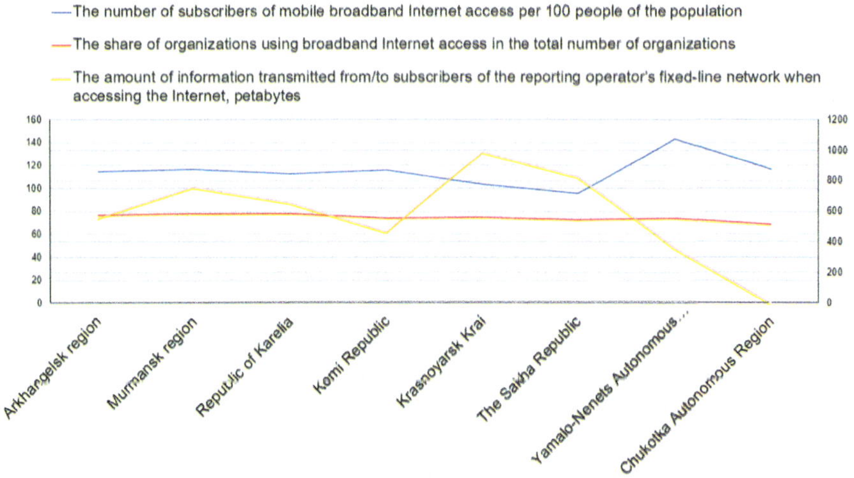

Fig. 2 Internet usage by people and organisations and the amount of information they transmit by 2022

In almost all oblasts, with the exception of the Nenets and Chukotka Autonomous Okruges, the indicators of Internet connectivity are at the level of the Russian average. There is a noticeable trend towards increasing penetration of high-speed Internet, which in turn indicates the emergence of the possibility of accumulating more data. On the one hand, this allows the government and businesses to use data to make management decisions, and on the other, it provides opportunities to increase the transfer of useful data. There is a noticeable imbalance in the regional distribution of activity, with Krasnoyarsk Krai, Murmansk Oblast and Sakha Republic leading the way, which is explained by the urbanisation of the population in regional centres.

4.2.2 Availability of Data Processing Centres

As mentioned earlier, the data centre is where this data is processed, stored and used. At the moment, the Arctic leads the world as a location for data centres. This is due to abundant electricity reserves, low price of clean energy, as well as cold climate, suitable fibre optic connectivity, reasonable land prices, low risk of natural disasters [20].

The construction of commercial data centres in the Arctic zone of the Russian Federation is a controversial issue. On the one hand, the climatic conditions allow to reduce the cost of their operation by up to 40%; on the other hand, there are high economic costs of construction, remoteness of territories and insufficient digital coverage of remote areas. Undoubtedly, data centre construction projects are complex, labour-intensive, expensive, and involve certain risks, but with proper planning, attracting specialists, and taking into account the benefits for the residents of the Arctic zone of the Russian Federation, this will attract large investments to the region, which can become a driver of its development and, in the future, the creation of server "cities" similar to IT valleys.

The availability of data centres in the Arctic zone of the Russian Federation can only be assessed on the basis of publications in official government sources; these statistics are not currently available. According to the available data, there are several corporate, commercial and government data centres operating in the Russian Arctic, built on the basis of global experience.

In 2022, Rostelecom commissioned a data centre in Murmansk, built on behalf of the Murmansk regional government. Its customers are both government and commercial. The centre offers rack rental or rack space for own equipment, data network, cloud services and modern digital services.

"Intelion Sever", as a resident of the Arctic zone of the Russian Federation, plans to build a computing data center in the Kola region with the support of the Far East and Arctic Development Corporation with investments of 1.1 billion rubles.

In 2024, Rosenergoatom Concern plans to launch a data center at the industrial site of the Kola NPP in the Murmansk Region, which will become the basic infrastructure for new high-load digital platforms and services of the Northern Sea Route and the Arctic zone, necessary for ensuring logistics, managing smart city services, supporting tourism and other important areas of regional development.

The corporate segment is represented by the largest companies, whose presence in the territory of the Arctic Zone of the Russian Federation is due to rich reserves of natural resources and special climatic conditions. Their main areas of activity are oil production and processing, mineral extraction, monitoring of environmental and climate changes, providing regions with telecommunications services.

Norilsk Nickel is placing data processing centers in Kola MMC. The data processing center's capacities ensure the operation of such digital projects of the Technological Breakthrough program as Production Dispatching, Metal Balance, the automated mining control system, and others.

The use of data center services by organizations indicates increasing volumes of information and the need to work with them (Table 1).

The Russian Arctic hosts 4% of all organisations that provided big data storage and processing services in Russia, and more than 5% of organisations used data centre services.

For example, the government of the Murmansk region is a user of Rostelecom's data centre services; four racks house the servers of the information technology

Table 1 Number of organisations that used data centre services in 2022

Region/ indicator	Number of organisations using data centre services by 2022, units				Number of organisations providing data centre services, units
	Total	Third-party	Proprietary	Proprietary and third-party	
Russian Federation	51 319	23 965	14 692	12 662	13 928
Murmansk Region	332	159	88	85	64
Arkhangelsk Region	346	151	99	96	18
Nenets Autonomous Okrug	31	15	15	1	3
Republic of Karelia	90	29	21	40	6
Komi Republic	260	117	85	58	26
Krasnoyarsk Territory	606	289	180	137	63
Sakha Republic	465	274	125	66	225
Yamalo-Nenets Autonomous Okrug	408	239	106	63	152
Chukotka Autonomous Okrug	100	53	37	10	48

centre, where the data received from video surveillance cameras are concentrated. Rostelecom's servers are used for the needs of the public security system of Murmansk and other cities of the Arctic region—they serve about 4000 surveillance devices located on streets, in schools and other socially important objects.

The data centre services are also used by companies located in the Arctic zone of the Russian Federation that generate arrays of information through the Industrial Internet of Things in the course of their activities.

4.2.3 Dynamics of Database and Computer Registration

As mentioned earlier, in order to collect data efficiently and use it for decision making, companies use proprietary or open source databases in conjunction with computer programmes.

Let us assess the level of innovation activity in the field of big data on the basis of data on the registration of databases and computer programs by residents of the Arctic zone.

Figure 3 illustrates the registration activity of the Arctic regions.

The leaders in the number of registered databases and computer programs are the Yamalo-Nenets Autonomous Okrug, Arkhangelsk and Murmansk regions. This can be explained by the presence of large mining companies. Thus, the key companies of the Russian oil and gas complex are represented on the territory of Yamal: Gazprom Neft, NK Rosneft and NOVATEK. Oil is produced by 22 enterprises, gas is produced by 36, gas condensate is produced by 23 enterprises, some of which have subsidiaries registered in the territory of the Arctic Zone of the Russian Federation. For example, Gazprom Dobycha Nadym, a subsidiary of Gazprom located in Yamal, registered

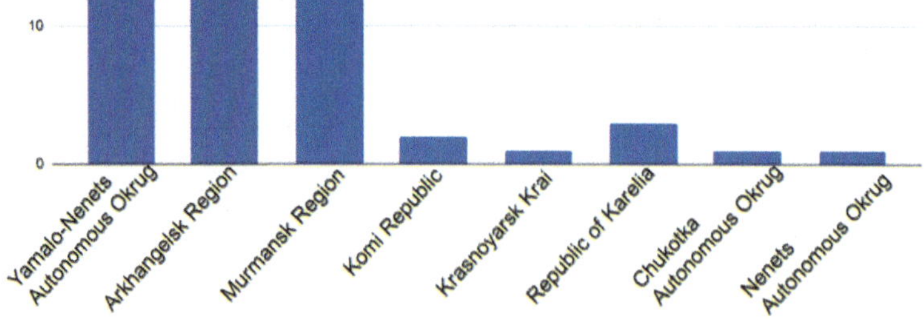

Fig. 3 Total number of databases and computer programs registered by resident companies in the regions, units

the database "Composition and Genesis of Mechanical Impurities in Well Products" and 2 computer programs "Servis + " and "InfSvod".

Let us consider the dynamics of registration of databases and computer programs by year in all regions of the Arctic Zone of the Russian Federation (Fig. 4).

According to Rospatent (Fig. 4), the peak of database and computer programme registrations occurred in 2022—more than 40 new intellectual property objects. This is largely due to the transition to domestic hardware and software, while the general trend of growth in registrations during the period under study remains.

Figure 5 is a diagram showing the concentration of registered databases and computer programmes in the Arctic regions.

The Yamalo-Nenets Autonomous Okrug is the leading region in terms of registration of the intellectual property objects in question in the commercial sector. For example, Gazprom Dobycha Yamburg registered 36 intellectual property objects in Yamal from 2016 to 2023, which indicates a high degree of concentration of intellectual property objects in large mining companies, since their activities are associated with specific natural conditions, data on which require unique approaches to working with them.

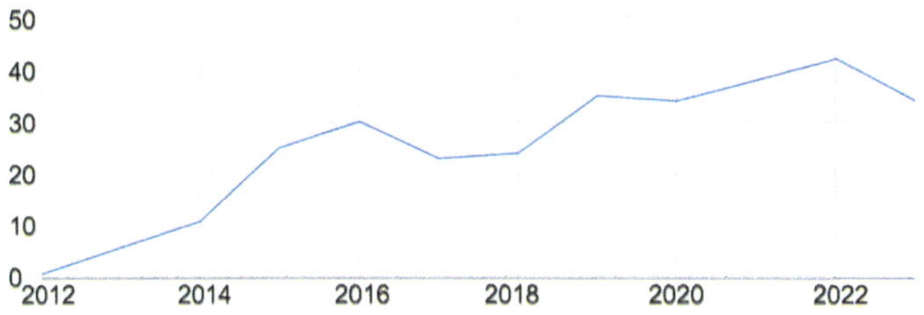

Fig. 4 Dynamics of the number of registered databases and computer programs in the territory of the Arctic zone of the Russian federation for 2012–2022, units

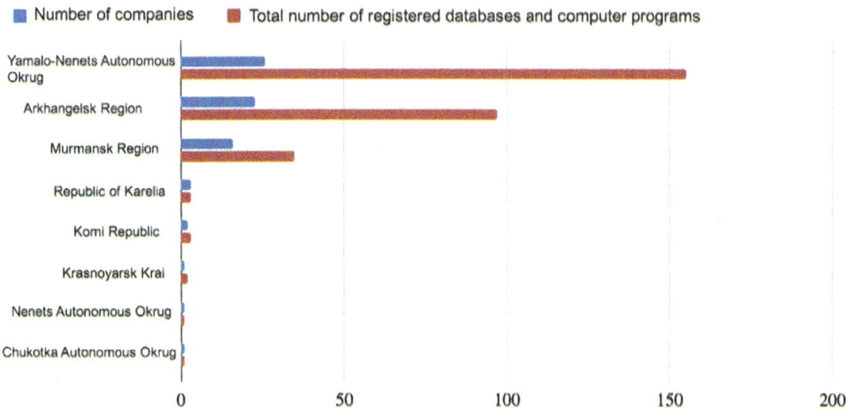

Fig. 5 Distribution of the total number of registered databases and computer programs in the regions of the Arctic Zone of the Russian federation

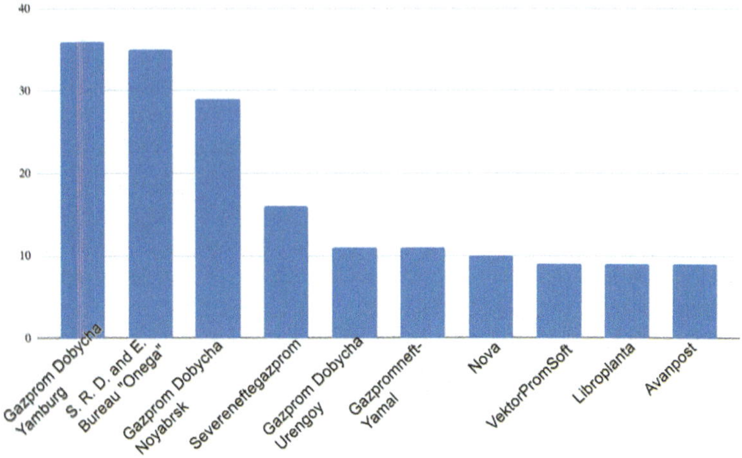

Fig. 6 Top 10 companies in the Arctic zone of the Russian federation by portfolio of registered databases and computer programs, in the period from 2016 to 2023

Figure 6 shows the volume of registered databases and computer programs of the ten largest companies operating in the territory of the Arctic Zone of the Russian Federation in the period from 2016 to 2023.

The use of IIoT, big data collection and analysis by oil and gas companies not only allows them to improve the efficiency of equipment use and monitor its condition in real time through predictive maintenance, but also to predict changes, simulate high-quality digital models to improve the quality and timeliness of management and investment decisions, and predict the development of trends in geology, engineering and production.

Active registration of databases and computer programs indicates the integration of digital technologies into the business processes of companies. Weak activity of database registration compared to computer programs may be evidence that the accumulated data arrays are not unique, but require special solutions for working with them (Fig. 7).

Databases and computer programs are assets of organizations that are registered in case of need for their constant use and confidence in bringing benefit to the company. Databases are characterized by a rapid rate of moral obsolescence, as a result of which companies selectively approach the solution of registration issues. However, the increasing regularity of requests related to data processing and expansion of technical capabilities for their use due to progress, as well as the increase in the market value of companies with a well-thought-out portfolio of intellectual property rights, ensure a stable growth of registration activity.

Thus, the presence of large companies in the regions contributes to the development of technologies, the formation of demand for high-tech products due to the ability to make large investments, attract highly qualified personnel and create

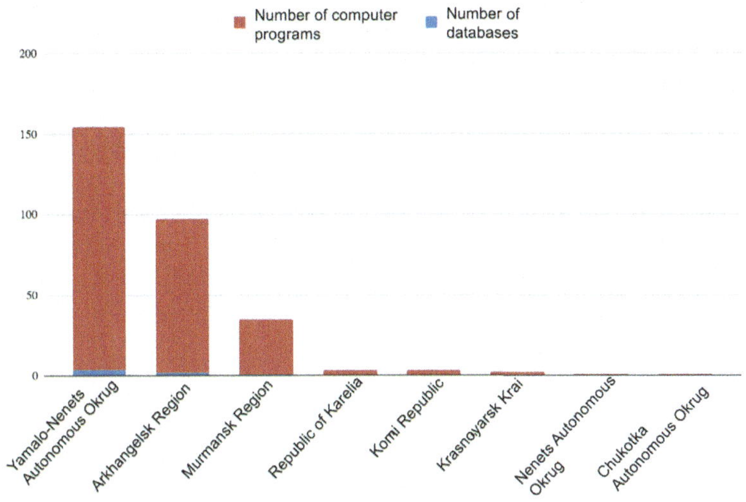

Fig. 7 Distribution of database and computer program registrations by region

the necessary infrastructure for their activities. Which, in turn, allows creating an infrastructure for the accumulation and analysis of big data.

5 Discussion

The infrastructure of the big data economy in Russia's Arctic zone is both an opportunity for the territories and a high-risk issue, involving high costs and structural changes, including the culture of business and consumption.

The high level of digital activity of the population and organisations in Krasnoyarsk Krai, Murmansk Oblast, Arkhangelsk Oblast, and Yamalo-Nenets Autonomous Okrug indicates that there are opportunities for data generation and the availability of data suitable for processing and analysis.

The deployment of data processing centers in the Russian Arctic, including with the help of state benefits, demonstrates the need to meet the digital needs of the regions and create a solid foundation for high rates of digital development. The use of data centers by commercial companies suggests that there is a growing need for cloud computing, the deployment of their own infrastructure for working with data, the benefits of which can be offered for the territory under study due to the nature of the data center.

The dynamics of registration of databases and computer programs by the largest companies in the Arctic zone shows that companies are moving towards not only using the results of analysis of existing data, but are also technologically capable of generating new arrays of information through the use of Industrial Internet of Things devices, and are also focused on creating unique computer programs for working with data and extracting value from them.

Thus, the indicators of readiness of the territory provided illustrate its disposition to change and development of the big data economy, however, these are only emerging trends, the development of which must be supported in order to continue their sustainable transformation.

The obtained research results can be detailed using indicators that are planned to be statistically collected within the framework of the national project "Data Economy and Digital Transformation of the State".

6 Conclusion

The authors of the article, based on the basic components of the big data economy, formulated 3 main indicators of the development of its infrastructure in the Arctic Zone of the Russian Federation: Internet coverage and its use by economic entities, the presence of data processing centers and the use of virtual infrastructure, and the registration of databases and computer programs by commercial organizations.

During the assessment of the indicators, a conclusion was made about the presence of prerequisites and the initial stage of the formation of a big data economy in the Arctic zone of the Russian Federation.

The Krasnoyarsk Territory has high digital development indicators: the largest percentage of the population and organizations are connected to broadband Internet, and the largest number of organizations used data centers for their purposes. Despite the high resource potential of the region and the location of large mining companies in it, databases and computer programs are practically not registered in this territory. The leader in this area is the Yamalo-Nenets Autonomous Okrug, Arkhangelsk and Murmansk regions.

Low levels of digitalization of the population are observed in the Nenets, Chukotka Autonomous Okrugs, and the Sakha Republic. The gap from others in the development of digitalization in these regions may be due to technical difficulties in creating a significant digital infrastructure for natural and social parameters, including population density and their type of employment. This may demonstrate the relevance of supporting entrepreneurial initiatives in these regions to reduce the gap, which will only increase with the transition to a big data economy.

The Big Data Economy is a natural consequence of the digitalization of most areas of activity. Its development depends not only on the degree of penetration of new technologies, but also on the activity of economic entities. The Arctic zone is on the way to a data-oriented economy due to the development of important technological indicators: the Internet, the use of cloud technologies, the registration of databases and computer programs. A broad demand from the market for ways to work with big data is beginning to form, which is one of the drivers of the development of the Big Data Economy. In this regard, the Arctic is becoming an object of discussion in terms of the possibilities of using its resources in the context of the development of the Big Data Economy.

The practical significance of the study lies in the possibility of using the results obtained to assess the infrastructural readiness of other regions of the Russian Federation for the big data economy as a stage following the digital economy, which can have a positive impact on the sustainable economic growth of the territories and the country as a whole.

The prospects for further research are determined by a deeper analysis of the indicators highlighted by the authors through the selection of variables and the assessment of the degree of their influence on the development of the economy of big data regions, the formation of proposals and recommendations for the development of infrastructure and big data technologies, taking into account the current international situation.

Acknowledgements The research was supported by Russian Science Foundation grant No. 23-78-10190, https://rscf.ru/project/23-78-10190/.

References

1. Zalta, E.N., Nodelman, U., Allen, C., Anderson, R.L.: Stanford Encyclopedia of Philosophy. http://plato.stanford.edu/
2. Ilin, I.V., Trifonova, N.V., Khusainov, B.D.: Digital transformation in Russian transport companies. In: International Scientific Conference "Digital Transformation on Manufacturing, Infrastructure & Service", pp. 945–954. Springer (2022)
3. Sagiroglu, S., Sinanc, D.: Big data: a review. In: 2013 International Conference on Collaboration Technologies and Systems (CTS), pp. 42–47. IEEE (2013)
4. Hu, X., Jiang, Y., Guo, P., Li, M.: How does China's big data policy affect the digital economy of cities? Evidence from national big data comprehensive pilot zones. Heliyon **10**(2) (2024)
5. Fadeev, A., Kalyazina, S., Levina, A., Dubgorn, A.: Requirements for transport support of offshore production in the Arctic zone. Transp. Res. Procedia **54**, 883–889 (2021). https://doi.org/10.1016/j.trpro.2021.02.143
6. Ruppert, E.: Big data economies and ecologies. In: An End to the Crisis of Empirical Sociology? pp. 13–28. Routledge (2015)
7. UNCTAD: Digital Economy Report 2021: Cross-border Data Flows and Development-for Whom the Data Flow. UN (2021)
8. Kokkinou, A., van Kollenburg, T., Mandemakers, A., Hopstaken, H., van Elderen, J.: The data analytic capability wheel: an implementation framework for digitalization. In: 36th Bled eConference: Digital Economy and Society: The Balancing Act for Digital Innovation in Times of Instability (2023)
9. Sergeev, V., Ilin, I., Fadeev, A.: Transport and Logistics Infrastructure of the Arctic zone of Russia. Transp. Res. Procedia **54**, 936–944 (2021)
10. Pichkov, O.B., Ulanov, A.A., Patrunina, K.A.: Digitalization of the Arctic. In: The Handbook of the Arctic: A Broad and Comprehensive Overview, pp. 441–461. Springer (2022)
11. Ilin, I., Borremans, A., Bakhaev, S.: The IoT and big data in the logistics development. Crude oil transportation in the Arctic zone case study. In: Internet of Things, Smart Spaces, and Next Generation Networks and Systems, pp. 148–154. Springer (2020)

12. Ilin, I., Levina, A., Gurzhiy, A., Borremans, A.: Architecture of the maritime logistics ecosystem of the northern sea route: vision and gap. In: Arctic Maritime Logistics: The Potentials and Challenges of the Northern Sea Route, pp. 63–80. Springer (2022)
13. Petrov, A.N., et al.: The russian arctic by 2050. Arctic **74**(3), 306–322 (2021)
14. Kuratova, L.A.: Features of digitalization of the Arctic regions of Russia. Soc. Econ. Dev. **50**, 125–143 (2023)
15. Babkin, A, Egorov, N., Kovrov, G.: Tsifrovizatsiya ekonomiki regionov i krupnykh neftegazovykh kompaniy v rossiyskoy Arktike. Vestn. Akad. Znaniy **3**(50), 49–56 (2022)
16. Osipova, E., Avagina, O., Cvetkov, A., Lupacheva, S., Sazanova, E. Digital transformation as a modern strategy for sustainable socio-economic development of the Arctic zone. In: E3S Web of Conferences, EDP Sciences, pp. 05007 (2024)
17. Bilal, K., et al.: A taxonomy and survey on green data center networks. Futur. Gener. Comput. Syst. **36**, 189–208 (2014)
18. Jerléus, K., Ibrahim, M.A., Augustsson, A.: Environmental footprints of the data center service sector in Sweden. Heliyon **10**(11) (2024)
19. Yang, Y., Guo, S., Liu, G., Yi, L.: Fine granularity resource allocation of virtual data center with consideration of virtual switches. J. Netw. Comput. Appl. **175**, 102916 (2021)
20. Saunavaara, J., Laine, A. Research, development, and education: laying foundations for Arctic and northern data centers. Soc. Econ. Dev. **125** (2021)

Evolution of Digital Systems in the Economy Through the Adoption of Multi-Agent Technologies

Aleksandr Antonov, Anastasia Levina, and Zhengwei Zhao

Abstract This paper raises the question of the history of digital systems development in the economy and analyzes it. The relevance of this topic lies in understanding the evolution of enterprises' and businesses' needs for digitalization. Analyzing the evolution technical support for these needs is also significant. The main purpose of this work is to analyze the reasons for information systems' development and the progress of their improvement to provide the required level of support. Another important task is to analyze what direction this development is heading towards in terms of information systems. At the beginning of the project, the first needs for digital support of the business are identified, the causes and consequences of these needs are analyzed, and the technical aspects of developing digital systems are also considered. Then, an analysis of existing solutions in information systems for business support is conducted. Based on this analysis, conclusions are drawn and a multi-agent approach to supporting business processes is described.

Keywords Industry 4.0 · Industry 5.0 · Multi-agent technologies · Society 5.0 · MRP · ERP

A. Antonov · A. Levina (✉)
Peter the Great St. Petersburg Polytechnic University, Saint-Petersburg, Russia
e-mail: levina_ai@spbstu.ru

Z. Zhao
Luoyang Normal University, Luoyang, China

I. Ilin and M. Youzhong (eds.), *Digital Systems and Information Technologies in the Energy Sector*, Lecture Notes in Networks and Systems 1244,
https://doi.org/10.1007/978-3-031-80710-7_14

1 The Impact of Digitalization on the Economy

1.1 The Emergence of the Need for Digitization

Information support has always been an essential factor in ensuring the quality of work in any enterprise or organization. In the early stages of development and establishment of complex and large-scale enterprises, the time taken to process various operations was significantly prolonged due to the lack of a systematic approach to storing and updating information after completing specific tasks. To address this issue, it became necessary to develop specialized information systems for enterprises.

The main development of information systems occurred during the period from the 20th to the twenty-first century. However, it's important to note that the active development of digital systems during this time was only possible due to a long and complex historical process influenced by various social, economic, scientific, and philosophical factors.

A key factor in the development of digitalization was the growth of information caused by rapid scientific and technological advancements. New sciences emerged, people's cognitive activities expanded, and production techniques evolved. These factors made manual data processing methods less and less efficient, leading to the need for new solutions. This confirmed the necessity to improve business processes and incorporate computers into people's daily lives.

Among the above-mentioned prerequisites, it is essential to consider the historical evolution of approaches to the study of digitalization in the economy from a philosophical perspective. One of the most significant events in human history is the shift from a mechanistic worldview to an informational one. This paradigm shift has had a profound impact on the way we perceive the world and has transformed the role of information in society, leaving a lasting impression on all aspects of life. One of the fundamental philosophical principles of digitalization is the concept of the information-based nature of reality. This idea can be traced back to the writings of ancient philosophers like Plato and Aristotle, who proposed the idea that the world around us is not made up of material objects but rather a collection of information structures. In the twentieth century, this concept was further developed by thinkers like Norbert Wiener, Claude Shannon, and Alain Turing, who laid the groundwork for information theory and cybernetics. They demonstrated that information is an essential aspect of the universe and plays a crucial role in managing and organizing systems [1].

New scientific revolutions have had a significant impact on the way of view information. It is worth noting two significant scientific revolutions: cybernetics and information theory. These approaches view information as an essential part of the fundamental characterization of the universe, rather than simply a reflection of material processes. Ursul Arkady Dmitrievich, a prominent Soviet and Russian philosopher and methodologist of science and technology, emphasizes the importance of understanding the role of information in evolutionary processes, both in nature and society. He argues that a modern worldview cannot exist without an understanding of information interaction. This highlights the crucial role of information in our understanding of ourselves and the world around us. Analyzing this idea, it is important to consider the world from the perspective of a complex information system. This system places information at the center of all tasks and processes [2].

It is important to note the technical prerequisites for digitalization. The key stage is the digital revolution caused by the creation of electronic computing machines (computers) in the middle of the XX century. Before that, the amount of information processed was limited by the capabilities of the human brain and mechanical devices. The first personal computer was created in 1981 by IBM with the name IBM PC. This computer was accessible to the public and had high speed and computing power, which made it possible to use it as a publicly available information processing tool. Later, analogues and new versions of "home" computers were created, which influenced the mass use of them and led to an increase in the need to use computer technology to process large amounts of information, including in manufacturing and business, to calculate various economic indicators.

It is important to note the main advantages of working with a computer in terms of new opportunities for their users. Such an analysis will help to better understand the importance of the transition from manual to automatic computing in terms of the technical prerequisites for digitalization and the development of digital systems in the economy. The main benefits of using a computer are outlined below.

Automation of information processing tasks. This change made it possible to remove routine tasks, which not only reduced the processing time of basic actions, but also allowed people to devote more time to creative tasks aimed at solving intellectual problems.

Processing large amounts of information. The emergence of the possibility of processing and analyzing previously inaccessible amounts of information has increased the number of options for solving various tasks.

Creation of new information systems: The ability to develop new software has not only influenced the speed of solving problems, but also the overall approach to tackling new challenges related to data analysis and resource planning.

Analyzing these advantages of computer usage, it should be noted that they have led not only to increased demand for computers, but have also created several new needs that have formed the basis for digitalization.

Need for information protection: The increasing amount of information has raised awareness of its significance, leading to the emergence of a need to safeguard data from deletion, theft, or alteration.

Demand for methods of storing and retrieving information: With the growing volume of data, new approaches to analysis are required. As a response, ideas have been proposed to establish unified digital systems that can organize and manage data in a unified format.

The need for trained personnel. Working with new information systems has led to the expansion of the job market by creating opportunities for specialists who can interact with these systems through their skills and knowledge.

Thus, several factors contribute to the need for digitalization, including social and technical requirements for increasing information volume and adopting a new way of thinking. To address these needs, various information systems have been developed to facilitate data processing.

1.2 The Evolution of Information Systems

The aforementioned digitalization needs have led to the development of digital information systems. The first definition of an information system was proposed by Joseph Weizenbaum in 1963 in his book "Computer Simulation of Human Reasoning". He defined an information system as a set of devices that collects, stores, processes, and distributes information [3].

The evolution of information systems can be seen through the development of automation and management systems for business processes. Several iterations of concepts have been developed that partially or completely cover key business processes and objectives. These concepts will be discussed in the following order, with an analysis of their innovations, benefits, drawbacks, and areas for improvement. The MRP (Material Requirements Planning) concept was developed in the 1950s. A definition of an MRP system is as follows: it is a system that uses a unique code to identify each component of a product and tracks its movement throughout the production process [4].

It is important to note that the MRP system in this form represents the automation of a specific aspect of the production planning process, namely, the business process of material requirement planning. Additionally, an important innovation of this approach is the scheduling process. The MRP concept only works if there is a comprehensive production plan in place, for which it is feasible to determine material and component needs. Furthermore, the initial ideas behind the basic systems did not include component analysis, which can be considered an additional innovation of the MRP information system. It is important to understand the MRP concept and its key components that are responsible for performing various actions related to analysis, processing, and inventory management.

Demand: The system allows us to not only assess and analyze current demand for finished products but also predict future demand.

Production plan: The MRP creates a plan that considers the quantity of materials needed and the dates by which these quantities should be produced.

Actual stock: The system reflects changes in stored stock and the behavior of inventory counting.

BOM (Bill of Materials): A record for each product that describes the composition of products, presence or absence of components, and their required quantities.

These components of the MRP (Material Requirements Planning) concept were essential to manage the complex business processes resulting from accelerated business development during the second half of the twentieth century, which led to a large-scale increase in production and sales. These factors contributed to the creation of MRP, which enables businesses to reduce production lead times, improve communication with customers, and establish a structured approach to production scheduling. It's important to acknowledge the role of technology in this development. Only through the creation of computers and supporting software was it possible to develop more sophisticated and complex systems that became MRP. However, it's also important to recognize the limitations of this approach, which contributed to the need for ongoing improvement and adaptation of information systems. The original MRP system focused solely on planning material requirements, without considering crucial production resources such as labor, equipment, and finances [5]. This limited its effectiveness in addressing the complexity of modern business operations.

To address these shortcomings, in the 1970s Oliver White proposed the idea of MRP II (Manufacturing Resource Planning). This concept extended the original MRP by including the analysis of inventory levels and production planning for required materials. Additionally, the concept included the planning of resources such as workers, machinery, and finances. These elements were not included in the original MRP model. By adding these resources to the planning and analysis system, a wider range of business processes became possible, increasing the overall usefulness and efficiency of information systems. At that time, it was no longer possible to ignore digital support for businesses.

It is also important to consider the MRP II concept from the perspective of key components that allow for analyzing and planning production. First, should consider the data used for planning, specifically the quantity and specification of materials needed for production, when they are needed, and when production, processing, and sales of finished goods occur. Secondly, it is essential to analyze the main components of MRP II.

Production Load Planning: This fundamental component was transferred from the MRP system and allows for taking into account the required number of materials in production by a specific deadline.

Resource Load Planning: A new component that considers the total number and capacity of resources, such as labor and equipment, in production. This innovation allows for analyzing and automating new business processes.

Production control: This component eliminates the need for real-time resource accounting, which arose as a result of the resource loading component described above.

Data analysis: With the increasing volume of information and its constant growth trend, it is necessary to effectively and promptly evaluate and analyze data obtained during production processes.

Thus, it can be concluded that the MRP II concept has evolved from the MRP concept, introducing a number of innovative features that have expanded the possibilities for planning and analyzing business processes. This has also influenced how businesses approach their operations, with a greater emphasis on improving product quality rather than simply reducing costs. While it is true that the MRP II approach has led to an increase in productivity, it is also important to note that it has allowed for the management of a larger number of planning processes and more diverse resources.

However, it is important to note the limitations of the MRP II approach. With the rapid growth of businesses, the number and complexity of business processes have increased dramatically, as well as the amount of information involved in these processes. The MRP II system's ability to support and analyze these processes has become inadequate. A new system that can manage all aspects of a company's operations is needed.

The evolution of information systems has led to the development of the Enterprise Resource Planning (ERP) system. This concept has no specific year of creation or creator, but the first mentions of it can be found in various sources from the 1980s. ERP has existed for several decades, combining many ideas, technologies, and concepts to process various business processes into a single system [6]. This integration allows companies to automate new business processes such as financial management, customer data management, and record-keeping and analysis. It's important to understand the transition from Material Requirements Planning (MRP) to ERP not just in terms of new components, but also in terms of historical factors that led to his shift.

Globalization: The expansion of markets and the growth of businesses has led to a need for more detailed accounting of business milestones. This is because it is important to accurately track all aspects of a business that enters the international market.

Increased competitiveness: To stand out on the international stage, businesses need to improve their competitiveness. This can be achieved through increased efficiency and automation of business processes.

Integration of business processes: As businesses grow, it becomes necessary to manage them in a unified system. This helps reduce time, information loss, and improves resource management.

The Need for Real-Time Response: In today's business environment, the ability to react quickly to changing situations is essential for success. As a result, there is an urgent need for solutions that can provide real-time information and support. This leads to the requirement for systems that can process and analyze data in real-time, providing the most current information available.

Having considered the various factors that have shaped the business environment during the latter half of the twentieth century, it is evident that the introduction of ERP systems represented a logical response to the challenges faced during that period. The increasing complexity of manufacturing processes, the necessity for integrating diverse data, as well as the escalating competition and global expansion of businesses, all necessitated fresh approaches to resource management [7, 8]. Under these conditions, ERP systems evolved as a response, providing a comprehensive information platform for crucial business operations.

An analysis of the critical components of Enterprise Resource Planning (ERP) systems permits a deeper comprehension of their role in assisting organizations in achieving their strategic objectives. A unified, integrated database encompassing all functional areas within a business guarantees the precision and consistency of data, as well as the capacity to monitor interactions between distinct procedures. The modular architecture of enterprise resource planning (ERP) solutions enables businesses to scale their systems according to changing needs and to integrate them with other information technologies if required. This flexibility makes ERP solutions an invaluable asset for organizations seeking to optimize their operations and achieve their strategic objectives successfully.

Considering all these factors, the ERP concept was developed to address the majority of business needs that emerged in the latter half of the twentieth century. This approach is also worth examining in terms of its key components.

Flexibility and Scalability: The ERP system was designed with flexibility in mind, allowing it to adapt and scale to meet the unique needs of each business [9]. This ensures that the system can evolve with the organization, rather than being a static solution.

Modularity: ERP systems consist of various modules, each responsible for a specific set of tasks. However, they can be combined or replaced as business needs change. This allows for customization to fit business requirements and eliminates unnecessary processes. Complex Process Automation: ERP systems use computational methods to plan complex and multi-step business processes [10]. This enables them to automate not just basic tasks, but high-level business processes as well (see Fig. 1).

Fig. 1 The main ERP modules

Automation of Complex Processes: ERP systems use computational methods to plan complex and multi-step business processes. This allows them to automate not only simple tasks, but also more complex business processes.

Data Management and Integration: ERP systems provide a centralized data storage system that allows easy access to analyzed data and quick output of results to system users. This helps make more informed decisions. Integration between the various modules of the system ensures that these decisions can be made quickly at all stages of a large business process.

Thus, it can be concluded that modern ERP systems, which have evolved from MRP systems to their current form, almost entirely meet the needs of businesses to automate standard business processes. This historical journey of changing business requirements and information system capabilities is a key indicator of the evolution of digital systems in the economy.

2 Modern Approaches to the Digitalization of the Economy

2.1 The Need for Digitalization in Modern Business

The development of information systems is a dynamic process. In today's world, enterprise resource planning (ERP) ideas are evolving in response to the growing needs of modern businesses. One of the primary objectives of any business is to maintain and improve its competitiveness. To achieve this, businesses must adapt to market changes and develop digital solutions.

These changes are evident in the context of digital transformation in business. This approach enables us to analyze market trends from a historical perspective. At present, the emphasis is on customer focus and the volume of data processed. Here are some key features of modern business requirements.

Increased Customer Focus: In order to expand the company and capture new markets, it is essential to provide high-quality customer service. Various technologies can be used for this purpose, as will be discussed later in this report. However, it should be noted that the CRM (customer relationship management) system is the primary approach used. This system includes various tools for analyzing customer data.

Enhanced Security: Given the value of information, it is vital to protect user data from modifications, deletions, and unauthorized access.

Cost Reduction: With the rapid growth of information technology infrastructure, costs associated with supporting information systems have increased. The complexity of these systems has also increased, affecting both the purchase price and the maintenance of these services. It is therefore crucial to optimize the utilization of IT infrastructure in order to reduce costs [11].

Increasing productivity and efficiency: The increase in information volume necessitates not only the digitization of business processes but also the maximum automation of their execution. Analyzing the requirements of modern businesses leads to the conclusion that the use of new technologies is essential. These technologies exist and can meet the described requirements, which characterize the core features of Industry 4.0. However, it is important to note that these technologies are not sufficient to go beyond these requirements. Industry 4.0 is a concept describing the enhancement of enterprise competitiveness through the application of advanced intelligent technologies [12, 13].

Digital transformation, a process of significant changes in an organization resulting from the introduction of digital technologies, has become an integral part of contemporary business. It extends beyond simple automation to encompass a radical restructuring of business models, procedures, and cultural norms [14].

The driving forces of digital transformation are the rapid development of technology, changing consumer expectations, increased global competition and the need to adapt to new regulatory requirements. Digital technologies such as artificial intelligence, big data, cloud computing and blockchain offer companies unprecedented opportunities to optimize processes, create new products and services, improve customer interaction and increase innovation.

The introduction of digital technologies has a significant impact on all aspects of companies' operations, from production and logistics to marketing and sales. Automating routine tasks, creating digital platforms for customer engagement, and using data analytics to inform decision-making are all becoming integral parts of modern business.

However, this digital transformation process presents its own set of obstacles. The high costs associated with adopting new technologies, the lack of skilled personnel, employee reluctance to change, and cybersecurity threats all necessitate a comprehensive approach to managing this process.

Successful digital transformation demands a clear vision, a thorough analysis of the current state of the business, the development of a well-defined plan for change, and the establishment of a flexible organizational framework that can adapt to evolving realities. It also requires fostering a culture of innovation and providing training for employees to acquire new digital skills. Digital transformation is not a one-off project but rather an ongoing process that necessitates continuous attention and investment. Organizations that effectively adapt to changing circumstances will gain a substantial competitive advantage. Digital transformation is an integral part of the evolution of modern organizations. It presents both opportunities and challenges for businesses, requiring a well-defined strategy and substantial investment. Companies that effectively leverage the potential of digital technologies have the potential to achieve a sustainable competitive edge and thrive in the digitally-driven landscape.

2.2 Existing Technologies in Information Systems

In order to meet the needs of modern businesses, a wide range of innovative technologies have been developed that allow for the processing of large volumes of data at remarkable speeds, enabling comprehensive analysis and complete fulfillment of the requirements of contemporary business operations. When considering the technical approach to addressing the challenges presented by the digital economy, it is evident that there is a tendency towards integrating different platforms into a unified system. This trend is driven by the need for effective support across various departments within a company, requiring a unified, centralized, and structured approach [15].

This requirement is driven by the exponential increase in data volumes and the need for businesses to improve their competitive advantage through increased productivity, while simultaneously maintaining or even reducing costs. In addition, it requires a stronger focus on customer-centricity and the ability to quickly scale operations, while maintaining the flexibility to adapt core processes. Furthermore, it is essential to emphasize the importance of information management within the context of Industry 4.0. Only with a large amount of data can we make informed decisions that are based on evidence rather than intuition. These requirements can be addressed by Enterprise Resource Planning (ERP) systems, which allow us to automate, monitor, and plan business processes across all stages of production, through centralized data storage and real-time analytics [16].

In the following, will explore several existing technologies that represent the current state-of-the-art in their respective fields. This analysis will provide insight into the historical evolution of digital systems and their current status.

The Internet of Things (IoT) is a technology that connects numerous devices in a single network for the purposes of transmitting, collecting, and processing data from various sources. This enables businesses to optimize operations, increase efficiency, and develop innovative products and services.

Artificial intelligence (AI) refers to a neural network that utilizes machine learning to process large volumes of data for prediction and other purposes. It assists companies in focusing on customers by creating more customized and intelligent chatbots.

Big Data is used to analyze huge amounts of information to increase customer orientation. This technology helps companies quickly scale their businesses if necessary [17].

Analyzing the above, it can be concluded that existing technologies meet the needs of businesses developing within the framework of Industry 4.0., however, they cannot go beyond them. These frameworks appear due to limitations in computing power caused by a strict, immutable and inflexible mathematical approach.

3 Emerging Trends in the Digital Economy

3.1 The New Demands of the Digital Economy

In the previous chapter, was discussed business requirements in today's world and explored technological solutions to address these demands. We examined the challenges that businesses face in the current market environment.

This chapter, will be focus on a more significant question: what is the future of digitalization in the economy? To answer this question, it is important to consider the difficulties that businesses may encounter even when utilizing the latest technological innovations.

The main obstacle is uncertainty. In a complex system with many participants, each with its own set of variables that influence task completion, uncertainty can arise at any time due to changes in plans, unexpected events such as force majeure, or other unpredictable occurrences. These unforeseen events often not only affect the plan but can also lead to its collapse [18].

It is essential to keep in mind that a minor alteration in any single element can lead to a series of changes in other elements, directly or indirectly. At the same time, there is an increasing need for a personalised approach, which also influences the development and implementation of plans.

Based on these requirements, it can be concluded that there is a gradual shift from Industry 4.0 towards Industry 5.0. While this shift has not yet been fully implemented, its foundations are evident. From the perspective of societal transformation, for example, the term "Society 5.0" was introduced in 2016 to describe a "super-intelligent society" that utilises and depends on digital services and becomes fully integrated with the system. This approach promotes greater adaptability in the system and has implications for everyday life [19].

It is also essential to consider the fundamental differences between Industry 4.0 and Industry 5.0. A significant distinction is the level of flexibility in planning and responding to unexpected situations. In Industry 5.0, planning should be agile and coordinated across all aspects of the system in order to enhance competitiveness. This also has implications for pricing, which is undergoing a transformation in Industry 5 due to the need for flexibility for all participants in the system.

In contrast, pricing in Industry 4 remains relatively rigid, with adjustments not being made quickly enough to adapt to changing market conditions. This can lead to a decrease in competitiveness for businesses operating in Industry 4, as they may struggle to keep up with market demands. When examining the aforementioned trends, it is essential to identify the critical stages in the transition from Industry 4.0 to Industry 5.0. The exponential increase in data and resources involved in business and societal processes makes it difficult to analyze such information using traditional methods within a single, centralized framework. This means that even minor changes in a process due to unexpected circumstances can only be corrected by increasing resources dedicated to solving the problem, which may involve a delay.

The emergence of Industry 5 introduces a paradigm shift in industrial operations, providing a more flexible and responsive approach to dealing with unforeseen challenges. Unlike Industry 4, which relies on static planning, Industry 5 embraces real-time adaptability, allowing for swift and effective responses to changing situations.

One of the main benefits of Industry 5 is its improved efficiency in resource management. By utilizing advanced predictive analytics, companies can anticipate potential issues and optimize their use of resources, leading to more streamlined operations and increased productivity. This not only enhances the competitiveness of businesses but also contributes to a better standard of living for employees in the industry. Furthermore, the focus on customer centricity and flexibility in Industry 5 promotes a culture of ongoing improvement, with data-driven insights driving strategic decision-making. With real-time access to data, prices become more responsive to market conditions and customer preferences.

Therefore, it can be inferred that the essential requirement for the digitalization of future economic systems is their adaptability. Current solutions fail to address this need, as they rely on a single, inflexible model rather than accepting the idea of responsive modification to each component of the system.

3.2 Multi-Agent Technologies

The above business needs can be addressed by new generation technologies. These technologies are based on the concept of multi-agent systems. A multi-agent system is not a single entity, but rather a collection of autonomous agents or groups of agents that work together to achieve a common goal in the most efficient way possible. Each agent in the system has its own objectives and is responsible for completing its assigned tasks. However, these agents also compete with each other to optimize their individual performance, which ultimately leads to the overall success of the system. This approach not only maximizes the benefits of each agent's efforts, but also ensures the achievement of the overall objective.

The implementation of this approach is only possible through the use of the idea of swarm intelligence, also known as emergent intelligence. Tasks such as planning and adaptation are accomplished through this intelligence within individual elements of the system, rather than the entire system as a whole. This can be compared to a swarm of bees or ants, where each individual solves its own small task while collectively achieving a larger goal [20].

Such systems consist of interconnected and competing software components, known as agents (see Fig. 2). These agents possess a certain level of awareness of their surroundings (which is typically stored in a centralized database), a defined function (typically divided into opportunity and need agents), and a goal-oriented approach that operates based on various criteria. Furthermore, these agents can interact with one another in a virtual marketplace, negotiating for the best possible outcome for themselves and the system as a whole.

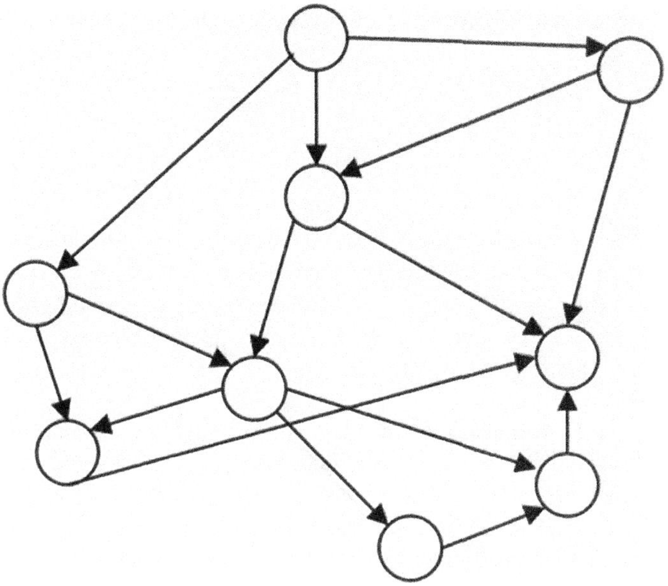

Fig. 2 The scheme of the agents' work

Agents can also be digital representations of real-world objects, such as an autonomous vehicle or a warehouse. They can also represent individuals, such as a driver or a delivery person. In their most advanced form, they are artificial intelligences with their own objectives, knowledge, and resources. When considering a system with multiple agents, each one being an AI, it is easy to draw parallels to human society. From a theoretical perspective, it is worth examining the Deming cycle, which encompasses planning, implementation, verification, and action [21]. If each individual adheres to this cycle, the outcome will be a flexible system that can respond to external changes.

Therefore, it can be concluded that the adaptability of decision-making processes in modern businesses can be facilitated by a new generation of digital systems, particularly a multi-agent system. This approach can be viewed as the latest phase in the evolution of the digitalization of the economy.

In summary, the history of information systems can be divided into distinct phases, starting with MRP systems and progressing to ERP class systems. Each phase has been supported by fundamental technologies that address the specific requirements of businesses [22]. Furthermore, the current approach to digital support for businesses in a rapidly changing world has been examined.

A key takeaway from this study is the recognition that information systems should not be viewed as a monolithic entity, but rather as a collection of agents each with a distinct role. This perspective is informed by an analysis of the historical evolution of information systems, which reveals a trend towards creating and developing such systems in a way that deviates from the current state and attempts to recreate it digitally.

4 Conclusion

An examination of the historical trajectory of information systems demonstrates their continuous evolution driven by changing business requirements and technological advancements. Starting with simple Material Requirements Planning (MRP) systems designed to optimize resource allocation, information systems have developed into comprehensive Enterprise Resource Planning (ERP) solutions covering all aspects of a business.

The key insight is that information systems are not static structures, but rather flexible systems that can adapt to new circumstances and challenges. A modern approach to business transformation entails creating flexible and scalable information systems capable of responding quickly to external changes. It should also be noted that viewing information systems as a collection of agents, each with its own specific role, enables a more accurate representation of their intricacy and diversity. This perspective opens up new avenues for the development and implementation of innovative information solutions that can improve business efficiency and help maintain competitiveness.

References

1. Sizo, R.S., Vorobyev, D.V.: Teoriya idey Platona i Aristotelya: sravnitel'nyy analiz. Gumanitarnye nauchnye issledovaniya 12 (2017)
2. Ursul, A.D. Priroda informatsii: filosofskiy ocherk. 2nd edn, Tsentr issled. glob. protsessov i ustoychivogo razvitiya (2010)
3. Weizenbaum, J.: How to make a computer appear intelligent. Datamation **8**(2), 22–24 (1962).
4. Stonis, R.A. "MRP-SISTEMA." Nauka, tekhnika i innovatsii: gipotezy, problemy, rezultaty, 74–77 (2019)
5. Öztürk, C., Örnek, A.M.: A MIP based heuristic for capacitated MRP systems. Comput. Ind. Eng. **63**(4), 926–942 (2012). https://doi.org/10.1016/j.cie.2012.06.005
6. Hencha, R., Verma, D.S.: Study of material requirement planning processes & its analysis and implementation (a case study of automobile industry). Int. J. Sci. Technol. Res. **8**(8), 648–653 (2019)
7. Ivanov, D., Frolov, K., Frolov, A.: Implementation of ILS methodology on the basis of ISO 55000 series standards. Lect. Notes Netw. Syst. **951**, 79–91 (2024)
8. Tick, A., Ilyina, I., Sheleyko, V.: Convention and exhibition industry in the context of digitalization trends and development problems. Technoeconomics 2. **2**(5). 54–65 (2023). https://doi.org/10.57809/2023.2.2.5.5
9. Frolov, K., Iliashenko, V., Ershova, A.: Strategic procurement of resources for transportation: architectural and methodological issues. Transp. Res. Procedia **54**, 945–952 (2021)
10. Schwab, K.: The fourth industrial revolution. Crown Publishing Group, New York (2017)
11. Aktaş, N., Demirel, N.: A hybrid framework for evaluating corporate sustainability using multi-criteria decision making. Environ. Dev. Sustain. **23**(10), 15591–15618 (2021)

12. Chen, H.J.: Clarifying the impact of surprise in e-learning system design based on university students with multiple learning goals orientation. Educ. Inf. Technol. **25**(6), 5873–5892 (2020)
13. Kalyazina, S.E., Balabneva, O.A.: Impact of the industry 4.0 paradigm on key software requirements. Technoeconomics **2**(2). 21–31 (2022). https://doi.org/10.57809/2022.2.2.3
14. Lee, J., Kao, H.A., Yang, S.: Service innovation and smart analytics for Industry 4.0 and big data environment. Procedia CIRP **16**, 3–8 (2014)
15. Lazareva, M.V., Mohinur, K.: Modern management and digitalization. Talqin Va Tadqiqotlar Ilmiy-Uslubiy J.I (Sci. Methodol. J. Interpret. Res) **1**(17), 301–307 (2023)
16. Kagermann, H., Lukas, W., Wahlster, W.: Industrie 4.0: Mit dem Internet der Dinge auf dem Weg zur 4. Industriellen Revolution. VDI nachrichten 13 (2011)
17. Preobrazhenskiy, Y.P., Myasnikov, O.A.: Analiz perspektiv informatsionnykh tekhnologiy v sfere internet veschey. Vestnik voronezhskogo instituta vysokikh tekhnologiy, 43–45 (2020)
18. Big Data. Tadviser. https://www.tadviser.ru/index.php/Big_Data. Accessed 10 Aug 2024
19. Nirmala, J.: Super Smart Society: Society 5.0. Robotics Tomorrow (2016). https://www.roboticstomorrow.com/article/2016/09/super-smart-society-society-50/8739. Accessed 7 Aug 2024
20. Bonabeau, E., Theraulaz, G.: Swarm smarts. What computers are learning from them? Sci. Am. **282**(3), 54–61 (2000)
21. Isniah, S., et al.: Plan do check action (PDCA) method: literature review and research issues. Jurnal Sistem dan Manajemen Industri **4**(1), 72–81 (2020)
22. Niv, G. Organizatsiya kak sistema. Printsipy postroeniya ustoychivogo biznesa Edvardsa Deminga = The Deming Dimension. Alpina Pablisher (2011)

Modern Approaches of Adaptive Planning and Management in Discrete Production

Aleksandr Antonov

Abstract This article discusses contemporary approaches to discrete production planning. The significance of this subject lies in the necessity to consider the planning process from the perspective of flexibility, as it entails adjustments to the system due to outside influences. It is also essential to analyze systems that can guarantee the proper operation and creation of a successful production plan. The primary objective of this study is to investigate contemporary approaches to adaptive scheduling for discrete manufacturing and propose an alternate methodology for implementing this process. At the outset of the paper, the fundamental characteristics of discrete manufacturing are outlined, followed by a review and comparison with continuous manufacturing. Key areas in which the use of discrete manufacturing with adaptive scheduling at various planning levels could be advantageous are identified. Next, we will discuss supporting systems for planning and controlling discrete manufacturing, such as Enterprise Resource Planning (ERP) and Manufacturing Execution Systems (MES), as well as possibilities for integrating them. In the final section, a novel multi-agent-based approach to adaptive scheduling for discrete production is presented and analysed, highlighting its key benefits.

Keywords Discrete production · Adaptability · Adaptive planning · Multi-agent technologies · MRP · ERP · PDCA

A. Antonov (✉)
Peter the Great St. Petersburg Polytechnic University, St. Petersburg, Russia
e-mail: antonov4.as@edu.spbstu.ru

I. Ilin and M. Youzhong (eds.), *Digital Systems and Information Technologies in the Energy Sector*, Lecture Notes in Networks and Systems 1244,
https://doi.org/10.1007/978-3-031-80710-7_15

1 Discrete Production Planning

1.1 Discrete Production

Discrete manufacturing is a crucial approach to organizing the manufacturing process. It entails performing individual operations in a discontinuous manner. This technique of production organization is commonly employed in industries such as mechanical engineering, computer production, electronics, and other related fields, where there is a need for controlling and accounting for individual components used in the manufacturing process. This allows the creation of products with specified characteristics. Due to the requirement for precise control and batch-based accounting, the planning process diverges from traditional continuous manufacturing and necessitates monitoring at each step. Later in the chapter, these differences will be discussed in more detail, including an analysis of the disparities between discrete and continuous manufacturing as well as the various stages of planning involved in discrete production.

For further analysis of discrete production planning, it is essential to examine in greater detail the characteristics of this approach as a whole. One of the most significant factors to consider is the capacity to create unique, customized products. This methodology cannot be categorized as a mass production technique, as it does not include standardization or uniformity [1].

Additionally, it is important to note that personalization is only feasible for specific components or aspects of a finished product, and this does not significantly extend production time, enabling mass production. This approach allows businesses to tailor their offerings to customer needs and remain competitive by expanding product specifications to satisfy individual customer requests.

An intermittent approach to production involves the sequential execution of individual operations, while allowing some operations to occur simultaneously. It is important to note that the duration of each operation may vary based on its complexity, and by breaking the production process down into smaller steps, we can ensure not only uniqueness, as described above, but also enhanced quality control. Through this approach, quality checks are performed after each stage in the production cycle, until the entire process is completed, ensuring a high standard of product quality. However, it is essential to acknowledge that the greater complexity and uniqueness of the operations may increase the difficulty of planning and logistics [2].

Another significant aspect of discrete manufacturing relates to the presence of a diverse range of technological processes. This is also viewed in terms of the availability and use of various equipment for specific tasks such as assembly, processing, and testing. It should be noted that equipment can be utilized for various purposes and processes.

However, it often requires specific configurations to change its function. These configurations often involve not only timing but also resource expenses, necessitating flexibility in managing production capacity. An example of this is a car manufacturing facility that produces both passenger cars and trucks. When switching product lines, the production lines need to be prepared for a swift transition between operations.

For further analysis, it is also necessary to consider continuous production. It is essential to understand the characteristics of this approach, analyze it, and identify the key differences from discrete production. This will enable us to take a new look not only at the characteristics and complexities of discrete manufacturing, but also highlight the main benefits of continuous production in a more prominent way.

Continuous production entails organizing the production process in a manner that avoids interruptions and operates in a continuous manner. As a result, this method frequently functions without human intervention and with maximum automation. Moreover, processes within continuous production are standardized, with all equipment performing the same function throughout the production cycle. This method minimizes the time and resources expended on reconfiguring equipment, ensuring consistent product quality.

Continuous production also requires fewer complex quality management systems, as control is implemented at the level of the overall process. For instance, in the chemical industry, quality assurance can be integrated into the production process and deviations can be identified during sampling. Therefore, it can be concluded that in this scenario, less emphasis is placed on individual products, as the entire production flow is viewed as a cohesive unit.

When comparing discrete and continuous production methods, several factors need to be taken into account in order to determine which method is most suitable. These factors include the type of product, quality standards, volume of production, and flexibility to adapt to market changes and variations in the product. For instance, when considering discrete production, the example of electrical appliance manufacturing can be used. This type of production typically requires a high degree of customization and a significant change in the production process. This can lead to improved quality control, while maintaining the flexibility needed to adapt to changes in production plans [3].

After identifying and analyzing the circumstances in which the use of individualized production techniques is appropriate, it is essential to consider whether such methods may not be applicable in certain areas under specific conditions. At present, I wish to progress with my work and examine a parameter such as adaptability. This has previously been recognized as a benefit of individualized planning, but has not been regarded as a crucial factor for application in a particular area.

In examining this matter, it is crucial to define what flexibility entails. Flexibility can be described as the capacity of a system to modify and adjust to alterations in the external surroundings, such as marketplace shifts or customer demands. These transformations may occur frequently, implying that there is a high likelihood of frequent adjustments to the production procedure.

Despite this, the analysis demonstrates that discrete production may still have relevance in certain instances. This leads to the conclusion that further consideration is possible in any fields where the application of discrete production holds significance.

1.2 The Complexity of Adaptive Discrete Production Planning

Discrete production planning is a complex and multilevel process that requires careful consideration of various factors affecting the efficiency and flexibility of production. An important aspect of this process is adaptability, which demonstrates the quality and effectiveness of planning, depending on the speed with which the overall plan can be adapted to changing circumstances. It should be noted that adaptability must be considered not only generally, but also at various levels of planning: strategic, tactical, and operational. Strategic planning entails making decisions regarding long-term objectives and developing production capacity. Tactical planning involves allocating resources for the medium term, while operational planning focuses on day-to-day operations and managing specific tasks [4].

In this context, these levels can be conceptualized in terms of a hierarchical structure. The highest level is the strategic level, followed by the tactical levels, and then the operational levels. This illustrates the interdependence between these levels and their dependencies (see Fig. 1). Thus, if there is a change occurs at the strategic level, it may affect the planning activities at the tactical and operational planning levels as well. Conversely, if changes are made only at the tactical level, they would have an impact on operational planning level. Similarly, alterations to the operational level would not affect tactical planning level will not be affected. Understanding this relationship gives an idea of how adjustments at various planning stages can influence one another. Furthermore, it emphasizes the significance of considering all aspects when making decisions.

The strategic level of planning is of utmost importance and influences all other levels of planning. It focuses on long-term objectives and goals for the overall growth and development of the organization. The primary objective of this level is to establish a shared vision and determine key priorities that can help the company achieve competitive advantage. In order to do so, companies engage in preliminary market research at this level to gather crucial strategic information [5].

This includes analyzing the market value represented by the company and determining how it can best position itself in front of its target audience. Market analysis, identification of the target audience, and assessment of the competitive landscape are all part of this process. Decisions are also taken regarding the direction of growth, new product development, and the markets the company intends to enter. Furthermore, it is essential to note that, at this stage, decisions regarding expanding production capacity, upgrading equipment, or implementing new technologies are made. A key objective is also to evaluate the necessary investments required to achieve the set objectives. This includes determining the financial, material, and human resources necessary to implement the strategy. In addition, companies at this stage analyze external and internal factors that may impact their business in order to identify potential risks and opportunities for growth and to develop adaptive strategies accordingly.

The tactical planning level is located at the average level and consists of planning for medium time intervals, approximately from several months to several years. This

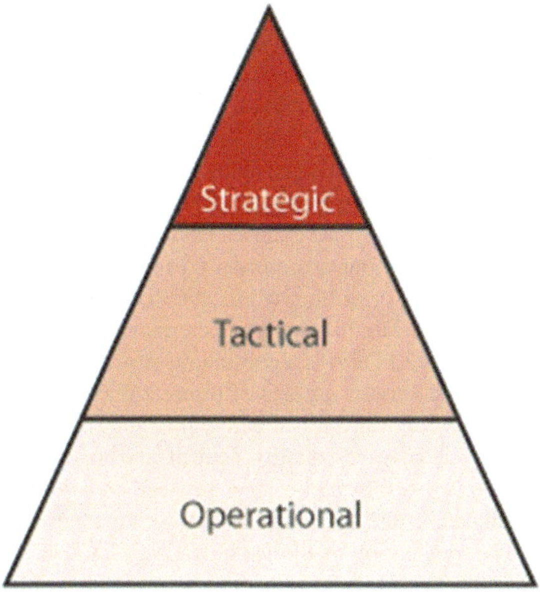

Fig. 1 Levels of production planning

level of planning operates with tasks and goals that concretize plans from the strategic level and translate them into more detailed plans. As an example, we can consider such a task of tactical planning as the development of a detailed production plan that determines which goods will be produced, in what volumes and in what time frame. Another important feature of tactical planning is taking into account the seasonality of demand and fluctuations in the market, which allows the company to flexibly respond to external changes. The tactical level also covers inventory management of raw materials, components and finished products, which allows for effective inventory planning and affects cost reduction. Also, this feature of tactical planning allows you to avoid a shortage of materials, which, in turn, contributes to uninterrupted production. It is worth noting that an important aspect of tactical planning is also the allocation of resources, including labor, equipment and materials, which allows you to optimize the production process.

The operational level of planning refers to the lowest level of the planning hierarchy and focuses on short-term objectives and tasks that are typically accomplished within a day or week. This form of planning is essential for managing day-to-day operations and achieving specific minimum production goals. Operational planning involves the allocation of tasks among team members, creation and management of work schedules, as well as ensuring that production standards are adhered to and necessary resources are available. This approach helps optimize the manufacturing process, reduce costs, and promptly address emerging issues. Additionally, quality control measures are implemented at this level to ensure that products conform to established standards. This is critical for maintaining the company's reputation and meeting customer expectations.

It can be inferred that strategic, tactical, and operational planning are interrelated and form an integrated system, in which each level affects the efficiency of the organization and is affected by the higher level. Strategic planning sets the overall direction and goals, tactical planning develops specific strategies for achieving these objectives, and operational planning manages day-to-day activities to ensure the implementation of plans. Collaborative work across all planning stages is essential for effective discrete production planning, which enables companies to effectively adapt to changes in the external environment and market requirements.

It is also important to consider a discrete production planning approach, such as the Deming cycle, or in other words, the PDCA (Plan-Do-Check-Act) cycle (see Fig. 2). This concept has its own characteristics and benefits for implementing adaptive planning. The PDCA is a continuous process improvement tool that can be tailored to optimize various aspects of discrete production. The PDCA cycle is crucial for enterprises not only for effective planning but also for enhancing product quality. Each cycle involves quality control and analysis to guarantee the quality of the process, as well as addressing the need to adjust the entire technological process, which can be due to changes in market demands or resource availability [6].

The application of the Deming Cycle in discrete manufacturing starts with the planning stage Plan. During this stage, goals, objectives, and resources are identified and established in order to plan and achieve future outcomes. The planning process begins with an analysis of current production processes and technologies. In the context of discrete manufacturing, this involves detailed planning of production activities, calculation of necessary resources, estimation of task completion times, and establishment of production priorities. It should be noted that small adjustments can still be made during this stage based on feedback from the Do stage, which will

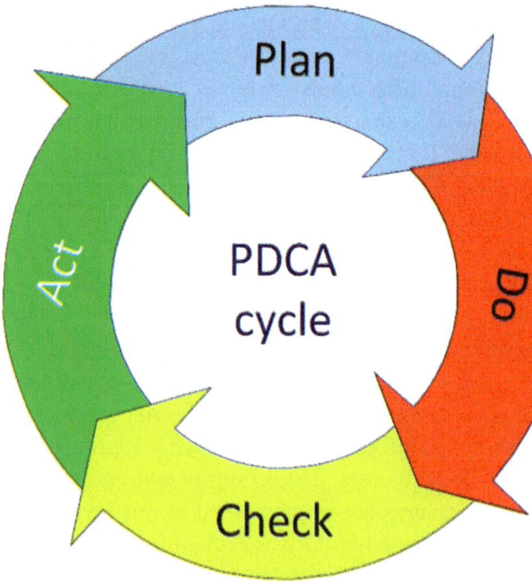

Fig. 2 Deming cycle, PDCA

be discussed later. However, it is important to understand that this approach may not always be completely accurate as the Plan stage works with only the available data at that point.

Next, comes the planning stage Do, which involves taking the plans developed during the previous phase and putting them into action. During this stage, specific production activities that were outlined during the planning phase are executed. In discrete manufacturing, this may involve activities such as launching a new production line, adjusting equipment, or implementing new technologies. A critical aspect of this phase is ensuring that all activities are carried out in accordance with established standards and quality criteria. These criteria are verified through rigorous production monitoring techniques to detect any deviations from the expected performance indicators at an early stage. Any deviations are addressed promptly to ensure that the production process remains on track [7].

The third phase of the process is known as Check. This stage involves reviewing the results of carrying out plans, comparing these results with the expected goals, and implementing quality control measures. During discrete production, this phase covers product quality assurance, analysis of performance indicators, and the identification of areas for improvement. For instance, if during the implementation phase it is noticed that certain activities are taking longer than anticipated or if product quality fails to meet standards, this triggers a review of the reasons behind these deviations. This step is crucial to identifying weaknesses in the organization's production process and determining what changes and enhancements are required to enhance efficiency and quality.

The final stage in the process is the Action. During this stage, specific actions are taken based on the results from the previous verification phase. These actions could include adjusting production schedules, modifying manufacturing processes, implementing new quality control measures, and reconfiguring assembly lines. For discrete manufacturing, companies might also choose to adjust equipment, revise employee schedules, or modify material management procedures. The main goal of this stage is to take specific measures based on information gathered during previous phases. These measures aim to improve the manufacturing process, reduce costs, minimize waste, and enhance customer satisfaction [8].

As a main conclusion regarding the application of the Deming Cycle in discrete manufacturing planning, it can be stated that its systematic approach to each process results in continuous process improvement. This approach allows for adaptation to changes in the external environment, effectively addressing issues related to fluctuating demand, limited resources, and the need for quality enhancement. By regularly implementing the Plan-Do-Check-Act (PDCA) cycle, businesses can enhance their production flexibility, minimize the risks of errors and downtimes, and optimize the utilization of equipment and resources. These advantages make the Deming Cycle an indispensable tool in discrete manufacturing, as success hinges on a company's capacity to continuously enhance its processes and swiftly adjust to alterations in external circumstances [9].

However, while taking all of the above considerations into account, it is also essential to address the issue of adaptive planning in the context of certain limitations. Some of the main limitations include the lack of available resources, the time needed for order fulfillment, and uncertainty regarding demand. These factors can pose significant challenges in terms of the flexibility and adaptability of production processes. For instance, fluctuations in product demand may necessitate a swift reconfiguration of manufacturing lines. Such actions require effective management and strategic planning, which may be challenging due to the unpredictable nature of demand. While the ability to execute such measures was discussed in a previous section, the additional challenge of unpredictable demand makes it difficult to generate accurate forecasts. As a result, this can adversely affect the planning process and lead to suboptimal resource allocation.

It is also important to separately and in greater detail analyze the issue of demand variability, as it significantly affects the change in order fulfillment time. In conditions of fluctuating demand, businesses may find themselves in a situation where it is not possible to fulfill customer orders in a timely manner, leading to customer dissatisfaction and, potentially, reducing the company's competitive position.

In order to minimize these risks, it is crucial to think ahead about flexibility and adaptability. Implementing procedures that enable easy change is essential, ensuring that it is not only feasible but also not crucial. This will enable businesses to react more promptly to market changes and adjust production plans in accordance with actual customer requirements [10].

Another significant consideration is the limited availability of resources and the necessity to optimize equipment utilization. Due to the intermittent nature of production processes, companies frequently encounter shortages of machinery, personnel, and raw materials. This constraint necessitates meticulous planning and allocation of resources in order to prevent downtime and ineffective equipment utilization. Consequently, optimizing machine load and labor costs becomes a critical priority. Even minor alterations in equipment load can result in substantial financial setbacks and diminished overall output.

In addition, changes to production conditions can result in downtime and losses. For example, if there is an unexpected need for equipment replacement or delays in material supplies, this can cause the production line to stop. Such situations have a negative impact on overall performance indicators and can make it difficult to meet production plans. To minimize these losses, it is important to implement flexible planning systems that account for potential changes and allow for flexibility in managing the production process [11].

Therefore, the key conclusion regarding the difficulty of adaptive production planning is that it requires balancing various factors such as multidimensional planning, uncertain demand, constrained resources, and potential downtime. To address these challenges effectively, it is essential to implement modern methods and techniques that enhance the flexibility and agility of production processes. These measures will enable businesses to compete successfully in a rapidly evolving market environment.

2 Discrete Production Planning Systems

2.1 MES Class Systems

The adoption of a Manufacturing Execution System (MES) for discrete production planning has become a fundamental component for improving productivity and streamlining manufacturing processes. A MES is a software solution that acts as a link between the planning and execution phases of production, empowering companies to more effectively and efficiently manage their operations. In the current dynamic and competitive business environment, adaptability, precision, and process control are paramount. The implementation of a MES significantly enhances management's capacity to plan and oversee production activities, resulting in optimized resource allocation and reduced waste. The system's real-time data collection and analysis capabilities empower managers to make well-informed decisions, allowing for adjustments to production as needed to maximize efficiency.

At the same time, the primary objective of MES is to integrate and coordinate all stages of the production cycle, from planning through to execution. This makes it a crucial tool for businesses engaged in discrete manufacturing. The use of a MES system enables the development of production plans at the planning phase, allocation of resources, and ensures timely completion of orders. This is particularly significant for discrete manufacturing, where products are frequently produced in individual pieces or small batches. As a result, meticulous control over each stage of the manufacturing procedure is essential. Thanks to a MES, it is feasible to create a comprehensive production plan for a facility that engages in discrete production, considering not only the present equipment load but also the availability of resources and logistical support [12].

It is also worth noting that MES provides real-time monitoring of production operations, allowing for the tracking of key indicators of the manufacturing process, such as equipment performance, actual operating times, defect levels, and deviations from planned targets. This analysis is essential in discrete manufacturing environments, where each production run may have unique characteristics. By providing this level of oversight, issues can be identified and resolved before they adversely affect product quality or order delivery dates. This results in a reduction of errors, downtime, and inefficient equipment usage, ultimately accelerating the production cycle and enhancing overall efficiency.

Another significant aspect of utilizing MES systems for discrete production planning involves quality management. These systems frequently incorporate features designed to monitor product quality metrics at each stage of the manufacturing process, which is critical in discrete manufacturing, where products undergo multiple processing steps. By monitoring quality at each step, defects can be prevented and compliance with standards ensured. The MES system automatically collects data on manufacturing operations, enabling real-time tracking of product quality and swift response to any problems.

MES also plays a critical role in optimizing resource management, which is a crucial aspect of discrete manufacturing planning. In this type of planning, there is often the challenge of limited resources such as equipment, materials, and labor. MES allows you to manage these resources more efficiently by providing accurate information about the current status of the production process and equipment utilization. This information allows for quick adjustments to production plans and avoidance of downtime or overutilization of individual production facilities. Furthermore, optimizing equipment utilization is especially important in discrete manufacturing environments, where adjusting production lines for different product types can be time-consuming and laborious, as in the case of mechanical engineering. It is also worth noting the reduction in production waste and defects thanks to the MES system. The system gathers and analyzes data throughout the manufacturing process, allowing for the rapid identification and correction of deviations from the standard. For example, if there is a recurring issue with defects at a specific stage of production, the MES can alert the operator or supervisor immediately, enabling them to implement corrective measures. This in turn reduces the number of faulty products manufactured and minimizes losses associated with the processing or disposal of defective items [13].

It is also worth mentioning that MES systems contribute to enhanced communication between various departments within an organization. This is accomplished by storing and managing data from the overall production process within a unified information system. Consequently, all departments involved in planning, logistics, and quality control have access to a variety of data and can swiftly obtain all relevant information. This enables them to make well-informed decisions based on such data, which enhances transparency in the manufacturing process and minimizes the risk of mistakes. Moreover, it expedites the decision-making process, resulting in more efficient operations.

In conclusion, it should be noted that the implementation of MES (Manufacturing Execution System) in specific production processes enables improved control over various aspects of production, ranging from planning to implementation and quality assurance. This results in reduced costs, enhanced productivity, and enhanced product quality. Moreover, the system offers increased flexibility and adaptability, which is crucial in a rapidly evolving market [14].

MES has become an essential tool for companies operating in discrete manufacturing, as it provides competitive advantages and ensures successful operations. Nevertheless, it is essential to acknowledge that frequent adjustments to the plan may not be feasible due to the absence of adaptive planning techniques in real time. Despite this challenge, the MES system is a high-quality software solution to support ongoing discrete production.

2.2 ERP Class Systems

The implementation of an enterprise resource planning (ERP) system for discrete manufacturing is essential to enhance the efficiency, consistency, and flexibility of production operations. This system integrates various business functions, such as production planning, resource management, procurement, logistics, finance, and sales, into a centralized information hub, facilitating the management of complex processes and coordination between departments. For businesses engaged in discrete manufacturing, this integrated approach is particularly beneficial, as it allows them to accurately schedule orders and optimize resource allocation. By consolidating all relevant information and processes within a unified system, businesses can ensure precise order fulfillment while reducing costs [15].

To initiate a study on the implementation of an Enterprise Resource Planning (ERP) system for discrete manufacturing scheduling, it is essential to examine the feasibility of integrating all business processes. In the context of discrete manufacturing, where products are assembled from individual parts, seamless coordination among various stages of production is crucial. The ERP system acts as a critical tool for automating and harmonizing data across different departments within an organization, significantly facilitating planning and resource allocation processes. For example, an ERP can be used to manage inventory levels and ensure timely delivery of materials and components, minimizing the risk of delays due to resource shortages and ensuring order fulfillment on schedule. Additionally, ERP allows businesses to monitor current order status, anticipate future material requirements, and adapt production plans accordingly.

Additionally, it is crucial to consider the capabilities of ERP systems, particularly with regard to resource management. This is a significant feature of these systems. In discrete manufacturing, there is often a high level of product variety and the need for swift response to changing customer demands. The ERP system can assist businesses in more accurately forecasting their material, financial, and human resource needs. For instance, it can provide insights into equipment utilization and work schedules, as well as determining whether additional staff are required to fulfill specific orders. This is essential for ensuring timely order fulfillment without excessive inventory accumulation or overloaded production facilities [16].

When considering the time frames and planning horizons discussed in the previous section, it is clear that the ERP system provides opportunities for both long-term and short-term planning at the strategic and operational levels. At the strategic level, the ERP facilitates the development of future plans that take into account the business requirements for manufacturing facilities and expansion, as well as modernization strategies. On the other hand, the system supports daily and weekly operational planning at the operational level, ensuring the timely completion of ongoing production tasks. This is made possible by the ability of the system to systematically collect data on order fulfillment and production progress in real time, enabling the company to more adaptively respond to fluctuations in demand, schedule deviations, and changes in production conditions.

Finally, a significant advantage of ERP systems is their capacity to adjust to specific business demands. Modern ERP solutions are modular and configurable, allowing them to be customized to the requirements of any organization, whether it manufactures complex technical products or produces standard items on a large scale. This adaptability makes ERP a versatile tool that can benefit both large corporations and smaller companies engaged in discrete production [17].

The ERP system is a powerful tool for designing and managing manufacturing processes. It integrates all business operations and improves resource management. This leads to more effective planning, improved product quality, and lower costs. By implementing the ERP system, companies can better respond to market changes, enhance internal processes, and ensure top-quality products, thereby increasing their competitiveness in the global market.

However, it is also worth considering the ERP system from a technical standpoint. These types of systems utilize NP-complete algorithms as mathematical planning techniques, which can result in an exponential increase in complexity of the planning challenge as more variables are incorporated. At the same time, enterprises need to be more adaptable in their planning to accommodate changing circumstances. Consequently, ERP systems might not be the optimal choice for the objectives discussed in this document. In the next chapter, an alternate solution will be presented.

3 Multi-Agent Systems

3.1 Application of Multi-Agent Systems in Discrete Production

Multi-agent systems (MAS) could be widely used in discrete production planning due to their flexibility, responsiveness, and ability for decentralized control. In the context of manufacturing, where each process comprises multiple stages, precise synchronization of resources and equipment is essential. The implementation of MAS allows for efficient planning and optimization of these processes, simplifying the management process and enhancing adaptability to changes in a dynamic market with fluctuating demand.

To understand the fundamental principles of MAS, it is important to define them as production management systems that use a novel approach to organizing and controlling production processes through a network of interconnected programmable agents. Each agent in the system is an independent component that can make decisions and perform specific tasks [18].

Discrete manufacturing is a production process in which products are created individually or in small batches. This requires a flexible and responsive approach to planning and execution. Multi-agent systems play a crucial role in this process as they allow tasks to be distributed among autonomous agents. Each agent has a specific role or responsibility within the production process. These agents work independently, but also collaborate with others to ensure the overall goal of producing products according to the plan. The agents' collaboration enables efficient production and ensures that each stage of the process is completed successfully.

One of the main advantages of using multi-agent systems for discrete production planning is their capacity to make decisions independently. Unlike conventional management systems such as Enterprise Resource Planning (ERP), which rely on central planning, multi-agent systems decentralize the decision-making process. This allows each agent to analyze local data and take decisions based on its unique circumstances, reducing the requirement for centralized planning and boosting its capacity to swiftly respond to changes in external and internal circumstances [19].

Another significant aspect of multi-agent systems is their ability to be flexible and adaptable in uncertain and dynamic environments. In the context of manufacturing, fluctuations in demand or unexpected disruptions may necessitate quick adjustments to production plans. Multi-agent systems are well-suited for this, as they allow agents to respond quickly to changes and reallocate tasks and resources, without the need for a full overhaul of the entire production process. For example, if a machine fails, the scheduling system can automatically rearrange orders to other machines, without human intervention. This reduces downtime and minimizes productivity losses, allowing for more efficient and agile production processes. Multi-agent systems make them an attractive option for industries that demand adaptability and responsiveness.

Coordination and interaction among agents are essential in the planning and execution of a production process. Agents can represent individual production units, such as machines, workstations, or logistics hubs. These entities must collaborate in order to maintain consistency and ensure the smooth operation of the overall system. Multi-agent systems provide the advantage of real-time coordination, allowing for synchronization of various stages of production in a timely manner. An example is an inventory management agent that interacts with a production line agent to ensure timely component delivery and coordinate supply schedules based on equipment load conditions [20].

Multi-agent systems are an efficient solution to the issue of optimizing resources in discrete manufacturing. Within this context, it is essential to optimally allocate resources, such as machinery, materials, and personnel, in order to maximize efficiency. This is achieved through the use of agents within a multi-agent system, which actively monitor resource utilization and propose solutions to minimize downtime and losses in productivity. These agents analyze the current production processes and suggest the most effective methods for reconfiguring equipment in order to minimize downtime during transitions between different product batches.

It is essential to focus on the issue of integration, which is a critical factor in this context. Multi-agent systems possess the ability to interact with various enterprise information systems such as ERP and MES. This integration ensures consistency in production processes and planning across all levels, leading to more accurate planning. Multi-agent systems act as an extension of ERP and MES systems by handling operational management and decision-making tasks. These systems provide strategic and tactical planning for the organization.

As a result, multi-agent systems hold the potential to become a crucial instrument for discrete production planning. These systems offer flexibility, adaptability, resource optimization, and process automation, allowing for more efficient coordination between various components of the production system. This reduces the need for centralized management and increases the speed of response to changes in production conditions, leading to increased competitiveness and productivity. In today's era of digital transformation and smart factory development, multi-agent systems provide new opportunities for companies to enhance their efficiency and competitiveness. By utilizing these systems, businesses can optimize their operations, respond more promptly to market demands, and ultimately achieve improved productivity and profitability.

4 Conclusion

An analysis of contemporary challenges in discrete manufacturing planning reveals that a major obstacle in today's world for effective planning is the rapidly changing environment, which results in an increase in the number of new and dynamic production factors. In order to maintain high levels of competitiveness, businesses must consider the adaptability of their operations. Discrete manufacturing is well-suited for this purpose, but the challenge of adaptable planning has become critical. Modern systems are not able to fully meet all demands. Therefore, the possibility of using a multi-agent system has been investigated. This approach provides a high degree of flexibility in planning and the ability to integrate with existing systems, thus enhancing the strengths of each system while minimizing their weaknesses.

References

1. Monostori, L., Kádár, B., Bauernhansl, T., Kondoh, S., Kumara, S., Reinhart, G., Sauer, O., Schuh, G., Sihn, W., Ueda, K.: Cyber-physical systems in manufacturing. CIRP Ann. **65**, 621–641 (2016)
2. Liu, Y., Xu, X.: Adaptive planning approaches for smart discrete manufacturing systems. J. Manuf. Syst. **60**, 130–145 (2021)
3. Liu, S., Wang, L., Wang, X.V., Wiktorsson, M.: A framework of data-driven dynamic optimization for smart production logistics. In: Advances in Production Management Systems. Towards Smart and Digital Manufacturing, pp. 213–221. Springer International Publishing (2020)
4. Zhang, L., Zhao, H.: Advances in discrete production system planning. Procedia CIRP **12**, 25–39 (2020)
5. Koren, Y., Shpitalni, M.: Evolvement of planning techniques in discrete manufacturing. CIRP Ann. **69**(2), 625–648 (2020)
6. Isniah, S., et al.: Plan do check action (PDCA) method: literature review and research issues. J. Sist. Dan Manaj. Ind. **4**(1), 72–81 (2020)
7. Realyvásquez Vargas, A., Arredondo Soto, K., Carrillo-Gutiérrez, T., Ravelo, G.: Applying the plan-do-check-act (PDCA) cycle to reduce the defects in the manufacturing industry: a case study. Appl. Sci. **8**(11), 2181 (2018). https://doi.org/10.3390/app8112181

8. Lu, Y., Morris, K.C., Frechette, S.: Current standards landscape for smart manufacturing systems (NISTIR 8107). Natl. Inst. Stand. Technol. 39 (2016)
9. Anggraini, W., Sari, D.K.: Lowering oil losses level on crude palm oil production using the Deming PDCA cycle. In: 1st Conference on Industrial Engineering and Halal Industries (CIEHIS), pp. 183–192 (2019)
10. Agostino, Í.R.S., Broda, E., Frazzon, E.M., Freitag, M.: Using a digital twin for production planning and control in industry 4.0. In: Scheduling in Industry 4.0 and Cloud Manufacturing, pp. 39–60. Springer (2020)
11. Arunarani, A.R., Manjula, D., Sugumaran, V.: Task scheduling techniques in cloud computing: a literature survey. Futur. Gener. Comput. Syst. **91**, 407–415 (2019)
12. Zhou, L., Chen, Z., Chen, S.: An effective detailed operation scheduling in MES based on hybrid genetic algorithm. J. Intell. Manuf. **29**, 135–153 (2018)
13. Chen, B., Wan, J., Shu, L., Li, P., Mukherjee, M., Yin, B.: Smart factory of industry 4.0: key technologies, application case, and challenges. IEEE Access **6**, 6505–6519 (2018)
14. Ivanov, D., Frolov, K., Frolov, A.: Implementation of ILS methodology on the basis of ISO 55000 series standards. Lect. Notes Netw. Syst. **951**, 79–91 (2024)
15. Frolov, K., Iliashenko, V., Ershova, A.: Strategic procurement of resources for transportation: architectural and methodological issues. Transp. Res. Procedia **54**, 945–952 (2021)
16. Almashaqba, Z., Sahandi, R.: Blockchain for enterprise resource planning systems: a literature review. Comput. Ind. Eng. **157**, 107253 (2021)
17. Abdulraheem, A.S., Abdulla, A.I., Mohammed, S.M.: Enterprise resource planning systems and challenges. Tech. Rep. Kansai Univ. **62**(4), 1885–1894 (2020)
18. AlMuhayfith, S., Shaiti, H.: The impact of enterprise resource planning on business performance: with the discussion on its relationship with open innovation. J. Open Innov. Technol. Mark. Complex. **6**(3), 87 (2020)
19. Bonabeau, E., Theraulaz, G.: Swarm smarts. What computers are learning from them? Sci. Am. **282**(3), 54–61 (2000)
20. Archimede, B., Letouzey, A., Memon, M.A., Xu, J.: Towards a distributed multi-agent framework for shared resources scheduling. J. Intell. Manuf. **25**, 1077–1087 (2014). https://doi.org/10.1007/s10845-013-0748-8

Operational Planning and Scheduling of Information Processes During the Interaction of Group of an Intelligent Transport Technology Vehicle in an Airport Digital Space

Vitaly Ushakov

Abstract The paper presents a solution to the program control (planning) task for information processes occurring during the interaction (functioning) of moving objects group. In general, information processes include the obtaining information processes about controlled objects by moving objects, its preliminary and secondary processing, storage (if it is not possible to transfer it to end users), as well as the transmitting and receiving information processes to end users. The purpose for solving the task under consideration is to increase the efficiency and control quality for these information processes in the interests for ensuring timely and reasonable decision-making on the controlled objects state. Usually, this task is solved in an automated (man–machine) mode using various models. Among them, traditionally, either static models are distinguished, which are characterized by large dimensions and difficulties in accounting for time, or logical-dynamic models, in which taking into account information process control technologies leads to the need to describe them using the appropriate phase constraints. To solve this task, models and algorithms for controlling information processes in the interaction for the considered moving objects have been developed. The developed method for researching the information processes interaction for moving objects is based on a multi-stage coordination-iterative procedure for the interaction for the constructed static and dynamic models that describe this subject area. In the static model, the operations that are part for the information processes are linked to moving objects, and in the dynamic model, the information operations already distributed over the moving objects are linked to time. The end result at this work is the models and algorithms developed and implemented at the software level using the Matlab language and Matlab App Designer, the performance which was tested on a test cases number.

V. Ushakov (✉)
Peter the Great St. Petersburg Polytechnic University, St. Petersburg, Russia
e-mail: mr.vitaly.ushakov@yandex.ru

Saint-Petersburg State University of Aerospace Instrumentation, St. Petersburg, Russia

I. Ilin and M. Youzhong (eds.), *Digital Systems and Information Technologies in the Energy Sector*, Lecture Notes in Networks and Systems 1244,
https://doi.org/10.1007/978-3-031-80710-7_16

Keywords Information process · Intelligent transport technology vehicle · Airport digital space · Reachability area · Moving objects grouping

1 Introduction

A modern airport is a complex transport facility, in which, in addition to aircraft, a large number of airfield technical facilities of various types and purposes operate simultaneously. The main task of airfield services is to ensure timely departure, a fully operational and prepared aircraft at exactly the appointed time, with the required level of passenger service in terms of ensuring their safety and comfort. To solve this task, the actions of all airport services must be coordinated on the basis of the aircraft servicing schedule and the airport's daily schedule.

Ground servicing of civil aviation aircraft at the airport involves the implementation of many technological processes. Aircraft servicing [1, 2] is one of such processes, including the preparation of the aircraft cabin for passenger transportation, unloading and loading of mobile equipment with food and items necessary for servicing passengers and ensuring their safety in flight. Servicing must be carried out in accordance with established regulations. The time for completing the entire process and individual operations is strictly regulated.

2 About Ground-Based Intelligent Transport Technology Vehicle

The task of managing [3–5] and coordinating service maintenance with other airport processes and services is extremely complex. For example, most Moscow airports perform on average 300–500 departures per day. Accordingly, the same number of aircraft arrivals per day occur at the corresponding airports. Only for service maintenance at a large airport, 100–200 units of ground-based intelligent transport technology vehicle (ITTV) are used. The total number of all mechanization means is 300–600 units or more. At the same time, airport services must constantly make control decisions, analyzing a large number of factors and taking into account the changing conditions of the airport's operation. The main factors and reasons influencing the operation of the airport include: changing weather conditions, delays in aircraft arrivals, delays in aircraft departures, technical malfunctions of aircraft and the facilities servicing them, social events occurring at the airport. According to established practice, at present, operational planning of the functioning of aerodrome services is carried out in manual or automated modes. The complexity of the tasks solved by aerodrome services, together with the human factor, leads to ineffective control decisions and errors, which, in turn, lead to abnormal situations, accidents and, ultimately, to financial losses.

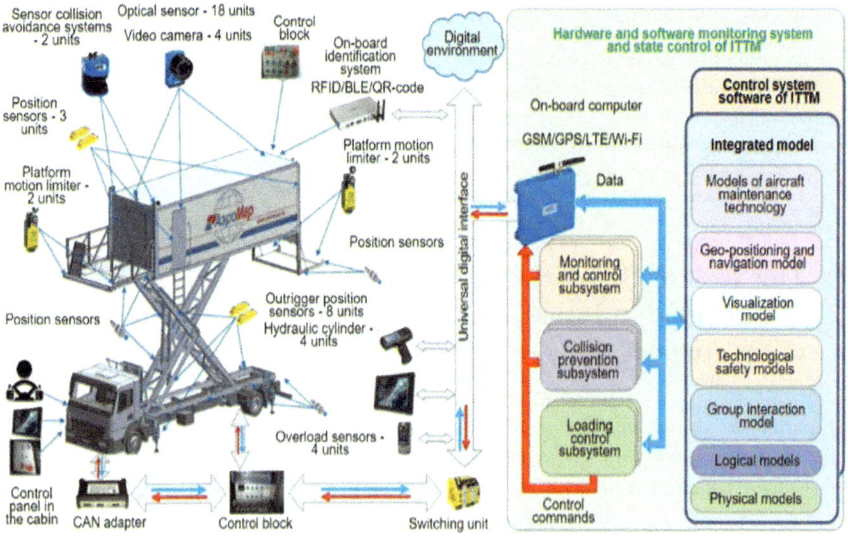

Fig. 1 Intelligent transport technology vehicle [11]

Increasing the level of passenger safety, reliability and efficiency of servicing civil aviation aircraft at the airport are extremely urgent tasks. To solve these tasks, a complex of ground-based intelligent transport technology vehicle of servicing civil aviation aircraft was developed and adopted for industrial use. Currently, each ground-based intelligent transport technology vehicle is a truck with a lifting mechanism of the body, equipped with various sensors, ensuring its functioning within the framework of a single digital space of the airport. Figure 1 shows the generalized structure of a ground-based intelligent transport technology vehicle. Each ground-based intelligent transport and technological vehicle includes a specialized distributed onboard information and computing complex, which includes the following subsystems:

- a loading and unloading control subsystem that collects data from specialized equipment and sensors;
- a collision prevention subsystem that receives and processes data from its own sensors;
- a subsystem for monitoring and proactive (anticipatory) control of the technical condition of the ground-based intelligent transport and technological vehicle, which collects and processes data from chassis and working equipment sensors;
- an information server (portal) that ensures the processing of incoming data and the exchange of data between the ground-based intelligent transport and technological vehicle and the airfield services control system.

Existing and prospective ground-based intelligent transport technology vehicle are intended:

- for transporting food and duty-free goods;
- for transporting bedridden patients;
- for transporting comfort items and additional equipment for servicing passengers in flight;
- for cleaning aircraft cabins and removing garbage;
- for loading and unloading cargo and baggage.

3 The Task of Optimal Operational Planning of Information Processes in a Single Digital Space of the Airport

The analysis shows that the solution of this class of tasks involves the synthesis of programs for managing information processes [6, 7] and flows in the considered ground-air transport and logistics system. From a formal point of view, this class of tasks refers to multi-criteria tasks of high-dimensional scheduling theory [8–10], with prohibitions on interrupting partially ordered work (determined by the operating regulations of specific airfield services), as well as taking into account special spatio-temporal, technical, technological, organizational transport and logistics restrictions that determine the specifics of the functioning of each specific airport separately. For the successful solution of all the above tasks, it is necessary to correctly organize the planning and control processes of the considered transport vehicles and their groupings, as well as the information processes underlying the corresponding control processes implemented by specific airfield services. Therefore, the task of optimal operational planning of information processes [11–13] was set and solved, ensuring the stable functioning of ground-based intelligent transport technology vehicle in a single digital airport space. The flight schedule presented on the online board of one of the Moscow airports, Zhukovsky, which is shown in Fig. 2, was chosen as the initial data for the example under consideration.

In the example under consideration, there are 8 sections of structure constancy, the planning interval for information processes is 1 day, and 9 mobile objects (in this case, ground-based intelligent transport technology vehicle) are used to perform aircraft maintenance. Structural state of the 5th group of ground-based intelligent transport technology vehicle is shown in Fig. 3.

A detailed description of the developed combined models and algorithms of information processes during the interaction of moving objects was given in [6, 11].

The following subsystems are distinguished in the information and computing network used to organize information processes that ensure the control of both aircraft and ground-based intelligent transport technology vehicle: the central server of the enterprise (airfield) under consideration, the on-board information and computing complex of each ground-based intelligent transport technology vehicle, computing means of the airfield service control system, as well as various communication channels based on modern telecommunication technologies (GSM, Wi-Fi, Bluetooth), and information is also provided on various scenarios of the structural dynamics of

Fig. 2 Online flight board for the day at Zhukovsky airport

a group of ground-based intelligent transport technology vehicle, which were used as the basis for modeling when calculating information process control programs [14–16].

The following values were used as initial data for modeling the group of ground-based intelligent transport technology vehicle and its hardware and software complex:

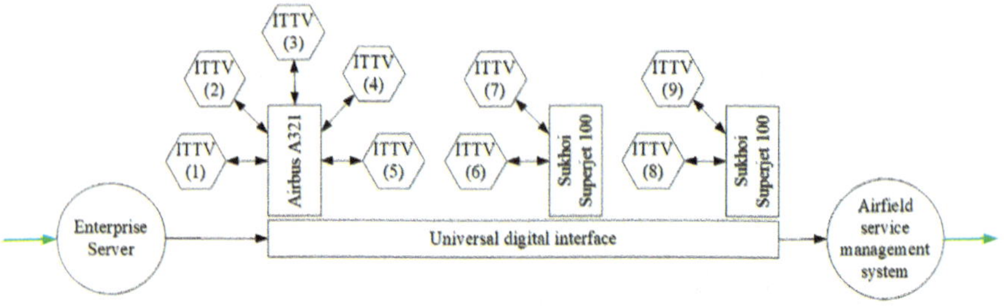

Fig. 3 Structural state of the 5th group of ground-based intelligent transport technology vehicle

- the maximum storage capacity of the enterprise server is 16 GB;
- the maximum storage capacity of the on-board information and computing complex of ground-based intelligent transport technology vehicle is 1 GB;
- the maximum storage capacity of the on-board information and computing complex of the airfield service control system is 16 GB;
- 12.5 GB of input data are received by the enterprise server at Sections 1, 3, 4, 5, 6, 8 of structure constancy;
- 6 GB of input data are received by the enterprise server at Sections 2, 7 of structure constancy;
- the maximum intensity (speed) of processing data received by ground-based intelligent transport technology vehicle is 300 Mbps;
- the maximum intensity (speed) of data and information transfer between the enterprise server and the on-board information and computing complex of ground-based intelligent transport technology vehicle is 80 Mbps;
- the maximum intensity (speed) of data and information transfer between the on-board information and computing complex of ground-based intelligent transport technology vehicle and the airfield service control system is 100 Mbps.

4 Solution of the Task of Operational Planning of Information Processes During the Interaction of a Group of Ground-Based Intelligent Transport Technology Vehicle in a Single Digital Space of the Airport

As a result of solving the task of operational planning of information processes during the interaction of a group of ground-based intelligent transport technology vehicle in a single digital space of the airport using a static model, 6.7 GB were processed and 2.4 GB of information were saved in the airfield service control system out of 87 GB received at the input of the enterprise server, which is 7.7% and 2.76%, respectively. At the same time, the volume of lost data was 6.2 GB (7.13%) and

0 MB of information. The remaining data and information were saved in the on-board information and computing complex of ground-based intelligent transport technology vehicle (an insignificant part) and on the enterprise server (a large part), which can be processed and then transferred to the airfield service control system in the future during the following 24 h. In summary, it can be noted that almost all data received by the on-board information and computing complex of ground-based intelligent transport technology vehicle were processed, and their losses are explained by the limitation of the maximum possible intensity (speed) of data and information transfer between the enterprise server and the on-board information and computing complex of ground-based intelligent transport technology vehicle, as well as the limitation of the maximum storage capacity of the on-board information and computing complex of ground-based intelligent transport technology vehicle, that is, technical limitations of the on-board information and computing complex of ground-based intelligent transport technology vehicle It can be noted that on the enterprise server, data losses begin to occur from the 5th section of structure constancy, which is explained by reaching the maximum storage capacity of the enterprise server.

Let us clarify that the aircraft crew receives information about the number of passengers and their weight, the number of pieces of baggage and its weight, fuel load, the amount of food for passengers and its weight in a generalized (aggregated) form, therefore not all data stored on the enterprise server should be transmitted to the aircraft. For example, information about the technical condition of ground-based intelligent transport technology vehicle remains with ground airfield services.

Due to the large number of ground-based intelligent transport technology vehicle and areas of structure constancy, Fig. 4 shows, as an example, a plan for software control of information processes during the interaction of a group of ground-based intelligent transport technology vehicle only for the first, second and third time sub-intervals.

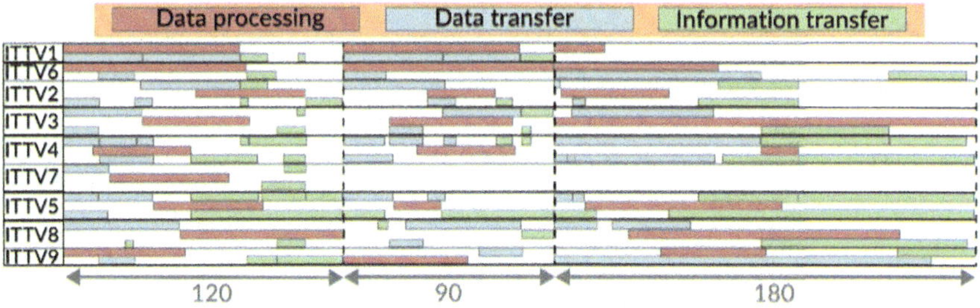

Fig. 4 Visualization of the plan obtained using the dynamic model

5 Evaluation of the Effectiveness of the Obtained Plan for the Implementation of Information Processes

To evaluate the efficiency of the obtained plan for the execution of information processes using a generalized indicator that evaluates the integral throughput of the considered information and computing network formed on the computing resources of airfield services.

For the initial data given above, we will set various disturbing effects from the interval [0, 1] for various plans for the execution of information processes. As a result of modeling, we will obtain three approximated reachability areas [17–19] for disturbances in the space of quality indicators of information process planning (see Fig. 5). As a result, we will obtain information process control program No. 1 (it corresponds to the green color of the approximated reachability area). Let us define two more plans for software control of information processes for other coefficients of importance of private indicators of planning quality (plan No. 2, which corresponds to the yellow color of the approximated reachability area [20, 21]; plan No. 3, which corresponds to the blue color of the approximated reachability area [22–24]).

Having analyzed Fig. 5, it can be determined that the most robust program for managing information processes is program No. 2, because this plan has the largest area of intersection of the approximated region of attainability for disturbances with the permissible zone.

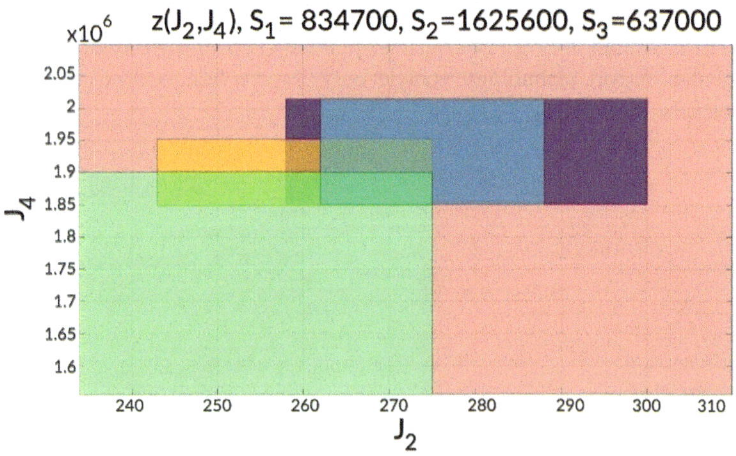

Fig. 5 Assessing the robustness of information interaction plans for a group of ground-based intelligent transport and technological vehicles

6 Conclusion

The practical implementation of the concept of complex (system) modeling in solving the tasks of operational planning of information processes during the interaction of a group of mobile objects in various subject areas has been carried out, which confirmed the positive integrative effect of the combined use of different types of models. The developed software prototype of a special modeling and algorithmic support for solving the tasks of operational planning of information processes during the interaction of mobile objects allows, unlike existing planning programs, to simultaneously synthesize the best technology (plan) for receiving and processing data and information during the interaction of mobile objects, and programs for optimal control of information control within the framework of the considered interaction of mobile objects.

Assessing the robustness of software control of information processes during the interaction of a group of ground-based intelligent transport technology vehicle allows the decision maker to determine the best plan in an interactive mode, essentially, as in an information system for decision support.

References

1. Zakharov, V.V., Salukhov, V.I., Mikoni, S.V., Zaytseva, A.A.: Corporate information system modernization during enterprise digital transformation. In: Lecture Notes in Networks and Systems, pp. 475–482. Springer International Publishing, Cham (2023). https://doi.org/10.1007/978-3-031-20875-1_44
2. Ilin, I., Levina, A., Iliashenko, V.: Innovation hub and its IT support: architecture model. In: Lecture Notes in Information Systems and Organisation, pp. 49–67. Springer International Publishing, Cham (2022). https://doi.org/10.1007/978-3-030-94252-6_4
3. Teslya, N., Potryasaev, S.: Execution plan control in dynamic coalition of robots with smart contracts and blockchain. Information **11**, 28 (2020). https://doi.org/10.3390/info11010028
4. Sokolov, B.V., Potryasaev, S.A, Korableva, O.N., Zakharov, V.V.: Methodology support and algorithms for dynamic synthesis of technologies and programs for proactive management of information processes in the industrial internet under a crisis. Int. J. Risk Assess. Manag. **25**, 194 (2022). https://doi.org/10.1504/ijram.2022.130524
5. Ilin, I.: Integration of information and management technologies (2022). https://doi.org/10.57809/2022.1.1.2
6. Ushakov, V.: Models and algorithms for controlling information processes in the moving objects interaction. Mar. Intellect. Technol. 235–247 (2022). https://doi.org/10.37220/mit.2022.57.3.031
7. Bakhtizin, A., Ilin, I., Nikitin, N., Ershova, A., Esser, M.: Multi-agent approach in planning and scheduling of production as part of a complex architectural solution at the enterprise. In: Lecture Notes in Networks and Systems, pp. 369–380. Springer International Publishing, Cham (2022). https://doi.org/10.1007/978-3-030-93872-7_30
8. Chauhan, S.S., Gordon, V., Proth, J.-M.: Scheduling in supply chain environment. Eur. J. Oper. Res. **183**, 961–970 (2007). https://doi.org/10.1016/j.ejor.2005.06.078
9. Ju, S., Woo, J.H.: Integration of long-term planning and mid-term scheduling of shipbuilding. Prod. Plan. Control. **34**, 524–542 (2023). https://doi.org/10.1080/09537287.2021.1940344

10. Ivanov, D.: Supply chain management and structural dynamics control. In: International Series in Operations Research & Management Science, pp. 1–18. Springer International Publishing, Cham (2018). https://doi.org/10.1007/978-3-319-69305-7_1
11. Sokolov, B., Ushakov, V., Zakharov, V.: Optimal planning and scheduling of information processes during interaction among mobile objects. Int. J. Prod. Res. **62**, 5905–5924 (2024). https://doi.org/10.1080/00207543.2024.2302388
12. Dolgui, A., Ivanov, D., Sethi, S.P., Sokolov, B.: Scheduling in production, supply chain and Industry 4.0 systems by optimal control: fundamentals, state-of-the-art and applications. Int. J. Prod. Res. **57**, 411–432 (2019). https://doi.org/10.1080/00207543.2018.1442948
13. Lohmer, J., Lasch, R.: Production planning and scheduling in multi-factory production networks: a systematic literature review. Int. J. Prod. Res. **59**, 2028–2054 (2021). https://doi.org/10.1080/00207543.2020.1797207
14. Ivanov, D., Dolgui, A., Sokolov, B., Werner, F., Ivanova, M.: A dynamic model and an algorithm for short-term supply chain scheduling in the smart factory industry 4.0. Int. J. Prod. Res. **54**, 386–402 (2016). https://doi.org/10.1080/00207543.2014.999958
15. Sokolov, B., Dolgui, A., Ivanov, D.: Optimal control algorithms and their analysis for short-term scheduling in manufacturing systems. Algorithms **11**, 57 (2018). https://doi.org/10.3390/a11050057
16. Ivanov, D., Sokolov, B., Kaeschel, J.: A multi-structural framework for adaptive supply chain planning and operations control with structure dynamics considerations. Eur. J. Oper. Res. **200**, 409–420 (2010). https://doi.org/10.1016/j.ejor.2009.01.002
17. Novikov, D.A.: Cybernetics: from Past to Future. Springer International Publishing, Cham (2016). https://doi.org/10.1007/978-3-319-27397-6
18. Sokolov B.V., Ushakov V.A.: Formation reachability area as a data vector using a dynamic model for controlling information processes in the automated control system for moving objects. In: Models and Methods for Researching Information Systems in Transport 2020 (MMRIST 2020), pp. 67–75. CEUR (2020)
19. Krylov, I.A., Chernous'ko, F.L.: An algorithm for the method of successive approximations in optimal control problems. U.S.S.R. Comput. Math. Math. Phys. **12**, 15–38 (1972). https://doi.org/10.1016/0041-5553(72)90063-8
20. van den Bosch, P.P.J., van der Klauw, A.C.: Modeling, identification and simulation of dynamical systems. CRC Press (2020). https://doi.org/10.1201/9780138746872
21. Sokolov, B., Ushakov, V.: Model-algorithmic support for abilities calculating of control system based on projection operators. In: Advances in Intelligent Systems and Computing. pp. 342–348. Springer International Publishing, Cham (2019). https://doi.org/10.1007/978-3-030-19813-8_35
22. Filippova, T.F.: Ellipsoidal estimates of reachable sets for nonlinear control systems with bilinear uncertainty. In: Studies in Computational Intelligence. pp. 121–131. Springer International Publishing, Cham (2020). https://doi.org/10.1007/978-3-030-22723-4_8
23. Engau, A., Sigler, D.: Pareto solutions in multicriteria optimization under uncertainty. Eur. J. Oper. Res. **281**, 357–368 (2020). https://doi.org/10.1016/j.ejor.2019.08.040
24. Lyubushin, A.A.: Modifications and convergence of successive approximations for optimal control problems. U.S.S.R. Comput. Math. Math. Phys. **19**, 53–61 (1979). https://doi.org/10.1016/0041-5553(79)90036-3

Meteorological Data Monitoring, Processing and Control Based on Internet of Things Technologies

Elizaveta Grigoreva and Vitaly Ushakov

Abstract In the article, within the framework of the "Smart City" concept, the task of meteorological monitoring for short-term forecasts for general use is solved using the synoptic method. The task is decomposed into subtasks of collecting, processing, analyzing and interpreting meteorological data. The purpose of the work is to observe and forecast weather and climate conditions in a certain area. To solve the task, a hardware and software complex (automated weather station system) was developed based on the Internet of Things technology, which is a distributed network of meteorological sensors. As an alternative way to solve the task of meteorological monitoring, small spacecraft were considered and compared with them. On the server side of the web application of the hardware and software complex, the Node.js platform with the Express framework is used, and the server interacts with the database via Object-Relational Mapping Sequelize. PostgreSQL is used as a database management system. Examples of the hardware and software complex operation are presented, and the results of the accuracy assessment of the forecasts predicted by the hardware and software complex based on weather monitoring in the Tosnensky District of the Leningrad Region are given. Rules have been developed in the hardware and software complex to ensure timely response to weather changes.

Keywords Internet of things · Hardware and software complex · Meteorological monitoring · Cloud computing · Wireless networks · Meteorological data · "Smart city" · Weather forecasting · Distributed network · Data monitoring task · Data processing task

V. Ushakov (✉)
Peter the Great St. Petersburg Polytechnic University, St. Petersburg, Russia
e-mail: mr.vitaly.ushakov@yandex.ru

E. Grigoreva · V. Ushakov
Saint-Petersburg State University of Aerospace Instrumentation, St. Petersburg, Russia

I. Ilin and M. Youzhong (eds.), *Digital Systems and Information Technologies in the Energy Sector*, Lecture Notes in Networks and Systems 1244,
https://doi.org/10.1007/978-3-031-80710-7_17

1 Introduction

"Smart city" [1–3] is a set of digital technologies that control the city and control information processes in it automatically. The goal of the "Smart city" is to improve the quality of life of people, resulting in increased productivity and interactivity of city services, reduced costs and resource consumption, and improved communication between city residents and the state. For example, "Smart manufacturing" [4–6], "Smart hospital" [7, 8], "Smart transport" etc.

The "Smart City" concept is based on several key technologies (see Fig. 1).

Firstly, these are wireless networks that transmit data between different devices without using wired communication channels. They provide communication between elements of the city infrastructure in real time, which allows for prompt response to changes and decision-making.

Secondly, cloud computing [9–11] is a technology that stores and controls data collected by urban "smart" systems. Any devices and software that have access to the Internet can connect to the data, which provides flexibility and scalability, and allows for the analysis of large amounts of data.

Third, the Internet of Things (IoT) [12–14] is a technology that connects various devices and allows them to exchange data with each other. IoT automates the operation of urban systems such as lighting, transportation, energy supply and security, which contributes to the creation of a more efficient and sustainable urban environment.

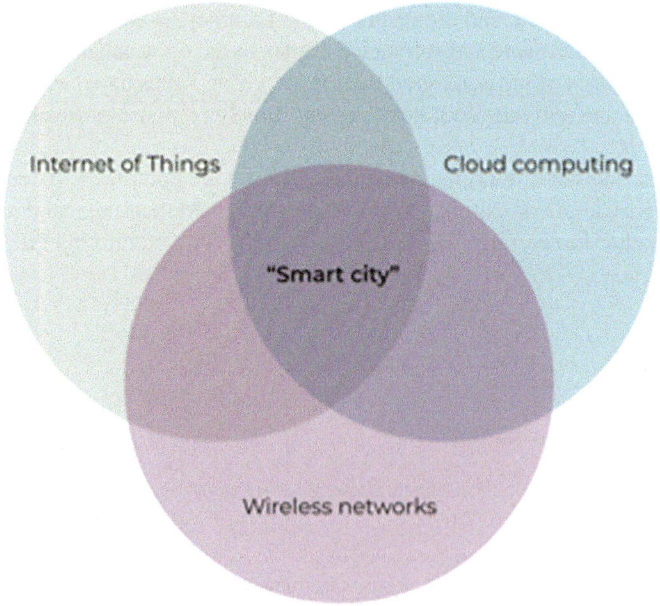

Fig. 1 A set of basic technologies of the "Smart City" concept

Examples of a "Smart City" are: "Smart Parking", which helps drivers find free spaces; "Smart Lighting", which controls the brightness level of street lighting in order to save electricity; "Smart Traffic Lights", which analyze the traffic situation and the flow of cars and, based on the information received, independently select the operating mode. In Australia, the "Smart Road System" is actively used automated dividing barriers ARMB [15], which help in accidents. They are installed along highways and at the entrances to tunnels. During accidents or other dangerous situations, the system closes roads and helps redirect the flow of cars.

One of the current areas in the "Smart City" is meteorological monitoring the process of collecting, processing, analyzing [16, 17] and interpreting meteorological data to observe and forecast weather and climate conditions in a certain area.

For use in the "Smart City" concept and to ensure timely response of citizens to changes in weather conditions, a hardware and software complex of a weather station will be developed that will allow collecting, processing and transmitting meteorological information without human intervention, as well as on its basis to generate local short-term forecasts for general use.

2 Monitoring Meteorological Data

Monitoring meteorological data is extremely important and necessary for modern society for a number of reasons.

First, monitoring meteorological data is necessary to study climate change and its impact on the environment. Analyzing long-term trends in temperature, precipitation, wind, and other parameters allows scientists to identify patterns and develop measures to adapt to climate change. This is especially relevant in light of global warming and associated risks such as melting glaciers, rising sea levels, and more frequent extreme weather events.

Secondly, meteorological monitoring data is used to assess and forecast the risks of weather-related emergencies. Timely warnings about impending hurricanes, floods, droughts and other dangerous phenomena allow for measures to be taken to protect the population and minimize damage. For example, meteorologists predicted heavy snowfalls and a cold snap known as the "Beast from the East" that hit Europe in the winter of 2017–2018. Authorities in many European countries were able to prepare for extreme weather conditions by setting up temporary shelters for the homeless, preparing emergency services, and warning the population about possible disruptions to transport and power supplies.

Thirdly, accurate and timely data on the state of the atmosphere allows for more reliable weather forecasts. This is of great importance for planning activities in various areas, such as agriculture [18], logistics, urban planning, construction and other industries where weather conditions play a key role. For example, forecasts warning of increased humidity or temperature fluctuations help farmers take timely measures against the spread of fungal diseases and other pests, which allows them to preserve their harvest, which in turn prevents fluctuations in food prices.

Synoptic meteorology [19–22] is the science of weather and methods of its prediction. It allows one to predict what the weather will be like at a certain time in a certain place.

Weather forecasts can be divided into categories: by time frame (ultra-short-term, short-term, medium-term and others); by territorial coverage (local, regional, country, global) and by purpose (general use, aviation, marine, river and agricultural). Short-term forecasts are calculated for 2–3 days. Their accuracy is from 85 to 95%. Such forecasts are the most common and in demand for the population, agricultural enterprises and various civil services. Calculating local forecasts for general use requires the least computing power and hardware support, and is also the most common.

When solving the task of meteorological monitoring, we will focus on short-term forecasts for general use, because they provide the most relevant and useful data for daily human activities, allowing us to track and quickly respond to changes in weather conditions.

3 Analysis and Justification of Possible Solutions to the Task of Meteorological Monitoring

To form a weather forecast, it is necessary to know the "current conditions" the current values of temperature, atmospheric pressure, humidity, wind speed and direction, precipitation and its amount. There are various sources of "current conditions", the main ones are weather stations, small spacecraft and weather stations with IoT technology.

Meteorological stations [20, 23, 24] are special sites where meteorological measurements of weather and climate are continuously carried out. For meteorological measurements, many devices are installed at the stations: thermometers, barometers, precipitation gauges and other devices, with the help of which meteorologists take measurements several times a day.

Weather stations provide accurate measurements, but they only collect data in their immediate vicinity and are often located far from each other, and the amount of precipitation cannot be determined from this data. The maintenance and operation of meteorological stations is quite expensive, since in addition to the affected resources of electricity and technical equipment, it is necessary to take into account the salaries of employees: technicians, engineers, meteorologists, necessary for the operation of each meteorological station.

Small spacecraft for collecting and, or processing meteorological data (weather satellites) are used to receive data on the planet's weather and climate to the information processing center. They can detect areas of intense precipitation and natural hazards, such as volcanic emissions and forest fire smoke, pollution, sand and dust storms, and ocean current boundaries. There are currently about 160 small spacecrafts in orbit, making about 80 million observations per day [25]. Examples of meteorological satellites are small spacecraft "The Khors" #1, "The Khors" #2, "The Khors"

Table 1 Weight coefficients of alternatives in terms of achieving the set goal

	Weighting factors in shares	Percents (%)
Weather stations	0.219	22
Weather stations with IoT	0.654	65
Small spacecraft	0.127	13

#3 and "The Khors" #4 spacecraft, created by scientists from Bauman Moscow State Technical University with the "GAMVEKI" instruments of the Fedorov Institute of Applied Geophysics (FGBU IPG). The spacecraft allow observing weather systems on a global scale and have high measurement accuracy. However, assembly, preparation for launch and launching the devices into orbit require large amounts of time and money.

Weather stations that use IoT technology are a distributed network of meteorological sensors that quickly and efficiently collect data from multiple locations. These sensors can be installed on the ground, in the air or in water, and, operating autonomously, transmit information about temperature, humidity, atmospheric pressure, wind speed and direction, precipitation levels and other parameters important for observation. The data is processed and transmitted using cloud computing and wireless technologies, which allows for instant analysis of information and operational decision making [26–28]. The data obtained using the meteorological sensors of the IoT are highly accurate, and also provide the system with flexibility and reliability, due to ease of maintenance and autonomous operation.

To select the optimal source of "current conditions", we will use the method of hierarchy analysis. The following criteria were identified for the analysis of the decision-making tsk: data accuracy, reliability, ease of maintenance, cost, and expandability. The criterion of "data accuracy" is of the greatest importance, since the forecast result depends on the accuracy of the received "current conditions". The criterion of "expandability" is of the least importance, since the hardware and software complex is initially developed to obtain a local weather forecast. Table 1 presents the final result of the weights of alternatives in terms of achieving the set goal.

Thus, the optimal source of "current conditions" was chosen to be a weather station with IW technology, which is best suited to the task at hand.

4 Monitoring and Processing of Meteorological Data

There are various methods of processing meteorological data and weather forecasting: synoptic method, hydrodynamic (numerical) methods and statistical methods.

Hydrodynamic (numerical) methods are based on solving systems of hydrothermodynamic equations. These equations functionally link the predicted weather characteristics with several initial atmospheric parameters. The main initial characteristics are: pressure, temperature, air density, and wind speed projections onto three coordinate axes. Any of the above characteristics can be used as the predicted value.

Statistical methods are based on empirically identified relationships obtained by processing a large volume of observational material. Based on the results of the processing, prognostic equations or diagrams are constructed that allow one to predict the future value of the forecast meteorological variable.

The synoptic method is currently the main method for developing short-term weather forecasts. The essence of the method is that based on the analysis of weather maps for several consecutive periods, a forecast of the synoptic position is made a forecast of the occurrence, movement and evolution of air masses, atmospheric fronts, baric systems. A map on which the expected position of synoptic objects is plotted is called a forecast map.

A forecast map is created for a specific time for the next day or for several days ahead. This map serves as the basis for making a weather forecast for a specific region. The weather forecast is constructed as a logical continuation of the forecast of the synoptic situation and is based on the principle that with the movement and change of synoptic objects, the corresponding weather conditions are also transferred and changed. Therefore, for the initial assessment of meteorological quantities, values are taken from the region from which the synoptic object is expected to move to the forecast area.

A conceptual representation of the weather forecast process for most meteorological elements is presented in formula 1.

$$\text{P\_element} = \text{a} + \text{t} + \text{w} + \text{s} \tag{1}$$

where a—Movement (Advection) is the movement of air masses and other meteorological elements in space;

t—Evolution (Transformation)—change in the characteristics of a meteorological element over time;

w—Diurnal cycle—cyclical changes in meteorological parameters during the day;

s—The influence of local factors is local features (relief, water bodies, vegetation, etc.) that can significantly change weather conditions in a particular area [29].

Since the task of forming local short-term forecasts is being solved, the synoptic method was used to predict the weather forecast during the development of the hardware and software complex.

5 Hardware and Software Complex for Monitoring and Processing Meteorological Data

The developed hardware and software complex for monitoring and processing meteorological data in a distributed network consists of five main modules: a module for obtaining initial data, a module for storing information, a weather forecasting module, a module for visualizing meteorological data, and a module for controlling the complex. Figure 2 shows the process of interaction between the modules of the hardware and software complex.

The source data module can receive data from several weather stations with IoT technology. Any set of sensors can be used to obtain data on "current conditions". Three types of sensors were selected for testing: the "AURA" meteorological parameter sensor, the "ALBION" meteorological visibility sensor, and the "ALBUS" wind speed and direction sensor. The meteorological parameter sensor "AURA" is designed to measure five meteorological parameters: wind speed and direction, atmospheric pressure, temperature and relative air humidity. The meteorological visibility range sensor "ALBION" is designed to measure the visibility range in the atmosphere and uses an optical method of measuring the visibility range based on the scattering of light in the atmosphere. The "ALBUS" wind speed and direction sensor allows you to measure the speed and direction of wind in the ground layers of the atmosphere, and is also capable of withstanding extreme temperature conditions.

In addition to sensors, the hardware and software complex uses devices equipment elements that have the ability to transmit data and signals on an event, as well as receive, store and process data. The devices collect data from several sensors on the state of the environment and transmit information to the cloud system, where further processing of meteorological data and forecasting takes place.

The initial data also includes forecast maps for the next three, current and three previous days, calculated advection data, transformations and daily variations of

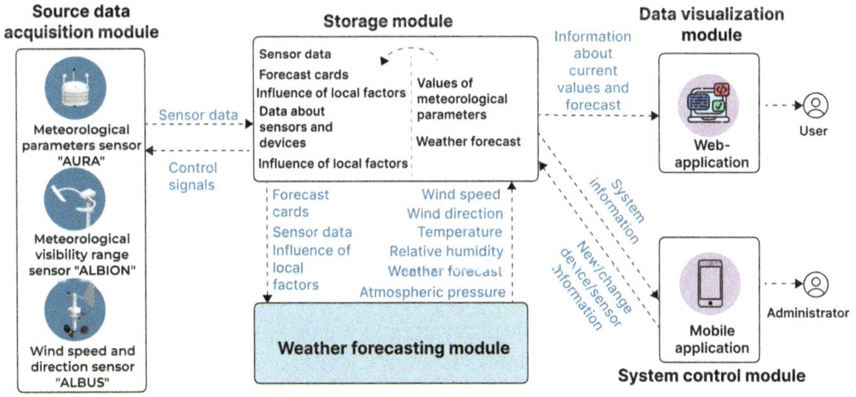

Fig. 2 Scheme of interaction of modules of the hardware and software complex

the meteorological element, information on local terrain features that affect weather conditions, which are stored in the cloud system.

Data received from sensors is aggregated on the device, processed and sent to the information storage module cloud storage, via the MQTT protocol, which allows for efficient message transmission between devices and cloud platforms. From the cloud storage, data is sent to the weather forecasting module, which was developed in the Python programming language. The weather forecast is generated based on new incoming data and information available in the system. Current sensor values are added to the forecast map, and the weather forecast is recalculated taking into account the new conditions. The generated forecast is sent to the cloud storage.

Forecast information is available for viewing in the end user interface the data visualization module. The developed web application allows displaying information in real time, and is also available for use on any device with an installed browser and Internet access. HTML, CSS, JavaScript, and the React.js framework were used in the development. Data is transferred to the web application via the HTTPs protocol. On the server side of the web application, the Node.js platform with the Express framework is used to process requests and control business logic. The server interacts with the database via ORM (Object-Relational Mapping) [30–32]. Sequelize, which simplifies working with data and makes the code more universal. PostgreSQL is used as a database management system, which provides reliable storage and fast access to large volumes of data.

The web application allows the user to view weather data at the current moment in time: precipitation, temperature, pressure, humidity, wind speed and direction, visibility and other parameters; as well as graphs reflecting the dynamics of changes in these values over time. In addition, the application displays brief information about the weather for the next 3 days, indicating the maximum daytime and minimum nighttime temperatures. The user interface of the hardware and software complex for monitoring and processing meteorological data is shown at Fig. 3.

The hardware and software control module allows operators to control and configure the operation of sensors and devices without writing software code. A mobile application for the Android operating system was developed. The application is intended for hardware and software system administrators and allows adding sensors: linking sensors to the control device via IP and linking them to a location on the map, as well as deleting and temporarily disabling sensors. The main page of the application and the sensor management page are shown at Figs. 4 and 5

Thus, the developed hardware and software complex for monitoring and processing meteorological data collects data from sensors and transmits it to devices, and then to cloud storage. Information on current weather conditions is processed, and a forecast is formed on its basis using the synoptic method. The received data is recorded in the cloud, from where it is sent to the user interface in a format convenient for human perception. The mobile application is a link between sensors, devices and the hardware and software complex. It allows you to view information about the location of sensors and devices, their readings and operability, and to disconnect and add components to the hardware and software complex.

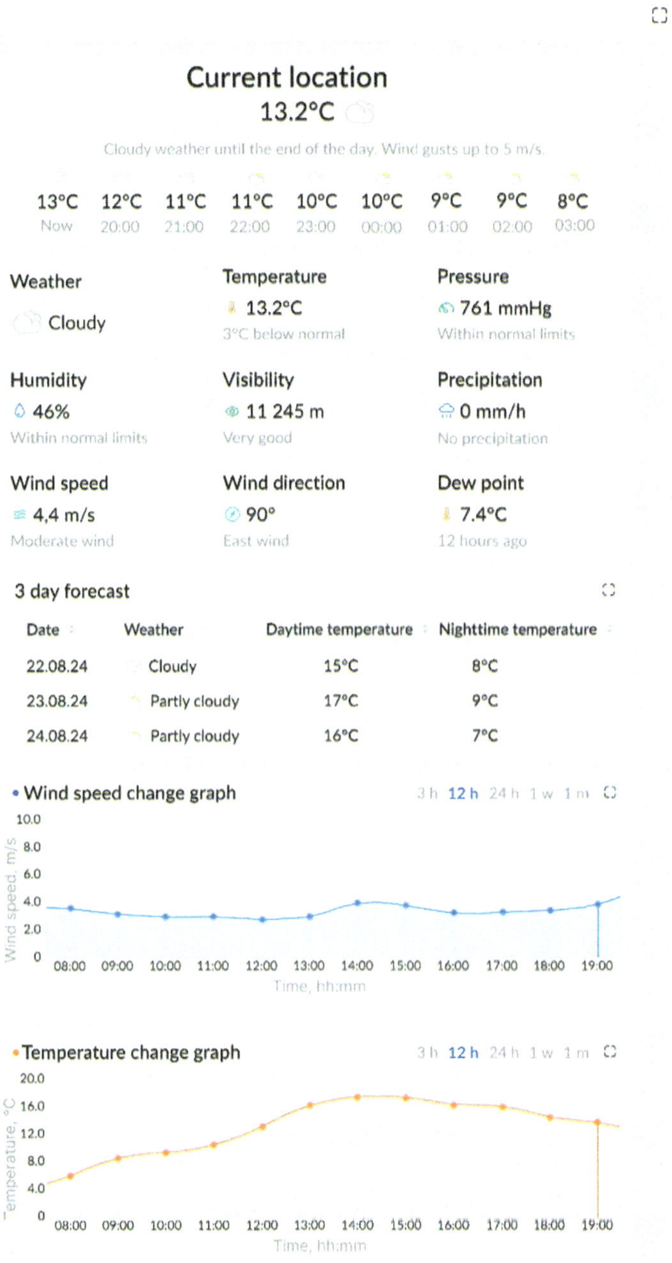

Fig. 3 User interface of the hardware and software complex for monitoring and processing meteorological data

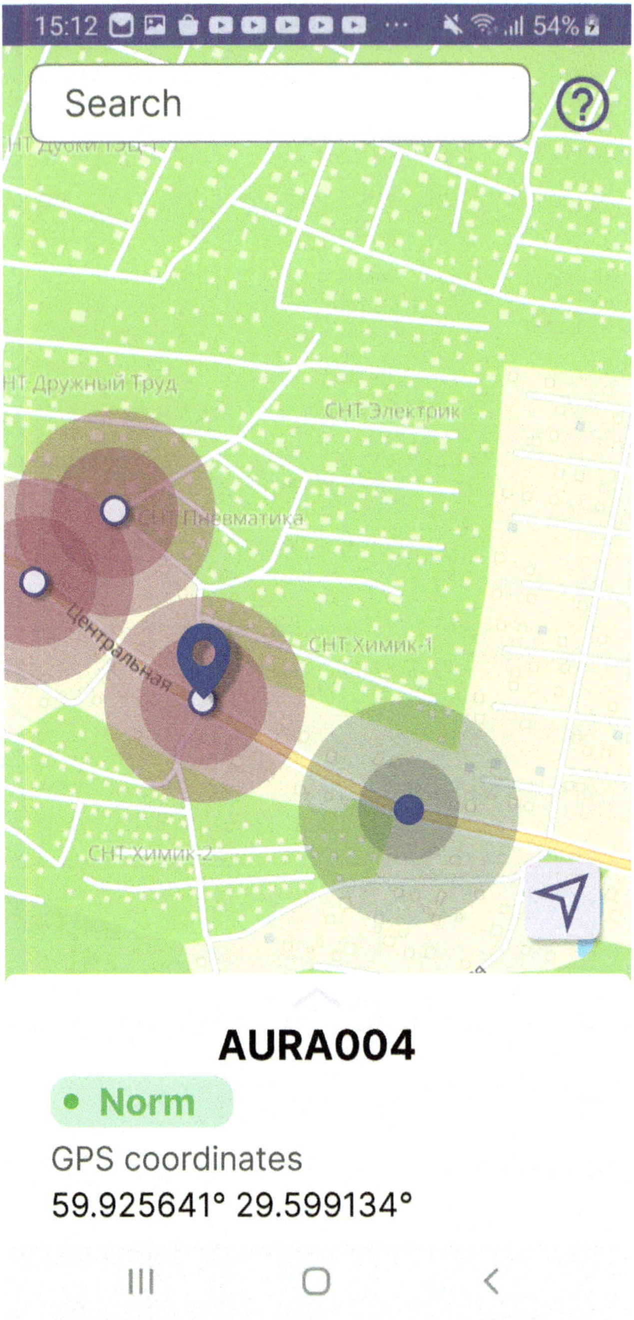

Fig. 4 Main page of the mobile application

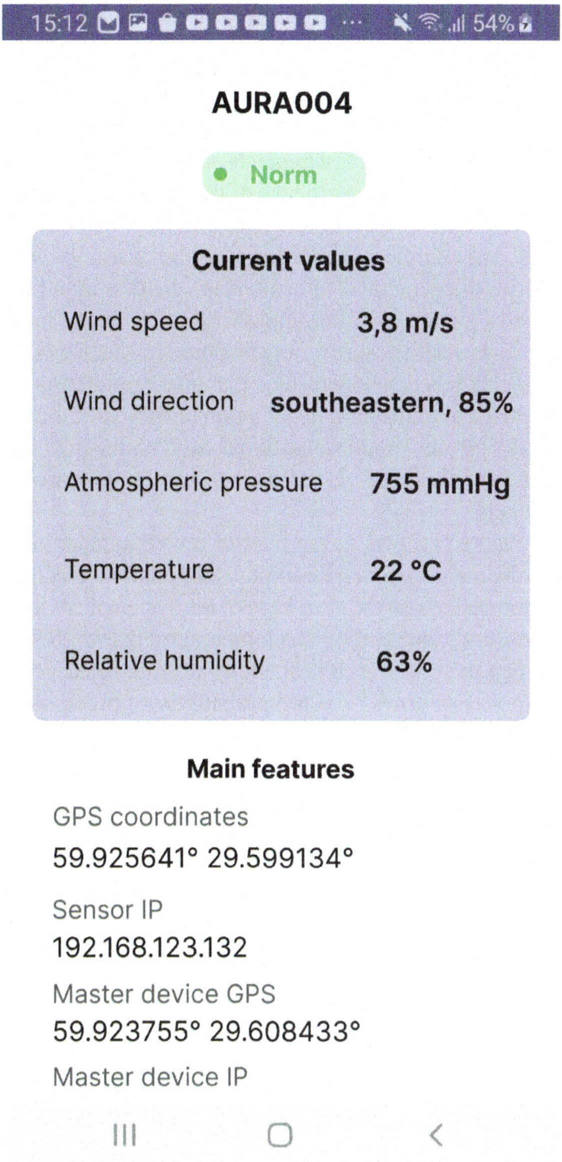

Fig. 5 Page for sensor control of the mobile application

6 Results of Monitoring and Processing of Meteorological Data

The forecasts are formed based on the synoptic method. As the study showed, the accuracy of the forecast depends on the area of coverage of the sensors, while the error of forecasts on the periphery of the territory covered by the sensors greatly exceeds the error of forecasts in its average value. This is due to the fact that the hardware and software complex cannot consider the overall picture of climate change around the world; the generated forecast maps cover only a limited area, and therefore cannot always predict cyclones approaching from outside.

However, to ensure timely response to weather changes, the hardware and software complex has developed rules that allow predicting climate changes without precise knowledge of the climate situation outside the area covered by the sensors. The rules were derived from observations based on the patterns identified. For example, if the wind speed exceeds 15 m/s with a sharp drop in pressure, then a storm or strong wind is likely.

Figure 6 shows a fragment of a web application with a forecast generated by a hardware and software complex that predicts precipitation. In this case, the readings of the sensors on the periphery of the system are within the range of the rule "if the air humidity exceeds 85% at a temperature below 20 °C and a wind speed of up to 5 m/s, then there is a high probability of rain. In accordance with this rule, the forecast map was reformed and the probability of precipitation appeared in the forecast for the next 24 h.

A fragment of the fog prediction web application is shown at Fig. 7. In this case, the temperature values drop to 5 °C in the next two hours, the humidity is 94%, and the wind speed is 0.5 m/s, which fits the rule "if the temperature drops below 6 °C with humidity above 90% and there is little or no wind, then fog is possible".

Based on the data on wind direction and speed, the hardware and software complex can predict clear weather (See Fig. 8). Having calculated the speed of cloud movement, as well as their area using readings from neighboring sensors and a forecast map, it was determined that the sky would become clear in 4 h, since by this time the entire mass of clouds would float over the observed point.

Thus, forecasts are generated based on rules defined in advance in the system, as well as through mathematical calculations and data from forecast maps, which allows for the prompt detection of weather changes.

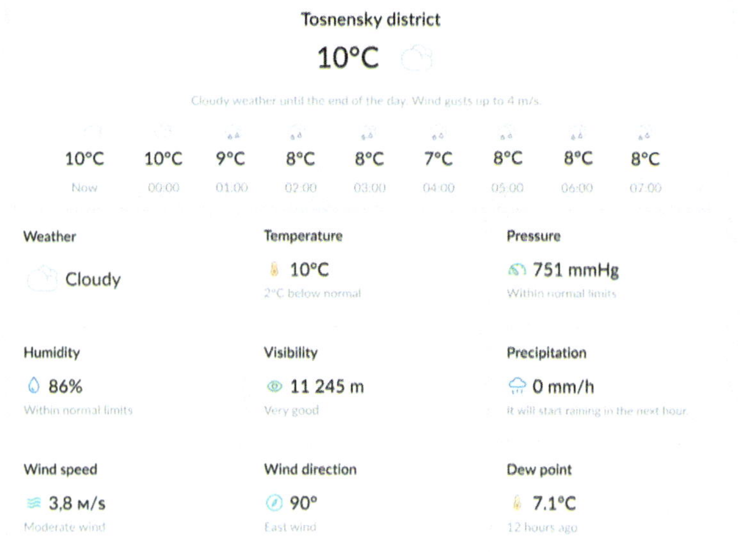

Fig. 6 Result of the precipitation forecast rule

Fig. 7 Result of the fog forecasting rule

Fig. 8 Result of the clear weather forecast rule

7 Accuracy Assessment of Monitoring Results and Processing of Meteorological Data

To assess the accuracy of weather forecasting in the developed hardware and software complex for monitoring and processing meteorological data during the week in the Tosnensky district of the Leningrad region from August 12 to 18, 2023, comparisons were made of the predicted and actual values of meteorological parameters: temperature, pressure, precipitation, humidity, wind speed and direction. A specific point was selected for which predictions were made and at which measurements were taken for the maximum values of the indicators. Predictions were made three days before the measurement. The data were entered into tables, on the basis of which graphs were constructed.

After all measurements were taken, the accuracy of the predicted values was assessed and compared with the actual readings. Figure 9 shows a graph of the maximum predicted and actual temperature values for the week in question.

The accuracy of the measurements was estimated based on the calculations of the average absolute percentage error. For temperature, it was 8.8%. The maximum value of the average absolute percentage error was found in the prediction of precipitation and was 12.43%.

Thus, the hardware and software complex solves the task. Increasing the density of the sensors will further improve the accuracy of weather forecasts.

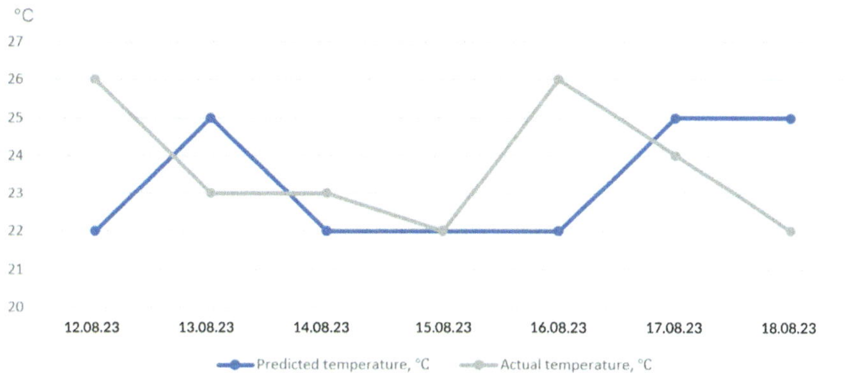

Fig. 9 Graph of maximum predicted and actual temperature values for the week

8 Conclusion

The task of meteorological monitoring for short-term forecasts for general use has been solved using the synoptic method, which allows for observation and forecasting of weather and climate conditions in a certain area. To solve the task, a hardware and software complex (automated weather station system) was developed based on the Internet of Things technology, which is a distributed network of meteorological sensors. The complex provides the ability to respond in a timely manner to changes in weather conditions thanks to automatic data collection and analysis, and helps to improve the accuracy of weather forecasts by collecting data from multiple sensors located in various geographic locations. The server side of the web application uses the Node.js platform with the Express framework to process requests and control business logic. The server interacts with the database via ORM (Object-Relational Mapping) Sequelize, which simplifies working with data and makes the code more universal. PostgreSQL is used as a database management system.

According to the results of the study, the accuracy of the forecast depends on the coverage area of the sensors, while the error in forecasts on the periphery of the area covered by the sensors greatly exceeds the error in forecasts in its average value. This is due to the fact that the hardware and software complex cannot consider the overall picture of climate change around the world; the generated forecast maps cover only a limited area, and therefore cannot always predict cyclones approaching from outside. To ensure timely response to weather changes, the hardware and software complex has developed rules that allow predicting climate changes without precise knowledge of the climate situation outside the area covered by the sensors. That is, forecasts are formed on the basis of rules defined in advance in the system, as well as through mathematical calculations and data from forecast maps, which allows for the prompt detection of weather changes.

To assess the accuracy of weather forecasting in the developed hardware and software complex for monitoring and processing meteorological data during the week in the Tosnensky district of the Leningrad region from August 12 to 18, 2023, comparisons were made of the predicted and actual values of meteorological parameters: temperature, pressure, precipitation, humidity, wind speed and direction. A specific point was selected for which predictions were made and at which measurements were taken for maximum values of the indicators.

The hardware and software complex for monitoring and processing meteorological data, based on the IoT technology, allows users to conveniently track weather changes and forecasts and can become an important component of the "Smart City" system.

References

1. Appio, F.P., Lima, M., Paroutis, S.: Understanding smart cities: innovation ecosystems, technological advancements, and societal challenges. Technol. Forecast. Soc. Change. **142**, 1–14 (2019). https://doi.org/10.1016/j.techfore.2018.12.018
2. Popov, E., Strielkowski, W., Vlasov, M.: Smart cities: knowledge generation effectiveness in the digital economy. In: Proceedings of the International Scientific and Practical Conference on Digital Economy (ISCDE 2019). Atlantis Press, Paris, France (2019). https://doi.org/10.2991/iscde-19.2019.2
3. Tzioutziou, A., Xenidis, Y.: A study on the integration of resilience and smart city concepts in urban systems. Infrastructures **6**, 24 (2021). https://doi.org/10.3390/infrastructures6020024
4. Grigoreva, E.D., Ushakov, V.A.: Research of fire detection speed in an industrial premises using the "smart manufactiring" system based on the internet of thing technology. In: 2024 XXVII International Conference on Soft Computing and Measurements (SCM), pp. 364–367. IEEE (2024). https://doi.org/10.1109/SCM62608.2024.10554274
5. Jeong, S., Na, W., Kim, J., Cho, S.: Internet of things for smart manufacturing system: trust issues in resource allocation. IEEE Internet Things J. **5**, 4418–4427 (2018). https://doi.org/10.1109/jiot.2018.2814063
6. Sunyaev, A.: The internet of things. In: Internet Computing, pp. 317–352. Springer Nature Switzerland, Cham (2024). https://doi.org/10.1007/978-3-031-61014-1_10
7. Levina, A., Iliashenko, V.M., Kalyazina, S., Overes, E.: Smart hospital architecture: IT and digital aspects. In: Lecture Notes in Networks and Systems, pp. 235–247. Springer International Publishing, Cham (2022). https://doi.org/10.1007/978-3-030-93872-7_20
8. Ilin, I., Rukina, P., Dubgorn, A.: Use of the data-driven concept in the management of a healthcare facility. In: Contributions to Management Science, pp. 151–165. Springer Nature Switzerland, Cham (2024). https://doi.org/10.1007/978-3-031-53614-4_9
9. Asore, O.B.: Introduction to cloud computing. J. Contrib. (2024). https://doi.org/10.6084/M9.FIGSHARE.26768029.V1
10. Sunyaev, A.: Cloud computing. In: Internet Computing, pp. 165–209. Springer Nature Switzerland, Cham (2024). https://doi.org/10.1007/978-3-031-61014-1_6
11. Sunyaev, A.: Fog and edge computing. In: Internet Computing, pp. 211–238. Springer Nature Switzerland, Cham (2024). https://doi.org/10.1007/978-3-031-61014-1_7
12. Mostafa, B.: Monitoring internet of things networks. In: 2019 IEEE 5th World Forum on Internet of Things (WF-IoT), pp. 295–298. IEEE (2019). https://doi.org/10.1109/WF-IoT.2019.8767203
13. Savelyeva, D.D., Tatarnikova, T.M.: Internet of things traffic consumption control system. In: 2022 Wave Electronics and its Application in Information and Telecommunication Systems (WECONF), pp. 1–5. IEEE (2022). https://doi.org/10.1109/WECONF55058.2022.9803422
14. Gunjal, P.R., Jondhale, S.R., Lloret Mauri, J., Agrawal, K.: Internet of Things: Theory to Practice. CRC Press, Boca Raton (2024). https://doi.org/10.1201/9781003282945

15. Startup Selfie Revolutionizing Road Safety with Automated Rapid Movable Barriers. https://www.startupselfie.net/2023/07/07/traffic-tech-road-safety-automated-rapid-movable-barriers/. Accessed 21 Aug 2024
16. Sokolov, B., Ushakov, V., Zakharov, V.: Optimal planning and scheduling of information processes during interaction among mobile objects. Int. J. Prod. Res. **62**, 5905–5924 (2024). https://doi.org/10.1080/00207543.2024.2302388
17. Ushakov, V.: Models and algorithms for controlling information processes in the moving objects interaction. Mar. Intellect. Technol. 235–247 (2022). https://doi.org/10.37220/mit.2022.57.3.031
18. Ilin, I., Laskin, M., Logacheva, I., Sarygulov, A., Tick, A.: Land plots evaluation for agriculture and green energy projects: how to overcome the conflict using mathematics. Mathematics **10**, 4376 (2022). https://doi.org/10.3390/math10224376
19. Colucci, S.J.: Synoptic meteorology|anticyclones. In: Encyclopedia of Atmospheric Sciences, pp. 273–279. Elsevier (2015). https://doi.org/10.1016/B978-0-12-382225-3.00071-2
20. Li, G., Wang, Y., Zhang, C., Xu, C., Zhan, L.: Study on the impact of building energy predictions considering weather errors of neighboring weather stations. Sensors **24**, 1157 (2024). https://doi.org/10.3390/s24041157
21. Mansfield, D.: Synoptic meteorology|forecasting. In: Encyclopedia of Atmospheric Sciences, pp. 280–288. Elsevier (2015). https://doi.org/10.1016/B978-0-12-382225-3.00399-6
22. Reynolds, R.: Synoptic meteorology|weather maps. In: Encyclopedia of Atmospheric Sciences. pp. 289–298. Elsevier (2015). https://doi.org/10.1016/B978-0-12-382225-3.00397-2
23. Alexe, M., Campineanu, A.: Weather station design using Arduino platform. Rom.Ian J. Pet. Gas Technol. **4** (75), 57–68 (2023). https://doi.org/10.51865/jpgt.2023.02.06
24. Ganesan, S., Lean, C.P., Chen, L., Yuan, K.F., Kiat, N.P., Khan, M.R.B.: IoT-enabled smart weather stations: innovations, challenges, and future directions. Malays. J. Sci. Adv. Tech. 180–190 (2024). https://doi.org/10.56532/mjsat.v4i2.293
25. Mayorova, V., Melnikova, V., Tenenbaum, S., et al.: Yareelo—group of nanosatellites for space weather monitoring. In: Proceedings of the International Astronautical Congress. IAC (2020)
26. Novikov, D.A.: Cybernetics: from Past to Future. Springer International Publishing, Cham (2016). https://doi.org/10.1007/978-3-319-27397-6
27. Sokolov, B.V., Ushakov, V.A.: Formation reachability area as a data vector using a dynamic model for controlling information processes in the automated control system for moving objects. In: Models and Methods for Researching Information Systems in Transport 2020 (MMRIST 2020), pp. 67–75. CEUR (2020)
28. Sokolov, B., Ushakov, V.: Model-algorithmic support for abilities calculating of control system based on projection operators. In: Advances in Intelligent Systems and Computing, pp. 342–348. Springer International Publishing, Cham (2019). https://doi.org/10.1007/978-3-030-19813-8_35
29. Saviche, A.I.: Synoptic Methods of Weather Forecasting. SPbPU, SPb (1982)
30. Juneau, J., Telang, T.: Object-relational mapping. In: Java EE to Jakarta EE 10 Recipes, pp. 333–377. Apress, Berkeley, CA (2022). https://doi.org/10.1007/978-1-4842-8079-9_7
31. Jungmann, L., Keith, M., Schincariol, M., Nardone, M.: Object-relational mapping. In: Pro Jakarta Persistence in Jakarta EE 10, pp. 99–153. Apress, Berkeley, CA (2022). https://doi.org/10.1007/978-1-4842-7443-9_4
32. Sutejo, A., Gernowo, R., Andreas Purwoadi, M.: Ontology learning from object-relational mapping metadata and relational database. Indones. J. Electr. Eng. Comput. Sci. **33**, 1116 (2024). https://doi.org/10.11591/ijeecs.v33.i2.pp1116-1125

Trends in the Development of BNPL Services as a Modern Digital Solution in Russian and Foreign Payment Industry

Evgenia Karlovskaya, Oksana Vaganova, Daria Gordya, Oksana Balabneva, and Lyazat Talimova

Abstract The article is devoted to the study of the rapidly growing interest-free installment service (BNPL) in the context of digital transformation of payment sector. The key features of BNPL in comparison with other similar methods of payment for the use of borrowed funds, in particular, the methods of installment cards and credit cards have been determined. An analysis of the current state of BNPL services in Russia and globally has been conducted. The advantages of their use for clients, providers and retail sellers have been identified. However, in addition to the positive effects of the introduction of the new payment model, the risks associated with the potential increase in the long-term burden on users of BNPL services have been highlighted. These circumstances make the issues of regulation of BNPL providers particularly urgent. The study considers approaches to regulation of the BNPL market in Russia and abroad.

Keywords Consumer lending · Installment · BNPL · BNPL services · BNPL providers · BNPL platforms · Banking ecosystems

1 Introduction

BNPL services provide the opportunity to pay for purchases in installments without paying interest. BNPL leverages digital technology to provide a seamless and convenient borrowing experience, making it a subset of the broader digital lending ecosystem.

E. Karlovskaya · O. Vaganova (✉) · D. Gordya
Belgorod State National Research University, Belgorod, Russia
e-mail: vaganova@bsuedu.ru

O. Balabneva
Peter the Great St.Petersburg Polytechnic University, St. Petersburg, Russia

L. Talimova
Karaganda University of Kazpotrebsoyuz, Karaganda, Kazakhstan

I. Ilin and M. Youzhong (eds.), *Digital Systems and Information Technologies in the Energy Sector*, Lecture Notes in Networks and Systems 1244,
https://doi.org/10.1007/978-3-031-80710-7_18

In recent years, such platforms have been developing more actively in the consumer market [1]. Digital transformation has a high impact on the payment segment. Every year, the demand for digital services in the banking sector is growing. This is a context in which credit and financial institutions are constantly expanding their capabilities and offering new digital products to customers to respond to the most relevant requests from service users and develop their customer experience. Users quickly get used to convenient banking decisions, and often their interest can only be maintained by introducing new options. Despite the fact that sometimes this process requires significant resources for financial companies, the organizations receive important benefits: in addition to attracting new clients, existing clients are becoming more loyal, and business competitiveness as a whole is improved. In these conditions, new technologies and opportunities are constantly appearing on the fintech market that change the interaction of a person with a bank. BNPL-services are among them. The growing popularity of BNPL services is mainly due to two global trends, which are the acceleration of digitalization processes, catalyzed by the coronavirus pandemic with subsequent lockdowns, and the following growth of online trading [2], on one hand, and the population's response to a decrease in real disposable income due to increased inflationary pressure in the global economy, on the other hand.

Against the background of economic instability, there is a tendency towards a transition to a savings model of consumer behavior, which affects both retail itself and retail lending, and even the consumer's attitude to holiday promotions and sales.

BNPL (buy now, pay later) is a payment model, which is characterized by the division of the purchase amount, usually by four equal parts (25% of the purchase price) that are paid every two weeks. Interest is not accrued on BNPL payments, and for registration of purchase user needs only the phone number and payment instrument details (card or account data).

2 Results and Discussion

BNPL service differs from standard banking products. For example, before opening a credit account or credit card for an installment purchase, the bank checks the solvency and credit history of the client, and only then, with the positive result of verification, decides to borrow the required amount for a certain time. This process is often called an "interest-free loan" in which banks receive their benefit. Interest to the bank for using its money may come from missed payments, often higher than on cards or regular loans. In addition, if there are no missed installment payments, there are hidden interests that are included in the price of the product or are paid to the bank by the trading company. For more credibility, let us compare the terms and conditions of credit cards, installments and BNPL services according to a number of key characteristics (Table 1) [3].

Using a BNPL service, the buyer does not sign a loan agreement, which distinguishes this service from lending, credit cards and microloans. BNPL providers make

Table 1 Comparative characteristics of the conditions for using credit cards, installment cards and BNPL services

Criteria	Credit cards	Installment cards	BNPL services
Usage	Widespread everywhere	Widespread, mostly offline	Limited, mostly online
Credit history	Need for a good credit history	Need for a good credit history	No interaction with credit history bureaus
Approval process	Requires passport data, takes up to several days	Quick approval using KYC ("Know your customer")	Instant, passport data is not required, a phone number and payment card can be used to identify the client
Interest-free period	Yes, usually 1–2 months	Yes, up to unlimited	Yes, usually 2–6 weeks
Repayment	User pays minimum payments during the repayment period. Interest is calculated on the balance	User pays a fixed payment on the appointed date. Delay leads to the accrual of increased interest in the form of a fine	User pays a fixed payment on the appointed date
Additional benefits	Rewards program, cashback, miles for purchases, etc.	No rewards program, cashback is optional	No rewards program or cashback

money from the commission they charge the seller or buyer under certain conditions. For example, customers are charged a commission for a larger number of payments.

It is obvious that BNPL services are an effective payment tool, which are primarily aimed at active and advanced Internet buyers [4]. According to the LendingTree [5] study, the distribution of the most popular categories when using BNPL service is as follows: "Clothing, shoes and accessories" (46%), "Household goods" (34%), "Electronics" (27%) and "Beauty, body and hair goods" (26%) (Fig. 1).

At the same time, the global sales market using BNPL services is growing annually by 90–100%. Figure 2 shows the volume of transactions made through BNPL providers for the period of 2019–2026. Data from 2024 to 2026 are represented by predicted values [6].

Considering the foreign sales market, we come to the conclusion that starting from the 2010s, the following companies began to actively use BPNL services: Klarna Inc.; Affirm, Inc.; AfterPay Limited; Splitit Payments, Ltd.; Sezzle; Perpay Inc.; Zip Co, Ltd; PayPal Holdings, Inc.; Openpay; LatitudePay Financial Services; HSBC Group. Table 2 shows the classification of the leading BNPL service providers [7].

The growing popularity of using these services is evidenced by the fact that in March 2023 Apple introduced a new "Apple Pay Later" feature for its Apple Pay e-wallet users. This innovation reflects Apple's desire to provide its customers with flexible and convenient payment solutions. Currently, the service is available only to US citizens in a limited list of states. Apple Pay Later allows users to apply for interest-free loans of $75-$1000 directly from within the wallet app, allowing for seamless online or in-app purchases without incurring fees. After making a purchase

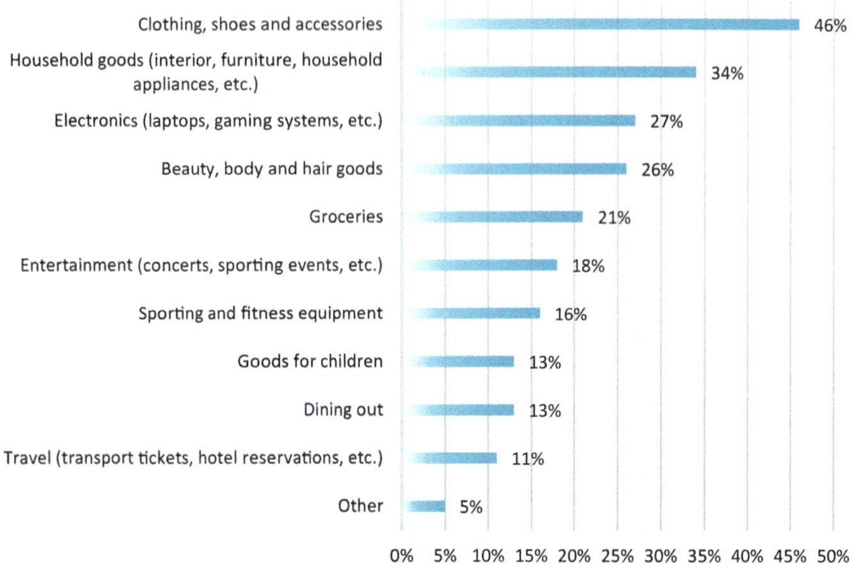

Fig. 1 The most popular areas for spending funds through BNPL services in 2024

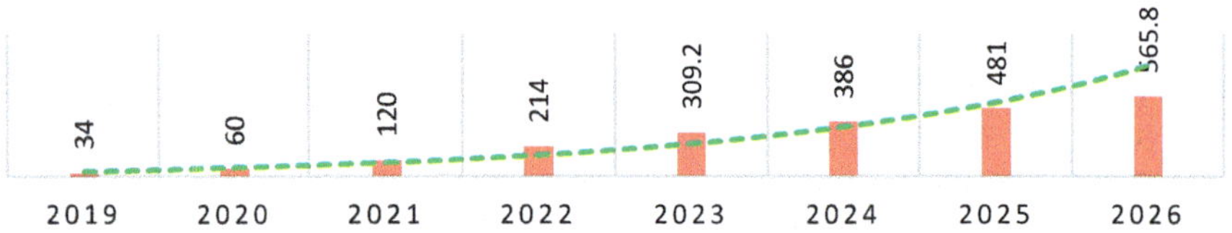

Fig. 2 Volume of transactions using BNPL services in the world in 2019–2026, billion US dollars

with Apple Pay Later, the loan details and payment history are transferred to one of the largest credit bureaus (Experian, TransUnion or Equifax). As stated on the company's website, at the moment the above information does not affect the user's credit rating, but may affect their credit history in the future [8].

The growth rate of BNPL popularity is so high, primarily due to the popularization of e-commerce as a way of purchasing goods or services. E-commerce gained momentum during the pandemic, as it allowed buyers to make transactions without leaving their homes, which was critical during the lockdown. After the pandemic, the online retail segment continued to grow. In 2022, the share of e-commerce in total retail sales in Russia was 11.6%. In 2023—already 13.8%. In 2023, the total volume of Internet trade amounted to 6.4 trillion rubles. This is 28% more than a year earlier, according to the Association of Internet Commerce Companies [9].

Such statistics create favorable conditions for BNPL development. This format of services became available for the first time on online platforms (marketplaces). The IT company ARTW analyzed the use of "payment by installments" services in Russia

Table 2 Comparative characteristics of leading BNPL service providers in the global market

Criteria	Affirm	After pay	Klarna	Sezzle	Zip co/ quadpay	PayPay "Pay in 4"
Home country	USA	Australia	Sweden	USA	Australia/ USA	USA
Number of total merchants	29,000	98,200	250,000	40,274	51,300	425,000
Number of active users (millions)	7.1	16.2	90	2.9	7.3	4.2
Merchant fee	up to 8% + $0.30 per transaction	up to 6% + $0.30 per transaction	"Installment"/ "Pay in 30 days" options: up to 6% + $0.30 per transaction "Financing" option: up to 3.3% + $0.30 per transaction	up to 6% + $0.30 per transaction	Varies by merchant	USA: 2.9% + $0.3 per transaction Other countries: 4.4% + currency transaction fee
BNPL terms	4 installments, "Financing" option is 3–36 month	4 installments, every 2 weeks	4 installments, "Financing" option is 3–36 months, or "Pay in 30 days" option	4 installments, every 2 weeks, 1st due at POS	4 installments, every 2 weeks, 1st due at POS	4 installments, every 2 weeks
Consumer interest	0–30% APR	no	"4 Installments" option: 0% APR, "Financing" option: 0–30% APR	no	no	no
Consumer fee	no	Late fees: $10 + $7 if not paid in one week; capped at 25% of order and never over $68	Late fees: "4 Installments" option – up to $7; capped at 25% of order; "Financing" option – up to $35 per month	Late fees: up to $10; rescheduling fee up to $5	Late fees: up to $21; Installment fees: $1 per installment	no

(continued)

Table 2 (continued)

Criteria	Affirm	After pay	Klarna	Sezzle	Zip co/ quadpay	PayPay "Pay in 4"
Soft credit check	yes	no	yes	yes	no	yes
Potential credit score impact	When making a big purchase or late payments	For missed or late payments	When using "Financing" option	Opt-in, "Sezzle Up" members can improve	For missed or late payments	For late payments over 30 days

[10]. According to the data obtained, 44% of online stores have already implemented BNPL services for paying for purchases, while 20% of them use several options for payment by installments services.

BNPL platforms in the Russian market have begun to develop relatively recently (in 2021). Figure 3 presents the chronology of the entry into the market of the largest domestic BNPL providers and key indicators of the BNPL market in Russia [11].

Thus, according to Fig. 3, it is possible to note the high growth rate of the BNPL segment in Russia. Market capacity in 2023 is estimated in the range of 90–100 billion rubles, which is more than three times higher than in 2022. According to the forecast, this indicator will double by the end of 2024. We can also conclude that the key players in the market at the moment are large banks and IT companies. The trend setter was the "Dolyame" service from T-Bank (formerly known as Tinkoff Bank). Soon others appeared, which were "Split" from Yandex, "Pay in Installments" from Sberbank, "Podeli" from Alfa-Bank (first offline BNPL service in Russia), "Plait"

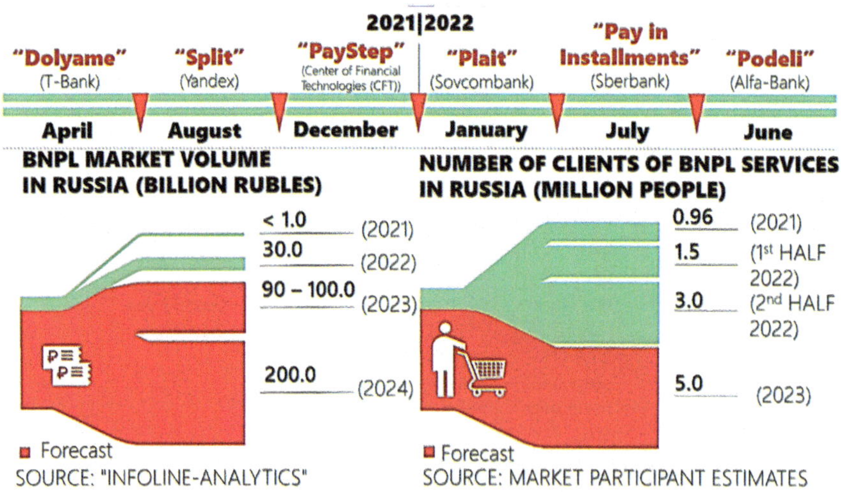

Fig. 3 Chronology of market entry of the largest BNPL providers in Russia

(formerly known as "Halva Chastyami") from Sovcombank. Let us consider the conditions for using these services in Table 3 [3].

Analyzing the activities of banks in the BNPL services market, we see its rapid development and the entry of an increasingly wider range of participants into it. Market participants who have already mastered the BNPL system and have seen its properties, expand their activities in this segment and use the mechanism of "free niches in the market". For example, T-Bank created DNPL (donate now, pay later) service for those who want to make donations to charitable foundations. Users will be able to sponsor non-profit organizations and donate to charities according to the standard operating principle of BNPL services, hence 25% of the donation amount is paid at first, and the remaining amount is paid over the next six weeks. The recipient will receive the entire donation amount immediately.

Several more Russian banks have announced their plans to introduce BNPL services in the near future. In March 2024, OTP Bank had established a new company (OTP Financial Technologies, LLC) and is planning to launch its own BNPL service, which will be focused on the development of the e-commerce segment (online stores), but the bank also plans to attract retail partners (offline stores). It should be noted that in the Russian market, BNPL payments offline at retail outlets are not yet widespread comparing to the Internet.

In April 2024, MTS announced its plans to enter the BPNL market with a new ecosystem product called "Flex". MTS Bank, which is part of the MTS Group, is

Table 3 Comparative characteristics of leading BNPL providers on the Russian market

Criteria	"Dolyame" (T-bank)	"Podeli" (Alfa-bank)	"Split" (Yandex)	"Pay in installments" (Sberbank)	"Plait" (Sovcombank)
Fee	Retailer pays 4.9%	Retailer pays 4.9%	Client pays no more than 299 rubles from every 5000 rubles	Retailer pays 4.9%	5–6%, discussed with partners individually
Number of payments	4	4	4–6	4	4
Late fee	0.1% of the unfulfilled obligation for each day of delay	0.5% of money to be paid for each calendar day of delay	no	5% of the payment amount regardless of the period of delay	3.9% of debt per month
Maximum purchase price	10,000–70,000 depending on the retailer	30,000 rubles	100,000 rubles	30,000 rubles	60,000 rubles
Number of service partners	> 300	30,000	> 300	≈ 100	≈ 50

considering a number of strategic initiatives to achieve growth goals in the bank's retail loan portfolio, including increasing operational efficiency, improving the value proposition of the products offered, increasing integration into the MTS ecosystem, as well as entering new segments to increase customer flow. In addition to basic credit and debit products, the customer acquisition funnel will be expanded to include non-banking products. "One of the important directions of the bank's development is entering the new segment of interest-free installments. In particular, the launch of a BNPL class product is currently being piloted to increase the volume of attracting new customers through commodity lending processes," notes MTS Bank [12]. Currently, users can extend their payment using "Flex" when making purchases in the travel service "MTS Travel", on the insurance marketplace "MTS All Insurance", and at the ticket aggregator "MTS Ticketland". In the near future, the option to pay with "Flex" will become available in the MTS online store and in all storefronts where MTS Pay is available. The second stage will allow users to extend payments in all offline stores when paying using QR of Faster Payments System (SBP). There are also plans of developing the service to provide users with the opportunity to divide the payment not only into four parts, but also to choose a longer period.

In addition to banks, large marketplaces are gradually starting to develop their BNPL services in Russia. The already mentioned BNPL service "Split" is actively developed by Yandex. It can be used for orders on "Yandex Market" and partner stores (Kari, Ecco, Holodilnik.ru, etc.). The amount of the order in this case is written off in equal parts every two weeks.

On "Megamarket" from Sber you can arrange some goods in instalments through Sberbank and its service "Pay in Instalments".

Currently, "Ozon" marketplace also provides the opportunity to use a BNPL service through Ozon Bank, using which you can divide the payment by six months, in some cases a commission is charged for this.

In April 2024, "Wildberries" marketplace began testing the BNPL-based payment of orders through its own Wildberries Bank. The service is now available to a limited sample of customers, but Wildberries is actively working on the development of fintech, creating convenient services for buyers and sellers of the platform.

Thus, we can say that BNPL services are currently actively developing. They successfully integrate with banking ecosystems and cover practically the entire solvent population of the country, which creates their competitive advantage [13, p. 92].

Based on the above, let us highlight the advantages that explain the growing popularity of BNPL services among buyers, providers and retailers who are service partners. These benefits are presented in Fig. 4.

BNPL services provide a wide range of advantages for all market subjects. For example, this payment model allows retailers to increase sales and attract new customers due to flexible payment terms. Financial and credit institutions use BNPL to develop and expand payment methods, which increases the conversion and turnover of transactions. Using installment payments across a wide partner network will attract new customers and increase interest in banking products.

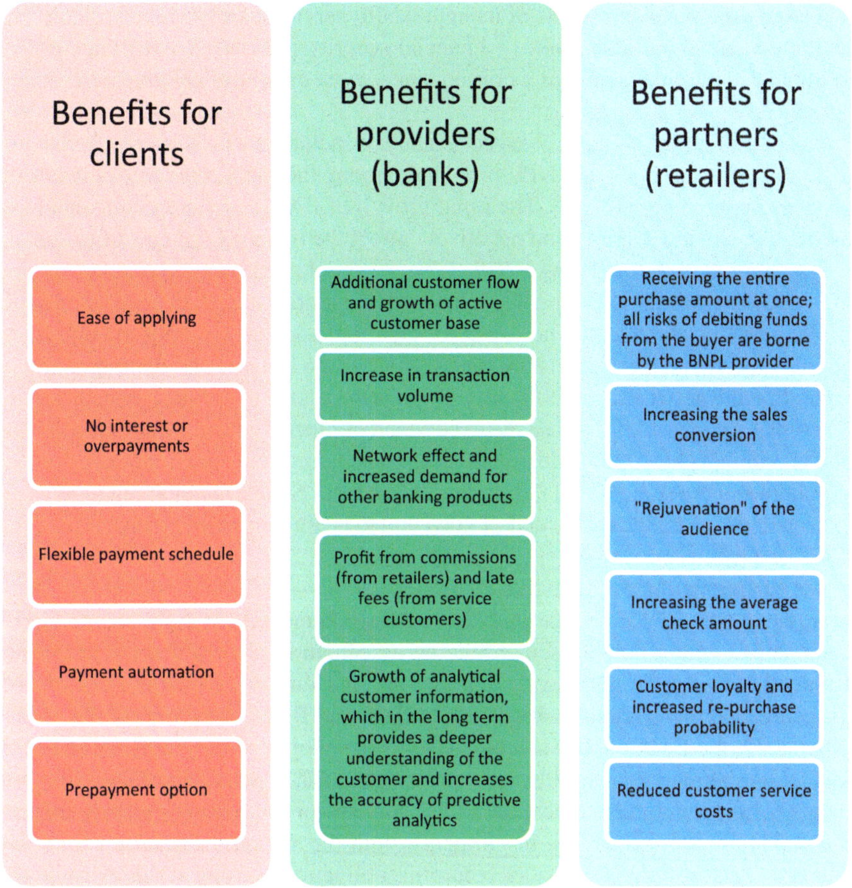

Fig. 4 Advantages of using BNPL services for various market participants

Many users of BNPL services prefer this payment method due to the lack of the required amount of money in full and to the ability to make payments in a more convenient time frame [14]. The use of loans and credit cards lowers the consumer's credit rating and limits the ability to attract large borrowed funds. As a result, BNPL services are characterized by simple forms of use and soft regulation, unlike banks, microfinance organizations, credit cooperatives, and even P2P platforms. Consequently, BNPL services become an effective instrument of household budget management.

The popularity of BNPL services among young people is especially high [15]. Young people are advanced users of the Internet. Thus, they are quickly adopting new digital services that allow them to make their desired purchases in installments without using credit banking products. At the same time, it is important to remember that behind every opportunity lies a threat [16, pp. 16]. For example, there is an unregulated amount of late fees and a low level of client awareness.

Despite the positive aspects of using a BNPL services, they can be criticized for giving consumers a false sense of financial security [17]. Such a feeling can lead to impulsive purchases and, as a consequence, to an uncontrollable increase in debt burden.

In addition to psychological aspects, analysts include the lack of clear legislation regulating the activities of BNPL providers among the main problems of using the services under study [18, 19]. The rapid growth of BNPL services could lead to an increase in shadow lending and negatively affect the financial security of citizens.

The approach to regulating BNPL services should reflect a sequence of actions to neutralize risks [20]. Taking into account the maturity and level of penetration of BNPL services abroad, it is necessary to take into account not only the efficiency, but also the social significance of BNPL [21]. To do this, when adopting a law on regulating the activities of BNPL services, it is necessary to publicly discuss reports on regulatory measures and use the experience of other countries. For example, soft law instruments have been adopted in Singapore, where there are codes of ethics for suppliers and consumers [22, 23]. In the EU member States, national regulations are expected to be adopted by the end of 2025 and to be implemented by the end of 2026 [24].

As for regulating the activities of BNPL services in Russia, the Central Bank, even when first BNPL service in 2021 had been launched, strongly recommended companies to enter into consumer lending agreements. The regulator is concerned about calculating the borrower's debt ceiling and limiting systemic risks of debt exposure, and therefore talks about the need to send BNPL data to credit bureaus.

In 2022, the Bank of Russia announced the development of a draft document on regulating the market of BNPL services. In particular, it plans to introduce requirements for informing consumers about the contents of the contract, clearly indicate the maximum number of free installments, limit the amount of penalties for delay, and provide for the possibility of recording obligations arising when using BNPL services in credit histories [25].

The danger of the lack of regulation and legislative designation of the term "BNPL service" lies in the fact that it could be used to describe any similar credit products that do not cover all the main components of BNPL service [26, 27]. The Central Bank of the Russian Federation is currently considering the establishment of a register of free installment operators (BNPL) as a regulatory measure.

3 Conclusion

To summarize all of the above, it can be noted that BNPL services have been developing abroad for quite a long time and managed to occupy a significant share in the e-commerce market. In Russia, on the contrary, BNPL services remain a fairly new payment method and their share of use in the market of goods and services is quite small.

However, it is necessary to take into account the specifics of the Russian financial sector, which manifests itself in the following. Firstly, the financial sector in Russia is characterized by high digitalization penetration. Secondly, in contrast to the greater spread of the high-risk model of fintech startups abroad, the key players who develop technological financial solutions in Russia are large banks with developed ecosystems and impressive revenues that they can spend on introducing new products and services [28, 29]. Hence, in the future it will be possible to observe a more rapid growth of implemented innovative solutions, in particular BNPL services.

Nevertheless, in addition to high convenience and a number of advantages for all BNPL entities, the rapid growth of the market also creates a basis for additional risks associated with the increase in the debt burden of the population. In this connection, the issue of the legislative regulation of BNPL services is particularly acute.

References

1. Grigoryan, G.: Fintech development degree in the RA banking system. Opportunities and challenges. Mess. ASUE **1**(76), 47–65 (2024). https://doi.org/10.52174/1829-0280_2024.1-47
2. Saydakova, V.A., Cherepanova, A.D.: BNPL-servis kak innovacionnyj trend v finansovoj sfere Rossii [BNPL Service as an Innovative Trend in the Financial Sector of Russia]. Vektor jekonomiki **11**(89) (2023). https://doi.org/10.51691/2500-3666_2023_11_5 (in Russian)
3. Pokupaj sejchas, plati potom: chem otlichajutsja servisy BNPL i SNPL ot rassrochek i kreditov [Buy Now, Pay Later: How Do BNPL and SNPL Services Differ from Installments and Loans]. https://rb.ru/opinion/bnpl-snpl-vs-credit-installment/. Accessed 3 June 2024 (in Russian)
4. Aalders, R.: Buy now, pay later: redefining indebted users as responsible consumers. Inf. Commun. Soc. **26**(5), 941–956 (2022). https://doi.org/10.1080/1369118X.2022.2161830
5. Buy Now, Pay Later Statistics for 2024. https://fitsmallbusiness.com/buy-now-pay-later-statistics/. Accessed 3 June 2024
6. Global Transaction Value of Buy Now, Pay Later (BNPL) in E-Commerce from 2019 to 2021, with Forecasts from 2022 to 2026. https://www.statista.com/statistics/1311122/global-bnpl-market-value-forecast/. Accessed 3 June 2024
7. Digital Payments Hit Super-Cycle, with More Action Yet to Come. https://www.bloomberg.com/professional/insights/data/digital-payments-hit-super-cycle-with-more-action-yet-to-come/. Accessed 3 June 2024
8. Apple Pay Later. https://www.apple.com/apple-pay/later/. Accessed 3 June 2024
9. Itogi goda: regiony peretjagivajut liderstvo v e-commerce [Year in Review: Regions Take Over Leadership in E-Commerce]. https://www.akit.ru/news/itogi-goda-regiony-peretyagivayut-liderstvo-v-e-commerce. Accessed 3 June 2024 (in Russian)
10. 44% rossijskih internet-magazinov ispol'zujut BNPL-servisy [44% of Russian online stores use BNPL services]. https://artw.ru/blog/archives/BNPL/. Accessed 3 June 2024 (in Russian)
11. Po rassrochnomu delu [On Installment Payments]. https://www.kommersant.ru/doc/6187011. Accessed 3 June 2024 (in Russian)
12. MTS-bank planiruet vyjti na rynok BNPL [MTS Bank Plans to Enter the BNPL market]. https://frankmedia.ru/161618. Accessed 3 June 2024 (in Russian)
13. Bykanova, N.I., Solovey, J., Gordya, D.V., Konshina, L.A.: Formirovanie jekosistem bankov v uslovijah cifrovizacii bankovskogo prostranstva [Formation of Banks Ecosystems in Conditions of Digitalization of Banking Space]. Nauchnye vedomosti Belgorodskogo gosudarstvennogo universiteta. Serija: Jekonomika. Informatika **47**(1), 91–100 (2020) (in Russian)

14. Singh, R., Kasana, E.: Digital lending dilemma: how BNPL impacts consumers' propensity towards debt. In: Singh, D., Malik, G., Aggarwal, S. (eds.) Transforming the Financial Landscape with ICTs, pp. 275–297. IGI Global (2024). https://doi.org/10.4018/979-8-3693-1503-3.ch013
15. Guttman-Kenney, B., Firth, C., Gathergood, J.: Buy now, pay later (BNPL) ...on your credit card. J. Behav. Exp. Financ. **37**, 100788 (2023). ISSN 2214–6350. https://doi.org/10.1016/j.jbef.2023.100788
16. Vaganova, O.V., Talimova, L.A., Gordya, D.V.: Cifrovoj banking: riski i vozmozhnosti dlja bankov i ih klientov [Digital Banking: Risks and Opportunities for Banks and Their Clients]. In: XI International Scientific and Practical Conference Proceedings on Current Problems in the Development of Economic, Financial and Credit Systems, pp. 14–19. Belgorod State National Research University, Belgorod (2023) (in Russian)
17. Hegawan, C.I.P., Faustine, T.M., Wijayanti, S., Manaf, P.A.: Exploring the influence of perceived usefulness and value of PayLater/BNPL on satisfaction, impulsive buying, and post-purchase intention. Int. J. Emerg. Trends Soc. Sci. **15**(1), 1–12 (2023). https://doi.org/10.55217/103.v15i1.667
18. Gerrans, P., Baur, D.G., Lavagna-Slater, S.: Fintech and responsibility: buy-now-pay-later arrangements. Aust. J. Manag. **47**(3), 474–502 (2022). https://doi.org/10.1177/03128962211032448
19. Johnson, M.E.: Regulate Now or Pay Later: A Late Start on Regulating the BNPL Industry Endangers Consumers. 27 N.C. Banking Inst 454 (2023)
20. Mukhtar, B., Binti Ahmad Tamrin, S.N., Kashif Shad, M., Woon Lai, F., Abdul Jalil, N.: A proposed framework for assessing BNPL (buy now, pay later) adoption and its impact on consumers' buying behavior. KnE Soc. Sci. **8**(20), 804–812 (2023). https://doi.org/10.18502/kss.v8i20.14639
21. Relja, R., Ward, P., Zhao, A.L.: Understanding the psychological determinants of buy-now-pay-later (BNPL) in the UK: a user perspective. Int. J. Bank Mark. **42**(1), 7–37 (2024). https://doi.org/10.1108/IJBM-07-2022-0324
22. Podhody k regulirovaniju BNPL (buy now, pay later) za rubezhom [Approaches to Regulating BNPL (Buy Now, Pay Later) Abroad]. https://asros.ru/upload/iblock/172/w1epmtpvenrqcmnexk84z3kscyont1g5/Issledovanie-_-Podkhody_k_regulirovaniyu_BNPL_za_rubezhom.docx. Accessed 3 June 2024 (in Russian)
23. Sng, A., Tan, C.M.: Buy Now Pay Later in Singapore: Regulatory Gaps and Reform. https://ssrn.com/abstract=3819058. Accessed 3 June 2024
24. Machura-Urbaniak, A., Lupinu, M.: Buy now, pay later' (BNPL) payment services: opportunities and legal challenges for EU consumers and businesses. J. Eur. Consum. Mark. Law **12**(5), 184–193 (2023)
25. Chast' rasplaty: CB uvidel riski rosta tenevogo kreditovanija v BNPL-servisah [Part of the Reckoning: the Central Bank Saw the Risks of Growing Shadow Lending in BNPL Services. https://iz.ru/1592618/iuliia-romanova/chast-rasplaty-tcb-uvidel-riski-rosta-tenevogo-kreditovaniia-v-bnpl-servisakh. Accessed 3 June 2024 (in Russian)
26. Servisy Buy Now Pay Later—novyj sposob pokupat' po-staromu [Buy Now Pay Later Services - a New Way to Buy the Old Way]. https://frankrg.com/wp-content/uploads/2023/04/3eb56042ccd8.pdf. Accessed 3 June 2024 (in Russian)
27. Faustov, A., Chemeris, O., Halada, K.: Using digital solutions in railway logistics: the experience of medium-sized businesses. In: Ilin, I., Jahn, C., Tick, A. (eds.) Digital Technologies in Logistics and Infrastructure. ICDT 2021. Lecture Notes on Data Engineering and Communications Technologies, vol 157. Springer, Cham (2023). https://doi.org/10.1007/978-3-031-24434-6_17
28. Vaganova, O., Bykanova, N., Gordya, D., Evdokimov, D.: Growth points of fintech industry in the perception of financial market transformation. Eur. Proc. Soc. Behav. Sci-Ences (EpSBS) **103**, 435–441 (2021)
29. Ilin, I.V., Levina, A.I., Dubgorn, A.S., Abran, A.: Investment models for enterprise architecture (EA) and IT architecture projects within the open innovation concept. J. Open Innov. Technol. Mark. Complex. **7**(1), 69 (2021)

Digital Aspects of Energy Sector

A Multi-Agent System in the IT Architecture of Project Portfolio Management of an Energy Company

Sofia Kalyazina, Igor Ilin, and Anastasia Levina

Abstract Project management in the energy sector has its own specificities related to the high complexity and scale of projects, as well as increased risk and uncertainty. These factors require the application of project management methodologies that ensure the consideration of multiple factors and the ability to respond flexibly to changes. This paper considers the classic stages of project management and project portfolio management, and analyses the applied project and project portfolio management methodologies. Taking into account the peculiarities and risks of project management in the energy sector, the use of multi-agent systems (MAS) in integrated project management is proposed. In this context, the principles of the operation of multi-agent systems and their advantages are considered. On this basis, a model of IT support for project activities at the portfolio management level is proposed, taking into account the relationship between IT systems, their services and relevant business processes, and the place of MAS in project portfolio management is outlined.

Keywords Project management · Project portfolio management · PPM · Energy sector · Enterprise architecture · IT architecture · Multi-agent system

1 Introduction

In today's world of rapidly evolving technology and changing market needs, organisations must constantly adapt their business processes and information systems. The synchronisation of these two key elements is a critical success factor, and Information Technology (IT) Architecture and Enterprise Architecture act as a central component to ensure that an organisation's goals are realised through information technology [1].

S. Kalyazina (✉) · I. Ilin · A. Levina
Peter the Great St. Petersburg Polytechnic University, Saint-Petersburg, Russia
e-mail: kalyazina_se@spbstu.ru

I. Ilin and M. Youzhong (eds.), *Digital Systems and Information Technologies in the Energy Sector*, Lecture Notes in Networks and Systems 1244,
https://doi.org/10.1007/978-3-031-80710-7_19

IT and enterprise architecture is a comprehensive framework that integrates business strategies and IT infrastructure into a single system, ensuring that resources are used efficiently to achieve objectives. Project management, as an integral part of this mechanism, plays an important role in successfully managing resources, optimising processes and preparing the business for transformation in line with market requirements [2, 3].

The key to success in this dynamic environment is agility and alignment of the IT architecture with business objectives. A mismatch between these two elements can lead to serious problems, including reduced productivity, increased risk and loss of competitive advantage.

The relevance of the research is to find an approach that enables agile project management in energy companies, taking into account the specifics of projects in this industry, which are characterised by their complexity and high level of risk. Determining the place of this approach in the enterprise architecture and IT infrastructure will allow companies in the energy sector to create an effective project management system capable of providing flexibility, sustainability and adaptability to changing conditions.

With the increasing complexity and risk of projects in the energy sector, agile project management is becoming a critical success factor. It makes it possible to respond effectively to unforeseen circumstances, minimise risks and ensure the successful implementation of projects.

The research aims to identify and analyse the approaches to agile project management that best suit the specificities of the energy sector. Determining the place of this approach in the enterprise architecture and IT infrastructure will allow companies to create integrated solutions that can ensure sustainable development and the achievement of strategic goals.

Accordingly, the objective of the study is to review the classical stages and methodologies of project management and project portfolio management and, on this basis, to identify an approach that potentially provides the required flexibility of project management with a focus on companies in the energy sector.

2 Material and Methods

As a theoretical and methodological basis of the study, the main current methodologies of project and project portfolio management (including MoP® portfolio management methodology, PRINCE2 project management method) were used. The architectural approach (TOGAF methodology using the Archimate language) was used to form the IT support model, which allows building flexible and scalable IT systems, contributing to improving the efficiency of business processes and achieving the strategic goals of the organisation. To achieve the set goals and objectives of the study, the method of economic and mathematical modelling was used; simulation modelling methods with a focus on agent-based modelling were considered. The mathematical apparatus of agent-based modelling is based on the model of agents'

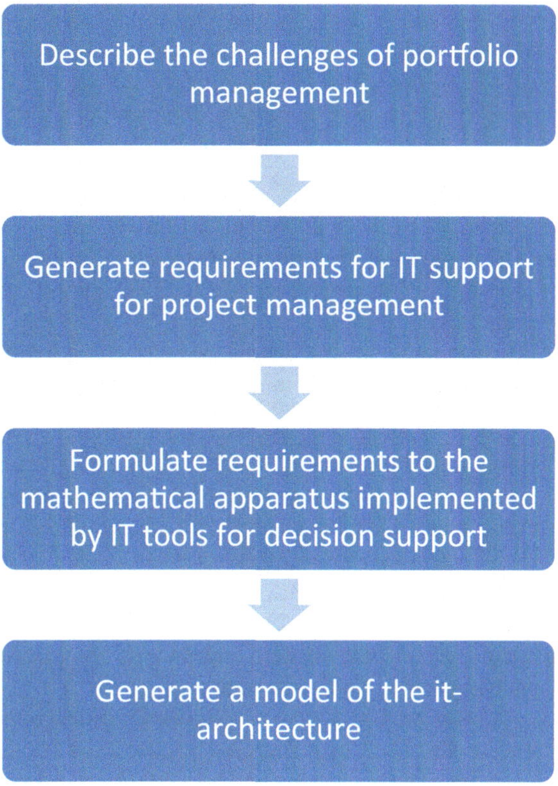

Fig. 1 Research algorithm

negotiation in various forms and allows to take into account the problems of project planning with limited time and resources under the conditions of increasing project complexity and complication of the planning task.

In order to achieve the objectives of the study, the research was conducted according to the following algorithm (Fig. 1):

Reorganisation on the basis of project management methods, which is a strict methodology of all stages and parts of the project cycle, allowing with the help of application of new techniques and technologies, to effectively achieve the set goals and results in the project in terms of composition and scope of work, time, cost and their quality, acts as a modern method of re-engineering of production activities.

3 Subject Area Overview

Project management is about solving problems and achieving project goals. In today's world, business is increasingly shifting from a regular managed process to a multitude of unique projects—challenges that arise to ensure the survival of the business or to adapt to a rapidly changing environment. The ability to turn these tasks into projects and then manage them is therefore highly valued [4].

At the macro level, organisations are motivated to implement project management tools for the effective delivery of their initiatives of any size. At the micro level, a corporate project management system, among other things, solves the following tasks:

- Reduce the cost of implementing initiatives.
- Create the conditions in the organisation for the project team to work.
- Inform top management about the status of the organisation's strategically important projects.
- Ensure sufficient and efficient project document flow.
- Meeting project implementation deadlines.

Project management focuses on communicating and managing stakeholder expectations to increase stakeholder satisfaction [5].

A project portfolio is a collection of projects managed together, all contributing towards a shared strategic goal. This collection is often united by a common theme. When a company has multiple strategic objectives, it may create separate project portfolios for each one [6].

A portfolio allows for a better allocation of company resources between ongoing projects, maximising effort in one place where necessary, even though some projects may be less successful than others.

A classic project portfolio minimises business risk, as the profits from successful projects can offset the resources spent on unsuccessful ones.

Project Portfolio Management (PPM) is a systematic approach for selecting, executing, and ensuring high-quality delivery of projects. This process considers the organization's business objectives, potential benefits, risks, and available resources to make informed decisions [7]. Figure 2 compares the project management and project portfolio management processes.

The decision to implement Project Portfolio Management (PPM) depends not only on a company's strategic goals but also on its financial standing, market demand for its products, and overall development. When considering PPM, it's crucial to analyze potential organizational consequences, risks, and opportunities [8]. Several methods are commonly used for project analysis and evaluation within portfolio management:

- Scoring: this method assigns points or percentages to predetermined criteria based on their importance. These criteria may relate to finance, strategy, risk, or labor costs. The total score reflects the overall value of the project and aligns it with the company's strategic objectives. Scoring is used for both project selection and dynamic analysis.
- Visualization: this approach facilitates project comparison and progress tracking within the portfolio.
- Cost and return on investment (ROI) estimation: analytical techniques are used to assess project costs, ROI, and profitability compared to previous revenues. These tools also determine the ratio of projected costs to actual costs within the portfolio.

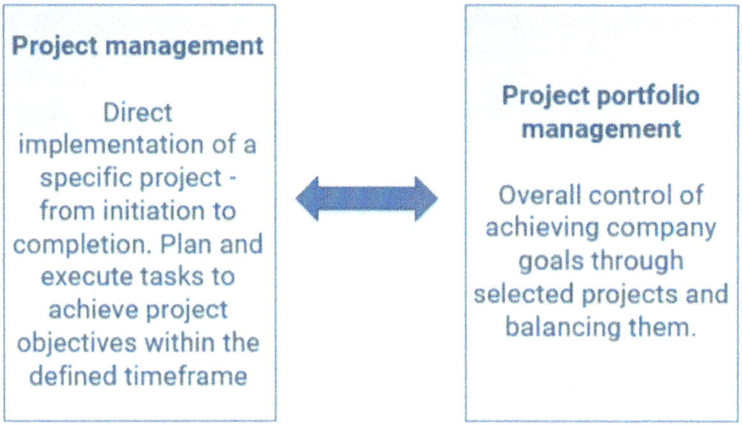

Fig. 2 Comparison of project management and project portfolio management processes

- Expert judgement: this method leverages the expertise and opinions of specialists to determine project relationships and comparisons. It helps identify interdependent, overlapping, or duplicative projects, excessive functionality, and mutually exclusive projects.

Several key roles can be identified in PPM. A project portfolio management team often consists of members from various departments and divisions [9]:

- Project portfolio manager: this individual leads the entire portfolio, overseeing the status of projects, ensuring timely updates, assessing risks, and responding promptly to project changes. They are responsible for portfolio management, receiving reports from individual project managers and communicating with them. The portfolio manager's skills are crucial to the effectiveness of the management process.
- Project manager: this team member manages the implementation of individual projects. Depending on the portfolio size, some projects may be directly managed by the Portfolio Manager.
- Project team member: these individuals are assigned to specific projects. They may belong to the same team or come from different departments and may not have worked together before. Project teams are temporary, formed for the duration of the project and disbanded upon completion.
- Interested managers: these employees hold management positions or are professionals who oversee the implementation of specific project functions.

The following stages of project portfolio management can be identified [10]:

Stage 1—Understanding. This stage develops strategic objectives and/or identifies the changes required to achieve those strategic objectives. The purpose of this stage is to gain a clear understanding of what is in the current portfolio and project development queue, past performance and projected costs, benefits and risks.

Stage 2—Categorisation. In this stage, initiatives are grouped into segments or sub-portfolios based on strategic objectives or other groupings as appropriate. The purpose of this stage is to help senior decision-makers understand the composition of the portfolio and thus make decisions about the balance and best use of available funding and other resources.

Stage 3—Prioritisation. Prioritisation prioritises change initiatives within a portfolio (or portfolio segment) based on one or more agreed indicators. The purpose of this step is to help the portfolio manager decide which initiatives to fund first, given the challenge of balancing risk and return.

Stage 4—Balancing. The purpose of this stage is to ensure that the final portfolio is balanced in terms of factors such as timing, coverage of all strategic objectives, impact across the business, overall risk, available resources, etc.

Stage 5—Planning. The purpose of this stage is to gather information from the portfolio definition cycle and create a portfolio strategy and implementation plan.

Stage 6—Portfolio Implementation. The purpose of this stage is to ensure that the planned change initiatives agreed in the portfolio strategy and implementation plan are successfully implemented and that the portfolio adapts to changes in strategic objectives, project and programme delivery and lessons learned.

Stage 7—Delivery. This stage is delivered through key practices: Management Control; Benefits, Financial, Resource, Risk Management; Stakeholder Engagement; Organisational Management.

Various methods and tools are used to evaluate projects in the selection phase and during project work.

IT architecture is a set of different technical and technological solutions that enable an enterprise to function efficiently. It is divided into Enterprise Information Architecture, Enterprise Solution Architecture and Enterprise Technical Architecture.

4 Results

Projects in the energy sector have a number of characteristics that increase the interest in implementing project management:

1. High complexity and scale: energy projects often involve complex infrastructure such as power plants, wind farms, solar farms, oil and gas pipelines, nuclear power plants, etc. They may cover large geographical areas and require interaction with various stakeholders (government, local communities, environmental organisations).
2. High risk and uncertainty: Energy projects are subject to various risks, such as.

- Technological risks: new technologies may not be fully understood, leading to problems in their implementation.
- Financial risks: investments in energy projects are substantial and may be subject to fluctuations in fuel prices, changes in government subsidies, etc.
- Environmental risks: projects can have an impact on the environment and require strict environmental standards.
- Political risks: regulation in the energy sector can be volatile, which can affect the viability of projects.

  In addition, effective risk management is required; risks need to be identified, assessed, planned and controlled to minimise their negative impact.

3. Long implementation periods, requiring long-term planning and effective risk management throughout the project life cycle. In addition, energy projects are designed for long-term operation, so project management must take into account not only the construction phase, but also the subsequent phases of maintenance, modernisation and decommissioning.
4. High safety standards: safety is a priority in the energy industry, so projects must meet strict safety standards and health and safety requirements.
5. Strict environmental sustainability requirements, including minimisation of greenhouse gas emissions, use of renewable energy sources, etc.
6. Engagement with various stakeholders such as
   - Government and regulators
   - Local residents
   - Environmental organisations
   - Financial institutions
   - Equipment and service providers
7. Technological innovation: the energy industry is constantly evolving with new technologies such as renewable energy, smart grids and energy storage systems. Project management in this area requires rapid adoption of new technologies and adaptation to change.
8. Addressing social impacts: Energy projects can have significant social impacts on local communities. Project management should take these into account and minimise negative impacts and create positive social change.

Accordingly, project management in the energy sector requires a thorough understanding of the industry, as well as high skills and experience in project management, finance, technology, safety and the environment [11].

Energy efficiency is a pressing issue in today's world. To address this challenge, companies can implement energy efficiency improvements at various levels, such as optimizing energy resource consumption, reducing the energy intensity of their products, and lowering utility and other expenses [12]. A key strategy for achieving these goals is the development and implementation of dedicated projects.

Energy supply projects, while varied, are complex, multi-faceted, and involve multiple stages. For instance, an energy audit can be considered a separate project, followed by the development of an energy efficiency model for the specific problem. The implementation of this model then becomes a third project [13].

Project management in this context also focuses on enhancing the reliability of electricity supply, which involves ensuring the energy system's ability to generate, transmit, distribute, and supply electricity to end users in the desired quantity and quality. This requires coordinated efforts among power plants, grids, and consumer electrical installations [14].

The relevance of using project management in the energy sector is confirmed by the following benefits:

- Identifying and analysing the efficiency of investments to increase the reliability of energy supply;
- Reduction of possible emergency situations due to the possibility of timely and quick analysis of the impact of changes in the work schedule, financing and energy supply on the implementation of projects;
- Possibility of creating a clear algorithm for the planning and operation of the energy supply system, with the help of mathematical methods for calculating various parameters of the project;
- Reduction of energy supply costs and improvement of the quality of the operations carried out, thanks to the control of all the works and the introduction of a risk management system.

Currently, the use of multi-agent systems has been developed in the area of modelling and improvement of business processes and project management, due to the fact that their structure is very close to the structures of the real world. Studies on the development of information systems for management decision support have shown that agent technologies can be effectively used in this area [15]. The application of multi-agent systems to technical tasks has been considered in the scientific literature for a long time. Currently, there are grounds for developing the application of multi-agent approach in the integrated management of project activities of the enterprise.

Multi-agent systems (MAS) have attracted the attention of scientists from various disciplines as a means of solving complex problems by dividing them into smaller tasks. The individual tasks are distributed among autonomous entities called agents [16, 17]. Each agent uses an ontology of interactions with other agents to decide the correct action to take to solve the task.

The ability of agents to learn and make independent decisions allows them to work together and be flexible. To learn new contexts and actions, agents use their interaction with the environment or with other agents. The next step is for agents to use their skills to make decisions and perform actions in the environment to solve the task they have been given [18]. When developing a MAS, many complex issues such as coordination between agents, training and security are addressed.

MAS can be defined as a loosely coupled network of problem solvers that interact to solve problems beyond the individual capabilities or knowledge of each problem solver. The agent research community has developed a number of interaction models, including coordination, cooperation and negotiation. The starting point for building a MAS is to define its architecture (agent level and system level), then use a methodology to design it, and finally implement it on the MAS platform.

One of the advantages of MAS is the cooperation between agents, which allows for better solutions to problems that cannot be easily solved by a centralised method. Due to disagreements between individual agents, conflicts are inevitable in the cooperation process. Negotiation is seen as a good way to resolve conflicts in MAS. Researchers study negotiation strategies. The utility function based on game theory is used in negotiation to make decisions. But the real applications are too complex to give the definition of utility in real application. Other solutions are rule-based reasoning and precedent-based reasoning [19].

The aim of negotiations is to reach a mutually beneficial solution. This is achieved through repeated bilateral interactions on concessions and the search for new alternatives. The negotiation process can take many different forms, such as auctions, contract-network style protocols and argumentation. Negotiation protocols are a set of rules that govern the interaction. These include the allowed types of participants (e.g., negotiators and relevant third parties), negotiation states (e.g., bids accepted, negotiation completed), events that cause state transitions (e.g., no more bidders, bid accepted), and the actual actions of participants in particular states (e.g., what can be sent by whom, to whom, and when). The objects of negotiation are the range of issues on which agreement is to be reached. These may be single issues such as price, or multiple issues related to price, quality, timing, etc. Agent reasoning models provide the decision-making mechanism by which participants attempt to achieve their goals. The complexity of the model is determined by the protocol used, the nature of the negotiation object and the set of operations that can be performed on it [17].

Since agents have autonomy and do not necessarily act in the interests of others, they need to influence other agents to direct their actions in a particular way. In this context, negotiation plays a key role in managing inter-agent relationships. Negotiation can take several forms, including auctions, protocols similar to a contract network, and argumentative discussions. In the case of auctions, two main interaction models can be distinguished: "one-to-many", where one agent organises the auction and many other agents participate in the bidding, and "many-to-many", where many agents initiate the auction and also many agents participate in the bidding. The first step in choosing an auction protocol is to determine the appropriate interaction model. Agents then need to devise a strategy for participating in the auction. There are several types of one-to-many auctions, such as the English auction, the closed first-price auction, the closed second-price auction (Vickrey auction), and the Dutch auction [20]. The aim is to identify ways in which agents can negotiate to reach an agreement, ways in which they can allocate their tasks in a way that is beneficial to all, ways in which they can find actions that enable change and serve the agents' goals, and ways in which they can decide according to the maximum utility that agents derive from the states.

The multi-agent system can be used as a virtual environment for selecting project management methods and tools. The system models management processes and roles in a project. Activity agents and resource agents represent activities and resources in the project. Service agents are automated experts for specific project management tasks. Action planning and resource conflict resolution methods are modelled by messaging and negotiation between agents. A project management tool uses a network of agents to provide services to distributed team members.

The agent-based approach replaces the central database and control computer with a network of agents, each with a local view of its environment and the ability and authority to respond to that environment locally. The overall performance of the system is not planned globally, but is achieved through the dynamic real-time interaction of the agents. Knowledge is built into the agent to deal with uncertainty and computational complexity. The architecture of the MAS system affects the patterns of information exchange and the relationships between individual agents. In most cases, physical objects such as a company, a workplace, a person are used as agents. An agent inherits all the functions of a physical object. A project schedule can be defined by interactively defining dependencies. This method provides a more flexible way of scheduling. The activity agent can select the most convenient predecessor to provide its input data.

For example, two agents may be competing for a resource, or outputs and inputs may not be fully matched. The agents then negotiate with each other to solve the problem. A resource agent is controlled by sending commands to it. Resource allocation is a typical project management problem. In traditional studies, resource conflicts arise whenever the total resource requirements for planned activities in a given period exceed the resources available in that period. In a mathematical formulation, the total resources are the constraint. A resource conflict occurs whenever a single source is needed for more than one activity at the same time. Resource conflicts are resolved through negotiation between agents. The goal of negotiation is to minimise the duration of the project.

MAS allows planning problems such as TCPSP (Time Constrained Project Scheduling Planning) and RCPSP (Resource Constrained Project Scheduling Planning) to be considered as project complexity increases and the planning problem becomes more complex.

Minimising project duration is often used as the objective function, taking into account other objectives such as minimising total project cost and balancing resource utilisation [18]. A trade-off between cost and project duration is usually required. An agent-based Time Constrained Project Scheduling Problem (TCPSP) approach augmented with cost optimisation criteria can help to solve this problem.

An agent that always tries to optimise the relevant performance metric (expected utility) is called a rational agent. This definition of a rational agent is quite general and can include human agents (with eyes as sensors, hands as actuators), robotic agents (with cameras as sensors, wheels as actuators), or software agents (with a graphical user interface as sensor and actuator). Agents coexist and interact with other agents in various ways. The relevant subfield of AI, which deals with the principles and design of multi-agent systems, is called Distributed Artificial Intelligence [21].

The knowledge base is very important for modelling the behaviour of the agent. Its behaviour is influenced by perception and belief. Perception uses a short-term memory that all agents can access to share their information about the environment.

The process of identifying beliefs becomes more complex by extracting information from perception using algorithms and matching it with knowledge patterns stored in the knowledge base. The filtering functionality is represented by constraints on the agent's goal and influences its behaviour as a constraint and guiding solution search space. An anticipation algorithm between intentions and actions is also considered, which uses knowledge templates to predict the environment and the possible course of action, thus providing a rational feature for the planning agent. Another link is established between perception and expectation, as the action-response mechanism is a feasible method to adjust the agent's behaviour and maintain feedback control.

In multi-agent control of technical systems, each component of the system is considered as an intelligent object with its own database and knowledge. This object can adapt to unpredictable situations and is able to operate in an environment with obstacles.

The multi-agent approach has a number of significant advantages that can make its application reasonable at all stages of project activity management. In Russian scientific sources there are examples of the use of a multi-agent system at the planning stage of enterprise management, while in foreign sources there are examples of the wider use of multi-agent systems (in enterprise resource management). It is assumed that MAS will be able to solve more effectively the problems currently solved by methods of mathematical modelling, linear programming, etc.

At the level of portfolio management, it is proposed to use agents at the stages of categorisation, prioritisation, balancing and planning to make flexible changes in the process of portfolio implementation, optimal reallocation of resources between portfolio projects, implementation of changes in the portfolio due to changes in strategy.

In the task under consideration, the dialogue between agents is organised through negotiation. Negotiations help to reach agreements between the agents of the system, when each defends its interests, which leads to a situation favourable to all. Two negotiation mechanisms are used—auction and contractual network.

The auction mechanism is used at the stage of selecting projects for inclusion in the project portfolio. The project with the maximum value of the multi-criteria evaluation wins.

At the portfolio formation stage:

- Candidate project agents submit project parameters to the Corporate Agent.
- The corporate agent analyses the parameters of the projects applying for inclusion in the portfolio, categorises and prioritises the projects and communicates decisions to the project agents as to whether the project will be included in the portfolio.

The prioritisation of projects is usually based on a single profit-oriented criterion rather than on multiple criteria, both quantitative and qualitative. Another limitation of current project prioritisation practices is the deterministic approach.

Figure 3 presents the proposed model of IT support for project activities at the portfolio management level, showing the relationship of IT systems, their services and related business processes, and the place of MAS in project portfolio management.

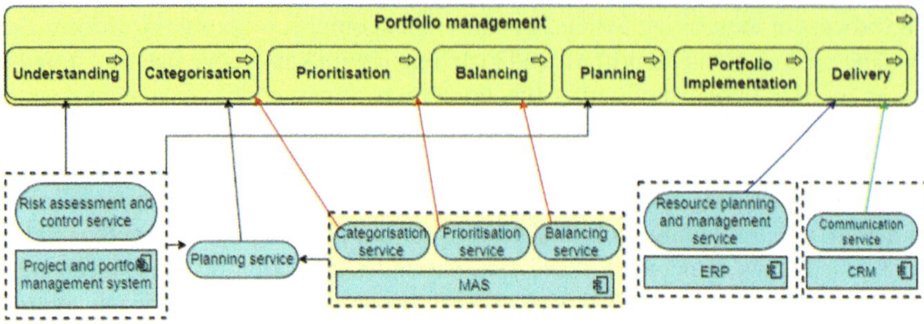

Fig. 3 A model of IT support for project activities at the portfolio management level

The proposed decision support apparatus, based on agent-based modelling and multi-agent systems, should ensure the formation of a portfolio of projects in the energy company that is optimal at each point in time, taking into account changes in the external environment.

5 Conclusion

This paper presents a variant of the implementation of energy project portfolio management using multi-agent systems (MAS). This approach, based on the interaction of independent software agents, can be used in the future for similar tasks in other related areas and at all levels of project management.

MAS show significant potential in the management of complex projects due to their ability to adapt to dynamic conditions and to optimise decision making. In the field of energy project management, the use of MAS promises significant improvements in efficiency, flexibility, planning and process transparency. The implementation of such a system will improve the competitiveness and success of project implementation. The agent-based approach allows many variables and conditions to be taken into account, which is particularly important in a dynamically changing energy market. One of the key aspects discussed in the paper is the use of multi-agent systems to optimise resource allocation and risk management in energy projects. In addition, MAS facilitate faster and better decision making by automating data analysis and forecasting processes.

Acknowledgements The research was supported by Russian Science Foundation grant No. 24-28-01796, https://rscf.ru/project/24-28-01796/.

References

1. Bente, S., Bombosch, U., Langade, S.: Collaborative enterprise architecture: enriching EA with lean, agile, and enterprise 2.0 practices. Newnes (2012)

2. Ilin, I., Levina, A., Borremans, A., Kalyazina, S.: Enterprise architecture modeling in digital transformation era', in Energy Management of Municipal Transportation Facilities and Transport, Springer, 124–142 (2019)
3. Wetering, R., Mikalef, P., Pateli, A.: A strategic alignment model for IT flexibility and dynamic capabilities: toward an assessment tool (2017)
4. Borremans, A.D., Zaychenko, I.M., Iliashenko, O.Y.: Digital economy. IT strategy of the company development. In: MATEC Web of Conferences, EDP Sciences, 01034 (2018)
5. Glukhov, V.V., Ilin, I.V., Levina, A.I. Project management team structure for internet providing companies. In: Internet of Things, Smart Spaces, and Next Generation Networks and Systems: 15th International Conference, NEW2AN 2015, and 8th Conference, ruSMART 2015, St. Petersburg, Russia, August 26–28, 2015, Proceedings 15, Springer, pp. 543–553 (2015)
6. Kaiser, M.G., El Arbi, F., Ahlemann, F.: Successful project portfolio management beyond project selection techniques: understanding the role of structural alignment. Int. J. Project Manage. **33**(1), 126–139 (2015)
7. Jordan, A.: Risk management for project driven organizations: a strategic guide to portfolio, program and PMO success. J. Ross Publ. (2013)
8. Perry, M.P., Perry M.: Business driven project portfolio management: conquering the top 10 risks that threaten success. J. Ross Publ. (2011)
9. Unger, B.N., Gemünden, H.G., Aubry, M.: The three roles of a project portfolio management office: their impact on portfolio management execution and success. Int. J. Project Manage. **30**(5), 608–620 (2012)
10. Wagner, R.: Best practice standards for portfolio management. In: The Handbook of Project Portfolio Management, Routledge, pp. 63–72 (2018)
11. Storch de Gracia, M.D., Moya Perrino, D., Llamas, B.: Multicriteria methodology and hierarchical innovation in the energy sector the project management institute approach. Manag. Decis. **57**(5), 1286–1303 (2019)
12. Ilin, I.V., Iliashenko, O.Y., Schenikov, E.M.: An approach to forecasting the structure of energy generation in the age of energy transition based on the automated determination of factor significance. Energies **17**(1), 68 (2023)
13. Fernando, Y., Bee, P.S., Jabbour, C.J.C., Thomé, A.M.T.: Understanding the effects of energy management practices on renewable energy supply chains: Implications for energy policy in emerging economies. Energy Policy **118**, 418–428 (2018)
14. Ilin, I., Laskin, M., Logacheva, I., Sarygulov, A., Tick, A.: Land plots evaluation for agriculture and green energy projects: how to overcome the conflict using mathematics. Mathematics **10**(22), 4376 (2022)
15. Halaška, M., Šperka, R.: Is there a need for agent-based modelling and simulation in business process management? Organizacija **51**(4), 255–269 (2018)
16. Dorri, A., Kanhere, S.S., Jurdak, R.: Multi-agent systems: a survey. IEEE Access **6**, 28573–28593 (2018)
17. Srinivasan, D.: Innovations in multi-agent systems and application—1, p. 310. (Springer 2010)
18. Bakhtizin, A., Ilin, I., Nikitin, N., Ershova, A., Esser, M.: Multi-agent approach in planning and scheduling of production as part of a complex architectural solution at the enterprise. In: Algorithms and Solutions Based on Computer Technology: 5th Scientific International Online Conference Algorithms and Solutions based on Computer Technology (ASBC 2021), Springer, pp. 369–380 (2022)
19. Vlassis, N.: A concise introduction to multiagent systems and distributed artificial intelligence. Springer Nature (2022)
20. Terán, J., Aguilar, J.L., Cerrada, M.: Mathematical models of coordination mechanisms in multi-agent systems. CLEI Electron. J. **16**(2), 5 (2013)
21. Bond, A.H., Gasser, L.: Readings in distributed artificial intelligence. Morgan Kaufmann (2014)

Energy System Architecture Incorporating the Internet of Energy Solution

Tatiana Tarasova, Anastasia Levina, Egor Shchenikov, and Manfred Esser

Abstract The article provides an analysis of the concept of the Internet of Energy: the structural elements of the Internet of Energy system, the main components of the architecture and the main distinctive features of this concept. The energy system created within the framework of the Internet of Energy concept makes it possible to successfully implement a distributed energy model in which two-way information exchange is carried out between users. Projects for the implementation of decentralized energy systems based on the architectural model of the Internet of Energy, created in Russia, the USA, and Singapore and others are considered, including the issues and prospects for their development.

Keywords Internet of energy · Digitalization · Microgrid · Distributed energy

1 Introduction

One of the urgent problems characteristic of traditional energy systems is the low efficiency of these systems in modern conditions. This problem is the driver for the emergence of new models of energy architecture. One of these models is the distributed energy model, which differs from the traditional one in that most of the generating, load control and storage capacities are located not in centralized facilities remote from end consumers receiving a unidirectional flow of electricity through trunk and distribution networks, but on the side of system users and distribution networks. However, when trying to integrate the distributed energy model into a centralized system, significant problems arise: a decrease in the overall reliability of the energy system, an increase in various costs, the emergence of technological

T. Tarasova · A. Levina (✉) · E. Shchenikov
Peter the Great St. Petersburg Polytechnic University, Saint-Petersburg, Russia
e-mail: levina_ai@spbstu.ru

M. Esser
GET Information Technology Group, Grevenbroich, Germany

I. Ilin and M. Youzhong (eds.), *Digital Systems and Information Technologies in the Energy Sector*, Lecture Notes in Networks and Systems 1244,
https://doi.org/10.1007/978-3-031-80710-7_20

barriers that impede the effective interaction of system entities. In view of this, for the successful implementation of the distributed energy model, in which there is a two-way energy and information exchange, it is necessary to create an appropriate architecture that assumes a decentralized version of the construction and management of energy systems. Digitalization plays a decisive role in the transition to distributed energy systems, making it possible to intelligently manage the energy system using various digital solutions. Thanks to digitalization, it is possible to reduce costs during the operation of the energy system and achieve its efficiency and stability. The purpose of this article is to analyze the architecture of a distributed energy system that meets modern energy requirements, created on the basis of the concept of the Internet of Energy. To achieve this goal, the following tasks were set: analysis of the prerequisites for the emergence of the Internet of Energy concept; analysis of the main elements of the Internet of Energy architecture; review of real solutions created by various companies based on the concept of the Internet of Energy.

2 Materials and Methods

To solve the set tasks, the following algorithm was defined: analysis of literary sources on the topic of distributed energy and the Internet of Energy; identification of requirements for modern energy systems; definition of elements of the Internet of Energy model and their functions; analysis of solutions implemented by various companies, created on the basis of the Internet of Energy concept.

- Analysis of literary sources. At this stage of the study, an analysis of the current state of theory and practice, achievements in the field of creating effective energy systems based on the concept of the Internet of energy was carried out.
- Identification of requirements for modern energy systems. Based on the analysis of the prerequisites for the transition to distributed energy models and the problems that these systems must solve, requirements for energy systems that meet the challenges of the world energy at present were identified.
- Definition of elements of the Internet of energy model and requirements for them. Based on the analysis of sources at this stage, the main elements (subsystems) of the Internet of energy and the functions they perform were determined. A model of the architecture of the Internet of energy based on the concept of enterprise architecture was compiled.
- Analysis of implemented solutions of various companies.

As part of the analysis of the materials, various sources describing the concept of the Internet of Energy, the prerequisites for its emergence and options for its implementation were considered.

Farhan in the work "Towards next generation Internet of Energy system: Framework and trends" considers the transition from the traditional energy system to the Internet of Energy by integrating intelligent infrastructure, smart metering and energy consumption management schemes into the energy system. This work examines in detail the concept of the Internet of Energy from the point of view of technical aspects and technologies. Security issues in energy management that can affect the performance of the Internet of Energy system are addressed, a structure for creating the architecture of the next generation of the Internet of Energy that meets high cybersecurity requirements is developed [1].

Ghiasi in the work "Evolution of smart grids towards the Internet of energy: Concept and essential components for deep decarbonization" the authors provide an overview of the evolution of smart grids in combination with the Internet of energy, and also conduct an analysis of the main components of the Internet of energy used for decarbonization. This work also presents mathematical models for assessing the role of the Internet of energy in the processes of reducing carbon dioxide (CO2) emissions into the environment [2].

Eisele in the work "Blockchains for Transactive Energy Systems: Opportunities, Challenges, and Approaches" consider a blockchain-based transaction energy system that ensures the security and confidentiality of smart contracts within the energy transaction market. Mishra P., Singh G. in the work "Energy Management of Sustainable Smart Cities Using Internet-of-Energy" consider various energy management strategies, analyze the possibility of extending the service life of devices that require low energy consumption, and also provide options for effective and sustainable solutions for smart cities based on the Internet of Energy [3].

Velmozhina et al. in the study "Review of modern strategies for the development of hydrogen bioenergy as key areas for achieving sustainable development goals" provide an overview of current trends and prospects in the field of hydrogen bioenergy, contributing to the creation of effective solutions for achieving sustainable development [4].

Ilin et al. in the article "Enterprise Architecture Analysis for Energy Efficiency of Saint Petersburg Underground" describe the project developed by the authors for reengineering the processes of the Saint Petersburg Metro, which is aimed at complying with the requirements of energy legislation in the company's activities. The work involves reengineering processes using enterprise architecture analysis [5].

Ilin et al. in the work "Use of renewable energy and tax burden on CO2 emissions in industrial enterprises" demonstrate the results of studies on the assessment of the specific share of renewable energy facilities in the energy balance of the Russian Federation. The work describes 3 options for implementing a carbon dioxide (CO2) emission taxation system for industrial enterprises [6].

Shemyakina et al. in the work "Architecture of the management system for hydrogen production at hydropplications" present various models of the architecture of the hydrogen production system at a hydroelectric power plant during a power outage, which are implemented through flexible management of reserve capacities using the Internet of Things and machine learning models [7].

Poljanskihh et al. in the study "Investment in renewable energy: Practical case in Estonia" describe an example of how the government can use various strategies to support "green" initiatives. The report considers the example of Estonia and the feed-in tariff model implemented in this country [8].

Anisiforov et al. in the article "Organizational and economic changes in the development of enterprise architecture" describe the synchronization of strategic and operational business tasks with the development of enterprise information systems. In the work, the authors touch upon the problems of developing the architecture of systems in the context of a constantly changing landscape of business processes [9].

3 Results

3.1 Definition of Requirements for Modern Energy Systems

Energy systems based on traditional centralized architecture have almost exhausted their efficiency potential. Such systems are not capable of responding to the current demands of global energy supply: increasing demand for electricity, changing qualitative characteristics of demand; compliance with environmental protection requirements [1, 10].

The following requirements can be identified for an energy system that meets the current situation in the energy sector:

- The distributed nature of the energy system, in which there are many producers and consumers of energy;
- The existence of information links that ensure interaction between electric power equipment;
- The ability to change the role of an entity in the energy system, thanks to which it can provide various services;
- Implementation of decentralized intelligent control of energy networks;
- Implementation of processes using transactions between entities, which are carried out within the framework of Internet of Energy services.

The architecture of the Internet of Energy meets these requirements and is aimed at reducing various types of costs that arise when integrating distributed energy elements into existing energy systems:

1. Reducing transaction costs. These costs arise in the process of transactions carried out by energy market participants. The most common transaction costs are the costs of concluding contracts, negotiating between the parties, and monitoring the fulfillment of the parties' obligations. The Internet of Energy makes it possible to reduce the costs of time and other resources for performing these functions by automating these processes.
2. Reducing the costs of information integration. This type of cost includes costs that arise in the process of exchanging information between elements of the energy system. This may include collecting, processing, and transmitting data between energy generation sources, consumers, and operators. The Internet of Energy makes it possible to significantly reduce these costs by standardizing

and unifying data exchange protocols, while simultaneously maintaining high accuracy of information exchange.
3. Costs of integrating physical equipment into the network. These include the costs of upgrading the technological infrastructure required to adapt the equipment for its integration into the distributed energy system: the costs of setting up the equipment, supporting the compatibility of different technologies. The Internet of Energy allows for the use of a flexible and adaptable infrastructure that facilitates the integration of new energy sources into the system and minimizes the associated costs [11].

Reducing the costs of distributed energy generation sources, including microgeneration and the use of renewable energy sources, turns consumers into active participants in energy exchange, who can both reduce their consumption of electricity from the grid and transfer excess energy to other network participants.

3.2 Architecture of the Internet of Energy

Users of the Internet of Energy can provide each other with various services: electricity sales, equipment rental, remote use, capacity reserve provision, etc. Interaction of users and provision of services is carried out through energy transactions. An energy transaction is understood as a set of operations performed by Internet of Energy entities and their equipment among themselves and with service operators, during which coordinated control of equipment operating parameters occurs, due to which one participant in the transaction receives the fulfillment of its technical demand, and the other receives payment [12, 13].

A structural unit of the energy system that is controlled using the Internet of Energy is usually called an energy cell. An energy cell is a group of users concentrated in a local area and a set of their energy equipment connected to distribution electric networks. A distinctive feature of the cell is the presence of a single interface for connecting to the Internet of energy, due to which it acts as a single subject of energy transactions.

There are different types of energy cells:

- Prosumer (can be both a consumer and a producer of electricity);
- Maneuverable generator (only generates electricity, i.e. the energy flow is only outgoing, uses traditional sources such as thermal power plants, diesel generators, etc.)
- Stochastic generator (only generates energy, using unstable sources—renewable energy sources);
- Active consumer (consumes energy and can manage its energy consumption);
- Passive consumer (consumes energy without the ability to manage energy consumption) [14, 15].

If the subject has the appropriate equipment, it can change its role, therefore, in the conditions of the Internet of Energy, the roles of energy cells can change chaotically.

The participation of an energy cell in the Internet of Energy is carried out by means of a software agent with which it is supplied. With the help of a software agent, the energy cell interacts with other cells, participates in the services of the Internet of Energy.

The user is also a subject of the Internet of Energy. It can be the owner of the energy cell or the operator (supplier) of services and applications. The user of the Internet of Energy can have several energy cells, and in this case, several software agents work for his needs [16, 17].

The architectural model based on the concept of the Internet of Energy is presented in Fig. 1. To build the model, the methodology of enterprise architecture was used, which allows reflecting both the landscape of business processes and the software and hardware components within the framework of one model [5, 9].

The Internet of Energy includes a multi-agent system of software agents of energy cells, as well as three subsystems, which will be discussed below [18].

Transactive Energy (TE) System: a system that provides data on energy transactions and contractual agreements of the parties. Software agents of energy cells

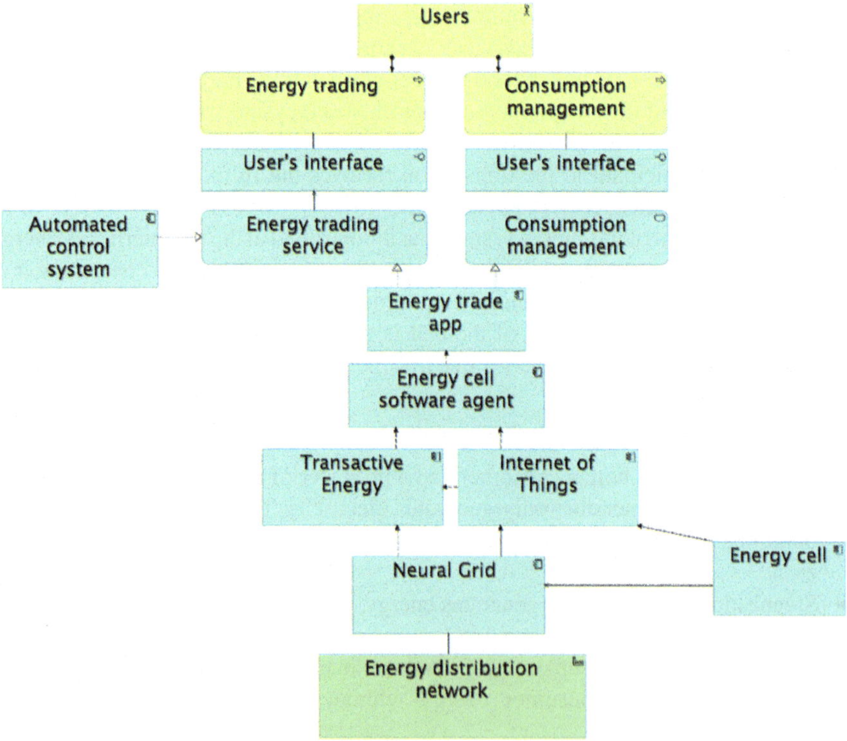

Fig. 1 Architectural model of the internet of energy

interact with TE to conclude smart contracts, verify their execution and financial settlements.

The transaction system manages the commercial parameters of energy transactions. This system provides automated creation and secure storage of smart contracts that record transaction parameters. TE also offers control over the execution of contract terms, financial settlement of fulfilled obligations of the parties, recording of disagreements, use of data for their resolution and performance of clearing functions.

A smart contract is a computer algorithm containing the terms of the contract and designed for automatic control and execution of contractual relations of the parties. The terms and obligations of the contract are described in a programming language, the program is launched on a trusted computing system. Also, for the implementation of transactions, blockchain technology is partly used, providing a high level of confidentiality and security [19].

The capabilities of the TE system are as follows:

1. Creation, recording and secure storage of data on energy transaction parameters.
2. Ensuring the operation of digital wallets of Internet of Energy users and secure storage of their personal data and financial information.
3. Verification of the fulfillment of obligations of the parties to energy transactions based on data obtained from the IoT system.
4. Ensuring invoicing and transaction processing, including instant transfer of funds when the recorded transaction conditions are met.

Internet of Things (IoT) system: a machine-to-machine interaction system that provides communication between software agents of energy cells and intelligent control systems. It also connects Neural Grid (NG) system components with energy cell agents and service applications [20].

The Internet of Things (IoT) is a system that implements the information component of energy transactions, in which machine-to-machine interaction and exchange of data and control signals required for energy transactions occurs between energy cells and their software. The Internet of Things provides software agents with access to local device automation systems and/or local energy cell control systems, as well as to the creation, storage, and deployment of digital twins and shadows of cells and their devices. In addition, the system supports the operation of auxiliary software for implementing machine-to-machine interaction [3, 12].

Distributed power regulation system—Neural Grid (NG). This component of the Internet of Energy controls the flow of electricity in the power grid, ensures static and dynamic stability of processes, and performs process protection functions. NG components are connected to software agents of energy cells and service applications via the IoT system to receive data on smart contract parameters and generate power flow values that need to be transmitted. Machine learning technologies make it possible to flexibly manage power reserves [7, 13].

NG is a system that implements the physical (electrical) component of energy transactions. It provides decentralized power balance management, manages power flows in the system to create the electricity transmission mode necessary to fulfill the terms of the smart contract, and maintains static and dynamic stability of the system during the simultaneous execution of several transactions. In addition, the system supports the integration of individual types of elements in the power grid.

The capabilities of the NG system include:

1. Maintaining static and dynamic stability of the power system during energy operations and simultaneous flow of a large number of energy transactions.
2. Application of necessary effects on the mode of electric power transmission to implement energy transactions.
3. Control of the balance of active and reactive power in the power system or power unit to ensure energy transactions.
4. Ensuring the quality of electric power required by the consumer.

The considered systems provide information and financial interaction between energy elements, service applications, operators and users, and also provide management of the energy system, maintaining efficient and sustainable energy supply.

The basis for the functioning of the Internet of Energy is energy transactions between energy cells, which occur through the interaction of software agents of cells with Internet of Energy service applications, as well as with each other.

Key elements of energy transactions:

1. Electricity—a product produced by energy equipment and supplied to consumers in certain volumes and with the quality characteristics required by a specific consumer within the framework of their request.
2. Power—the ability of energy elements to provide a given power level under certain conditions.
3. Flexibility—the ability of energy cells to influence the parameters of the electricity transmission mode upon request or under certain conditions provided as a service or product.

The Internet of Energy provides two main types of energy transactions that differ in their participants and are the main scenarios performed within the architecture of the Internet of Energy:

1. Peer-to-peer (P2P) transactions—transactions carried out between two energy cells and involving a coordinated and synchronized change in the parameters of electricity generation and consumption. The application of the corresponding service acts as an electronic platform and provides a set of rules and algorithms for the interaction of software agents of the cells.
2. Operator (P2O) transactions—occur between the energy cell and the Internet of Energy service operator, which can act as a service provider for the cell or as an aggregator of services provided by cells to energy system entities that are not part of the Internet of Energy (e.g. large generators, grid companies, system operator). The service operator acts as an intermediary between the centralized energy grid and the Internet of Energy, and the service application interacts with the cell's software agent to generate and execute the transaction.

First, this is the Transactive Energy (TE) system: a system that provides data on energy transactions and financial contracts. Software agents of energy cells interact with TE to conclude digital contracts, verify their execution and financial settlements.

Second, the Internet of Things (IoT): a machine-to-machine communication system that provides communication between software agents of energy cells and control systems. It also connects the Neural Grid (NG) system components to the energy cell agents and service applications. Third, the distributed frequency and power balancing system—Neural Grid (NG). It controls the power flow in the power grid, ensures static and dynamic stability, and performs process protection functions. NG components connect to the energy cell software agents and service applications via the Internet of Things system to receive data on contract parameters and generate aggregate power flow values.

These systems provide information and financial interaction between energy elements, service applications, operators and users and ensure optimal management of the energy system, thereby ensuring the efficiency and sustainability of energy supply.

3.3 Projects Review

The architecture of the Internet of Energy has been implemented in various projects launched in real conditions by large companies in different countries of the world. Let's consider some of them.

One of the first companies to provide a real solution for implementing the Internet of Energy was General Electric (GE). The company launched a startup called Current, aimed at transforming the energy sector. Despite the fact that Current was declared a startup in the company's official press release, it is an association of several companies producing energy storage devices, LEDs, and electric vehicles.

Current's goal was to provide the company's clients (commercial and industrial enterprises, municipalities, universities, hospitals, and others) with energy that has individual, necessary quantitative and qualitative characteristics. Current gives users the opportunity to manage their energy consumption: reduce and optimize it, as well as redistribute surplus generated electricity [21].

To achieve this goal, the company developed the Predix cloud platform. This platform is designed for analytical processing of large data sets, which enables the analysis of indicators on an industrial scale.

The Predix platform allows you to manage all stages of electricity production: from managing various applications of the energy system to equipment and control devices used by the power plant personnel. This platform allows you to automate many processes that were previously carried out manually. The main competitive advantage of Current is the continuous optimization of energy consumption. Predix is used to find new ways to effectively redistribute energy, optimize energy consumption, generate electricity, monitor power, and, as a result, receive new sources of income.

Another project that implements the architecture of the Internet of Energy is Transactive Grid (USA). It is a joint development of Brooklyn Microgrid, LO3 Energy and ConsenSys. This project is interesting because it uses a software platform based on blockchain technologies to implement energy transactions.

Blockchain technology has the following main characteristics: decentralization, transparency, automation of smart contracts and transaction tracking. As a result, this technology eliminates the need to rely on third parties when conducting energy transactions, guarantees the authenticity and reliability of data, and reduces transaction costs in general. The technical characteristics of the blockchain are very consistent with the idea of the energy Internet, which will potentially make blockchain one of the important technical solutions in the energy Internet in the future [6, 22].

Transactive Grid is 5 houses located on one side of the street and generating electricity, and 5 houses on the opposite side that buy this surplus electricity that is not consumed by the energy-producing houses. This project also uses generation based on renewable energy sources, so the producing houses are equipped with solar panels, and the houses of all project participants are equipped with sensors linked to a software platform to monitor the amount of energy produced and consumed and manage energy transactions between houses. In general, this model of interaction between houses is one of the possible variations of building smart cities [23].

To participate in the Brooklyn Microgrid, one must register as a consumer in the app. In order for the app user to purchase and buy local electricity, they must install a TransActiveGrid Generation 2 (TAG-e G 2) smart electricity meter and computing device. The TAG-e G 2 connects to the internet via Wi-Fi or Ethernet and allows data to be transmitted to the marketplace blockchain. The meters measure both energy consumption and the difference between the energy consumption and the generation of electricity by a specific power cell. Another platform that operates on blockchain technology is GridExchange in Canada. This platform facilitates the exchange of energy between utilities and their customers. The platform allows utilities to leverage customer-owned distributed energy resources to meet grid needs and reduce greenhouse gas emissions. The aim of the project was to demonstrate the value of distributed energy resources, allowing energy exchange participants to provide energy services to the electric grid using a blockchain-based transaction software platform that enables near real-time management of distributed energy resources. Such a platform allows for the transformation of energy markets and the reduction of greenhouse gas emissions [22].

The pilot project of the Internet of Energy, launched by Nanyang Technological University in Singapore, is called REIDS. This project is a platform for the design, implementation and testing of cost-effective energy solutions for countries in Southeast Asia. This platform implements elements of the Internet of Energy architecture and monitors the operation of microgrids in conditions as close to real as possible. This project uses generation based on renewable energy sources, and microgrids are connected to each other by a single network in a cluster, thanks to which they can exchange energy with each other as part of energy transactions [24].

The partners of the REIDS project, who place their microgrids on it, are major players in the global technology market, such as Rolls Royce (Great Britain), General

Electric (USA), EDF (France) and others. The advanced digital technologies used at the REIDS demonstration site are attracting the attention of large corporations and make the possibility of implementing an Internet of Energy architecture extremely interesting for them.

One of the problems that this project faced, which served as the beginning of the study of the possibility of implementing various energy transaction scenarios, was the imbalance between energy production and consumption. During the daytime, the energy production of solar panels installed on the roof of the workshop was almost twice the consumption. The installed storage systems were not enough to store excess energy, which could be used at night. As a result, at night the workshop was forced to use a diesel generator to ensure its operation. To solve this problem, a service for sharing equipment (solar panels and storage devices) was created, functioning due to the implementation of energy links between microgrids and the possibility of implementing energy transactions. The first project in Russia, which represents a unique complex cyber-physical model for testing the Internet of Energy technology, is the "Testbed "EnergyNet" from the NTI Competence Center based at the National Research University "MPEI". The capabilities of this testing ground are significantly greater than those of other Russian laboratories due to the implementation of Power-Hardware-in-the-Loop technology, which allows for control based on mathematical modeling in real time [25].

The unique equipment presented at the site allows for the creation of cyber-physical models of microgrids, one part of which is simulated in real time in a digital simulator, while the other is a real electrical network.

A solution based on the concept of the Internet of Energy was proposed by the US startup Leap, focused on distributed energy resources. Leap provides a platform to which the user can balance power within the network and access wholesale electricity markets via API. Thus, users can make transactions on global energy markets and manage their consumption, in particular, reduce it during peak hours. The platform unites energy market participants in a virtual power plant. The modular design of the solution allows for easy scaling both in terms of geographic location and by expanding the range of services provided (generation and various ancillary services) [26].

A startup from Slovenia, 2GG, is developing smart meters necessary for the implementation of the Internet of Energy. The startup aims to create a Smart-Metering-as-a-Service (SMaaS) solution that provides detailed energy consumption data to companies supplying electricity, utilities, and commercial customers. Through a secure communication node, the data is transmitted to a cloud server, which provides analysis and storage. Specialized analytical software allows for demand forecasting, leakage monitoring and prevention, supply optimization, and production efficiency management [27, 28].

A comparative analysis of the solutions considered by various criteria is given in Table 1.

Thus, the projects considered provide prospects for improving current and searching for new options for the effective implementation of the energy internet architecture: elaboration of various scenarios of energy transactions, creation of new

Table 1 Comparative analysis of projects for the implementation of the internet of energy

No	Name	Country	Technologies	Software	Partners
1	Current	USA	Cloud technologies, big data analysis	Predix	Walgreens, Simon property group, Hilton worldwide, Jpmorgan chase, hospital corporation of America (HCA), intel, trane
2	Transactive Grid	USA	Blockchain, smart meters	Brooklyn microgrid	Brooklyn microgrid, LO3 Energy и ConsenSys
3	GridExchange	Canada	Blockchain, smart meters	–	Sunverge, savage data, FLO
4	REIDS	Singapore	Microgrids	–	Rolls Royce, general electric, EDF
5	Testbed	Russian Federation	Control based on mathematical modeling	–	INBRES LLC, JSC Gidroelectromontazh, PELSi technology LLC
6	Leap	USA	API	Leap	Stem, Gridpoint, Sunnova and 80 + others
7	2GG	Slovenia	Cloud technologies, big data analysis	2GG	Public utility providers and commercial customers-you

services within microgrids, search for new technologies for connecting users and equipment. Also, based on the analysis, it can be concluded that most companies that have implemented the energy internet use their own platform and software developments, since at the moment there are no standard solutions for these purposes on the market. Startups such as Leap, for example, are able to cover only part of the tasks necessary for the successful implementation of the energy internet architecture.

4 Discussion

The concept of the Internet of Energy emerged in the wake of the energy transition, which is defined by three key aspects: decarbonization (transition to carbon–neutral energy), decentralization (distribution of energy production), and digitalization (implementation of digital technologies in the energy sector) [2, 4].

In particular, decarbonization promotes decentralization, since many types of renewable energy sources (RES), such as solar and wind power plants, do not require the concentration of energy, which reduces network costs [6].

Those aspects of energy transitions are also important for integrate Internet of Energy in smart cities. Although fossil fuels and renewable energy sources play a crucial role, environmental considerations take precedence in the development of smart cities, making renewable energy sources the primary and most viable choice to meet future energy needs. For policymakers and urban planners involved in the development of smart cities, the key priority should be to promote abundant renewable energy sources [29].

Despite the fact that the architecture of the Internet of Energy is designed to reduce the harm caused to the environment by reducing carbon dioxide (CO2) emissions and other greenhouse gases into the atmosphere, the question of the negative impact of the Internet of Energy on the environment remains open. This is due to the fact that the operation of data centers is also detrimental to the environment. Data centers are an integral part of the Internet of Energy, as they process and store huge amounts of data generated by smart devices and services. Data centers require significant amounts of energy to power servers, cooling systems, backups, and other components. With the growth of data volumes produced by various components of the Internet of Energy infrastructure, energy consumption by these centers is only increasing. According to expert estimates, the amount of energy consumed by data centers reaches 3% of the total global electricity production, and this figure continues to grow [30]. In this case, if data centers operate on traditional energy sources, this leads to an increase in greenhouse gas emissions into the Earth's atmosphere and the acceleration of climate change. Many countries solve this problem through government support for "green" initiatives, stimulating companies to reduce the level of negative impact on the environment.

Moreover, data center equipment requires regular updating. This leads to the formation of a large amount of waste, potentially containing hazardous substances and chemicals. Failure to comply with the conditions for the proper disposal of this waste leads to pollution of water bodies and soil, subsequently negatively affecting the state of the environment and the health of people living in the regions where waste dumps are formed. To solve the problem of high levels of CO2 emissions into the atmosphere, various options for a state taxation system can be proposed, as well as the development of an energy legislation system that will allow influencing enterprises that pollute the environment, however, not all countries in which large data centers are concentrated are considering these tax policy options [3, 8].

To build the architecture of the Internet of Energy, a high level of technological development of the company is required both in the field of energy equipment and in the software part. Therefore, the Internet of Energy projects implemented to date most often use their own science-intensive developments of platforms and IT solutions.

Thus, the environmental issue requires detailed study at the stage of designing and implementing Internet of Energy systems to ensure a balance between rapidly developing technologies and environmental protection. It is also difficult to implement the Internet of Energy everywhere because there are practically no standard solutions on the market that would allow any company to start using the Internet of Energy. At the moment, it can be observed that companies implementing projects for using the Internet of Energy resort to their own developments of platforms and other software, and standard solutions are able to cover only a small part of the requirements necessary for the successful construction of the Internet of Energy architecture.

5 Conclusion

The concept of the Internet of Energy allows solving many problems typical for traditional energy systems due to a fundamentally new approach to architecture. Carrying out energy transactions in the corresponding decentralized and intelligently controlled architecture allows implementing stable and efficient functioning of distributed energy systems.

Within the framework of this study, the concept of the Internet of Energy was analyzed: distinctive features, general principles of construction and functioning, its structural and architectural elements, the main operations performed in such energy systems. A model of the architecture of the Internet of Energy was compiled, described using the concept of enterprise architecture, allowing us to determine the main components that make up the Internet of Energy. Several of the most well-known projects were also considered, within the framework of which microgrids operating on the basis of the concept of the Internet of Energy were successfully implemented. The analysis showed that the sphere of Internet of Energy projects is promising for finding the most effective options for implementing energy systems, and large global corporations are interested in using these technologies.

Acknowledgements The research was supported by Russian Science Foundation grant No. 24-28-01796, https://rscf.ru/project/24-28-01796/

References

1. Farhan, M., et al.: Towards next generation internet of energy system: framework and trends. Energy AI. Elsevier **14**, 100306 (2023)
2. Ghiasi, M., et al.: Evolution of smart grids towards the Internet of energy: concept and essential components for deep decarbonisation. IET smart grid. Inst. Eng. Technol. **6**(1), 86–102 (2023)
3. Eisele, S., et al.: Blockchains for transactive energy systems: opportunities, challenges, and approaches. Computer **53**(9), 66–76 (2020). https://doi.org/10.1109/MC.2020.3002997
4. Velmozhina, K.A., Politaeva, N.A., Ilin, I.V., Shinkevich, P.S.: Review of modern strategies for the development of hydrogen bioenergy as key areas for achieving sustainable development goals. Int. J. Hydrogen Energy **70**, 729–736 (2024)
5. Ilin, I., Levina, A., Iliashenko, O.: Enterprise architecture analysis for energy efficiency of saint-Petersburg underground. Adv. Intell. Syst. Comput. **692**, 1214–1223 (2018)
6. Ilin, I., Vasilenok, V., Marchenko, R.: Use of renewable energy and tax burden on CO2 emissions in industrial enterprises. E3S Web Conf. **110**, 02102 (2019)
7. Shemyakina, A.A., Levina, A.I., Korablev, V.V., Lepekhin, A.A.: Architecture of the management system for hydrogen production at hydropplications. Int. J. Hydrogen Energy **69**, 1227–1235 (2024)

8. Poljanskihh, A., Levina, A., Dubgorn, A.: Investment in renewable energy: practical case in Estonia. MATEC Web Conf. **193**, 05065 (2018)
9. Anisiforov, A., Dubgorn, A., Lepekhin, A.: Organizational and economic changes in the development of enterprise architecture. Web ES3 Conf. **110**, 02051 (2019)
10. Hussain, H.M., Narayanan, A., Nardelli, P.H., Yang, Y.: What is energy internet? Concepts technologies and future directions. IEEE Access Inst. Elect. Electron. Eng. Inc. **7**(8), 183127–183145 (2020)
11. Bui, N., Castellani, A.P., Casari, P., Zorzi, M.: The internet of energy a web-enabled smart grid system. IEEE Netw. **26**(4), 39–45 (2012). https://doi.org/10.1109/MNET.2012.6246751
12. Thakur, S., Rani, R.: A comparative study of operating system for internet of things. Int. J. Sci. Res. Eng. Manag. **8**(5), 1–5 (2024). https://doi.org/10.55041/IJSREM34917
13. Korablev, V.V., Ilin, I.V., Levina, A.I., Shemyakina, A.A.: Economic assessment of the life cycle of hydrogen production at hydroapplications. Int. J. Hydrogen Energy **68**, 696–704 (2024)
14. Khokhlin, D.: Infrastructure center EnergyNet. Expert and analytical report architecture of the internet of energy IDEA. Version 2.0. Moscow (2021)
15. Chi, L., Zhao, Z., Xia, C., Chang, X.: A case study of developing an intelligent management system for energy internet. In: 2020 IEEE International Conference on Energy Internet (ICEI), Sydney, NSW, Australia, pp. 69–73 (2020). https://doi.org/10.1109/ICEI49372.2020.00021
16. Joseph, A., Balachandra, P.: Energy internet, the future electricity system: overview, concept, model structure, and mechanism. Energies **13**, 4242 (2020). https://doi.org/10.3390/en1316 4242
17. Shahzad, Y., et al.: Internet of energy: opportunities, applications, architectures and challenges in smart industries. Comput. Elect. Eng. Pergamon. **86**, 106739 (2020)
18. Balachandra, A.: Energy internet, the future electricity system: overview, concept, model structure, and mechanism (2020). https://doi.org/10.3390/en13164242
19. Meicheng, L.: Application and prospect of blockchain technology in the energy internet march 2024, n book: China's e-Science Blue Book (2023). https://doi.org/10.1007/978-981-99-8270-7_10
20. Marco, A., Diaz-Martinez, Q.: The internet of things as an essential element for the benefit of the energy sector: a review of the literature. South Afr. J. Ind. Eng. 35(2) (2024). https://doi.org/10.7166/35-2-3004
21. General Electric Homepage, https://www.ge.com/news/press-releases/ge-introduces-startup-current-transform-energy-sector. Last accessed 2024/11/09
22. Lin, G.: Some real-world examples of internet of things (iot) applications in renewable energy projects, Bachelor's Programme in Electrical Engineering (Double Degree, in co-operation with Hebut), Bachelor's thesis (2024)
23. Government of Canada, Current investmernts, https://natural-resources.canada.ca/science-and-data/funding-partnerships/opportunities/current-investments/the-transactive-grid-ena bling-end-end-market-services-framework-using-blockchain/22137, Last accessed 2024/11/09
24. Official website of Nanyang Technological University, https://www.ntu.edu.sg/erian/research-capabilities/renewable-energy-integration-demonstrator---singapore, Last accessed 2024/11/09
25. Testbed homepage, http://nti.mpei.ru/home/testbed/, Last accessed 2024/11/09.
26. Leap Homepage, https://www.leap.energy/, Last accessed 2024/29/09
27. Startup Insights, Top 5 internet of energy startups impacting the idndustry, https://www.startus-insights.com/innovators-guide/5-top-internet-of-energy-startups-impacting-the ind ustry/, Last accessed 2024/29/09
28. Mishra, P., Singh, G.: Energy management systems in sustainable smart cities based on the internet of energy: a technical review. Energies **16**, 6903 (2023)
29. Barinova, V., Devyatova, A., Lomov, D.: The role of digitalization in the global energy transition. Int. Organ. Res. J. Nat. Res. Univ. Higher School Econ. **16**(4), 1–18 (2021)
30. Energy efficiency predictions for data centres in 2023, Data centre Magazine, https://dat acentremagazine.com/articles/efficiency-to-loom-large-for-data-centre-industry-in-2023, Last accessed 2024/11/09

Impact of Digital Systems on Production Efficiency, Cost, Energy Consumption, and CO_2 Emission in a Heat-Treatment Production Center

Valerie Woimbee and Ilya Peshekhodov

Abstract Forvia Faurecia ranked at the 7th worldwide position as automotive supplier is accelerating its digitalization for the whole life cycle of its products. Part of automotive components require heat-treatment at high temperature under controlled atmosphere and regulated thermochemical reactions. Besides high amount of energy, high temperature and cycle times are involved in the context of fluctuating production. Most of process parameters are monitored, but nowadays evolution of information technology enables to escalate monitoring of process parameters to a new level thanks to a large range of sensors capable of communicating with each other. The present heat-treatment process has been re-modelled by taking opportunities to connect those sensors for a full digitalization. The purpose is to show how the new connected architecture can improve the heat-treatment process, extend the functionalities of each existing tool and impact positively on cost, CO_2 emission, quality, and process efficiency. From the current infrastructure to the new, comparative tables are given to quantify cost, energy, and CO_2 emission saving.

Keywords Heat-treatment · Digitalization · Energy · Cost · CO_2 · Process · Material

V. Woimbee (✉)
Scientific Expert Dr-Ing, Forvia Faurecia, Brieres-Les-Scellés, France
e-mail: valerie.woimbee@forvia.com

I. Peshekhodov
Material Data and Project Leader, Forvia Faurecia, Hanover, Germany

I. Ilin and M. Youzhong (eds.), *Digital Systems and Information Technologies in the Energy Sector*, Lecture Notes in Networks and Systems 1244,
https://doi.org/10.1007/978-3-031-80710-7_21

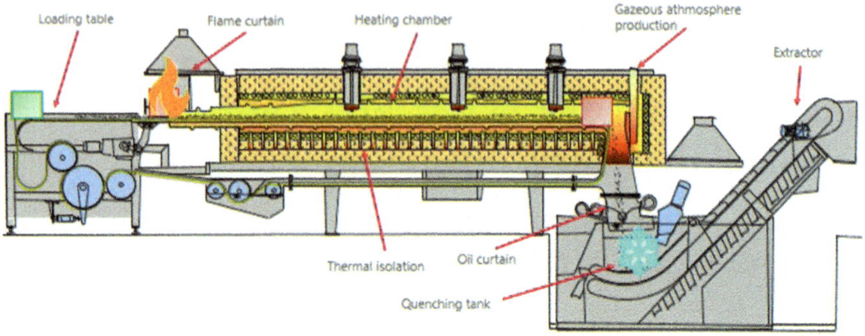

Fig. 1 Process of heat-treatment at air pressure (continuous): parts are loaded on the left and exit after quenching on the right (Safed Aichelin Furnaces)

1 General Situation on Heat-Treatment and Expectations of 4.0 Industry

1.1 Description of the Heat-Treatment Process

Steel parts are hardened by a heat-treatment process in a furnace, i.e., carbo-nitriding, single quenching, carburizing, ect. The process consists of heating at high temperature with phase transformation in the steel austenitic domain, followed by a thermo-chemical treatment with a carbon/nitrogen enrichment or only temperature maintenance in a neutral atmosphere (i.e., single quenching), then a final quick cooling to achieve final hardening. There are different technologies: either at air pressure (see Fig. 1) or at low pressure (from 1990) (see Fig. 2). The latter is often more automatized because the process has been developed in an environment of multi-cells treatment. Moreover, it has integrated latest technology of gas quenching which is typical, for example, for precipitation hardened ferrite-pearlite steels [1], replacing oil quenching, usual for quenched and tempered steels [2] for the final step of cooling. Both processes exhibit pros and cons and for current new investment, each gets approximately half of the market global share.

1.2 State of the Art for Process Information Technologies

For most heat-treatment processes, a supervising system pilots the different furnaces and process parameters (Temperature, cycle time, gas flow, measures of the atmosphere and its monitoring, maintenance, ect). The system enables to envision in real time the status of the equipment and, functioning state of different sensors, alerts maintenance and, production planning as well as follows the treatment online (see Fig. 3 for the integration of a supervising system on the shopfloor and Fig. 4 for an

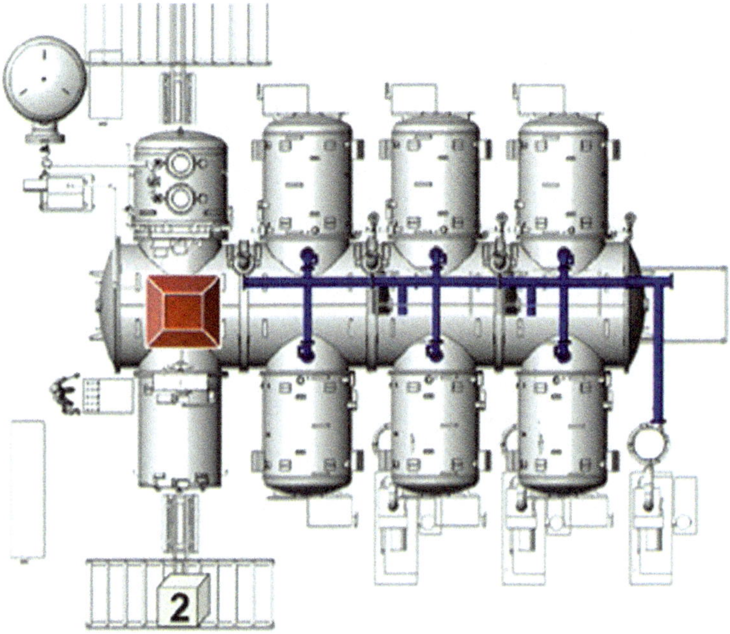

Fig. 2 Process at low pressure in multi-cells (ECM Furnaces)

example of system visualization accessible remotely) while recording and storing datas in a sql database, to ensure traceability.

1.3 Expectations of Industry 4.0

Energy consumption and CO_2 emission, and so cost depend on the process nature (at air pressure or at low pressure), most of the time favorable to low pressure because of lower gas used and lower cycle times at higher temperature. Regarding energy efficiency, an experiment has been launched to compare efficiency of both process types on an identical reference, a part produced in a Forvia Faurecia plant (both electrified processes) [3]. Experiments show a benefit of minimum 10% for the low-pressure process if each cell is full, but overall cost is slightly higher for this type of process. Both processes stay in the same order of magnitude with respect to cost. For the following, focus is on the classic heat treatment at air pressure, which is the current Forvia Faurecia process.

The energy shared for one line of heat-treatment process has been measured (see Fig. 5). Heat-efficiency is calculated by comparing theorical values of heat required by the part ($Q = m\, Cp\, \Delta T$) to total input energy measured on a machine site. Likewise,

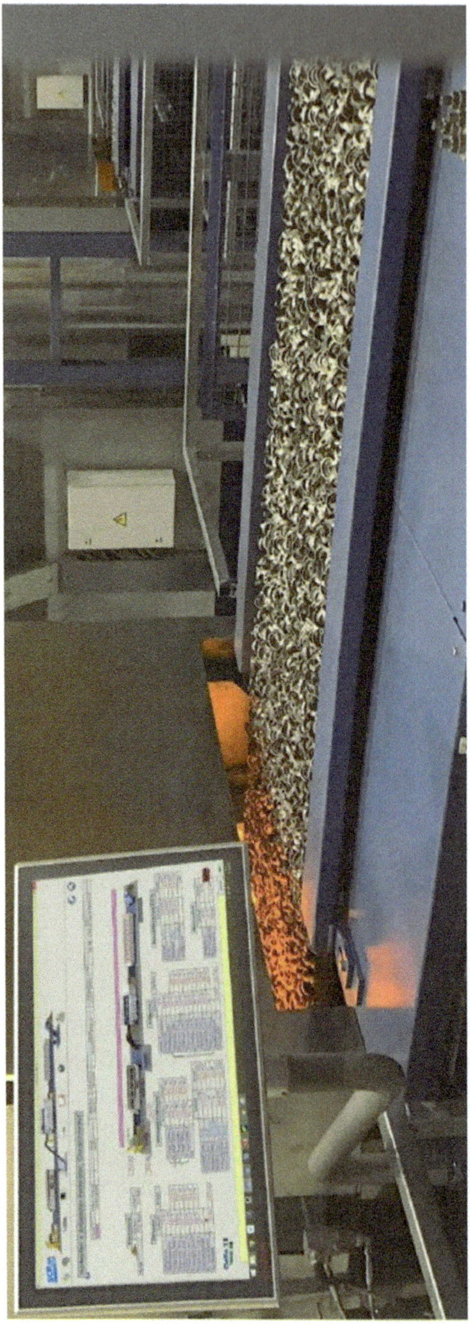

Fig. 3 State of the art of a supervising system for heat-treatment processes (continuous at air pressure) (SCR Crevoiserat SA): direct visualization of process parameters

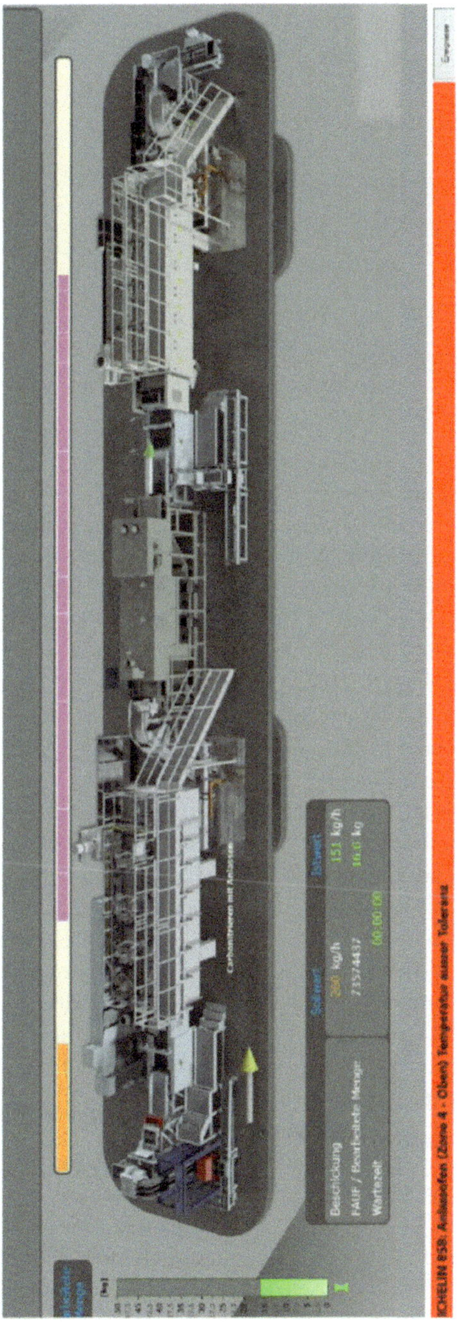

Fig. 4 Example of one of multiple view of load advancement and alert on process, SCR Crevoiserat SA

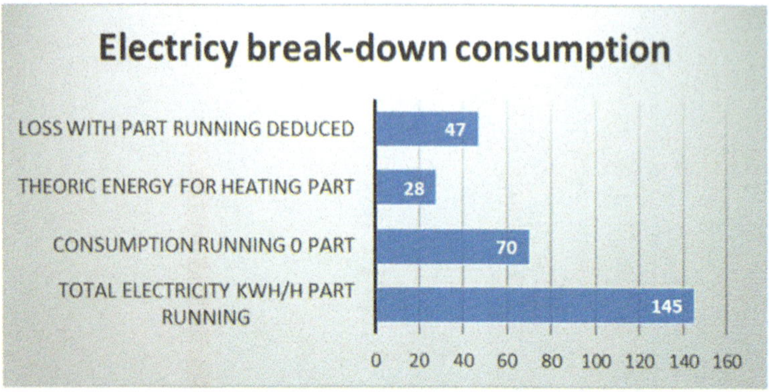

Fig. 5 Measures on plant site of energy sharing in Kwh for one hour, showing efficiency on the whole heat-treatment at air pressure (Safed Aichelin continuous electric furnace shown in Fig. 1)—numbers refer to a part cadency of 214 kg/h

additional configurations are measured: for example, configuration without any part running enables to differentiate between energy loss per parts and energy loss in the furnace itself. Energy absorbed by part to reach temperature during heat-treatment is 28 KWh, while 145 KWh is input. It shows a global heat efficiency of 19%, considered as one of the best, only induction heat-treatment technology can reveal higher efficiency, a simulation on the same parts estimates heat efficiency reaching 30% for induction hardening. The energy deduced for the charge material itself is the total energy subtracting the energy to run an empty furnace (see Fig. 5) (i.e. 145–70 kWh, which is equal to 28 + 47 KWh, or 75 KWh). 75/145 represents 52% of the total energy involved.

Sandkey diagram [4] in Fig. 6 illustrates the break-down of energy for a pusher furnace which is a batch furnace and has a closed hearth in comparison to our case. Difference between 28 and 52% of charge material loss, roots from the character of the furnace: a batch furnace is only open for the loading of parts and closed during the treatment, enabling to minimize energy loss by the parts, 28% charge material loss (see Fig. 6), while a continuous furnace remains constantly open because parts are loaded continuously, although efforts to limit the aperture, charge material represents 52% of total energy input (calculation from Fig. 5 see paragraph above).

The pusher furnace (see Fig. 6) is gas powered, as a result, flue gas accounts for the major energy loss with 36.2% of total energy, followed by charge load 28.2% and water cooling 15.4%, the latter relates to temperature regulation. Flue gas loss cannot be the same in case of an electric furnace, it is related to gas heater efficiency. Electric resistance efficiency in our case is expected to achieve above 80%.

At air pressure, in the case of a continuous open furnace where atmosphere is constantly renewed, gas flow drains energy and it is not accounted in the above diagram (Fig. 6) which considers a closed hearth. Nevertheless, it is likely to be the major source of energy loss for an open hearth furnace, as gas flow acts like an internal turbine exhausting energy out of the hearth, through exhaust fumes.

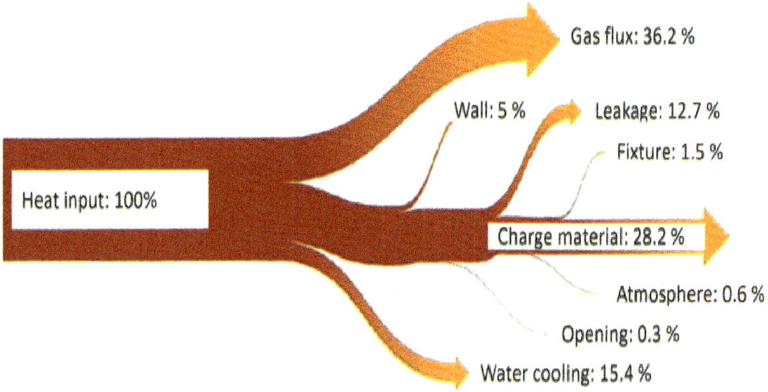

Fig. 6 Sandkey diagram for the base line heat balance of industrial furnace, pusher type, natural gas powered and closed hearth [4]

The water cooling energy loss in Fig. 6 relates to temperature regulation: any system of regulation. Temperature equilibrium is off balance each time, cycle time varies, parts cadency fluctuates, incident occurs ect. The return to balance is an energy expense. For instance, the quenching tank regulates its temperature by cooling.

The other sources of energy loss (see Fig. 6: wall loss, leakages, and fixtures account for a neglectable part of energy loss, meaning further insulation and equipment improvement shows nowadays a weak margin in energy loss minimization. The focus on energy saving should primarly deals therefore with the 'running organization' of the furnace.

To conclude, the main levers to minimize energy consumption for heat-treatment industrial furnaces relate to instantaneous program adjustments, it appears that gas flow and temperature regulation should constantly be adjusted, instead of a fixed programming, directly correlated to parts cadency and cycle advancement in order to avoid up and down of regulations which is a waste of energy. Only digitalization can enable these functionalities in real time. Additionally, an algorithm is to be created to minimize energy loss by tuning atmosphere and temperature management according to actual production flow, integration on real time of incidents and environmental data. The context is more or less usual in industry for material fabrication and heat-treatment. The iron industry evaluates this margin of improvement to reach 20% [5–10] by applying the best technologies, that means mainly digitalized solutions. For instance, heat treatment processes are some of the most energy-expensive processes of cast steel production and usually account for almost 70% of the total process energy consumption.

Regarding CO_2 emission [11], assessment has been done for two Forvia Faurecia locations (see Fig. 7), one in France and another in Poland and for three scopes: scope 1 for the gas used during the process, scope 2 for the electricity consumed during the process, and scope 3 for the carbon footprint of consumables [12–17]. Gas has a huge impact compared to electricity. Due to the dramatic impact of electricity emission

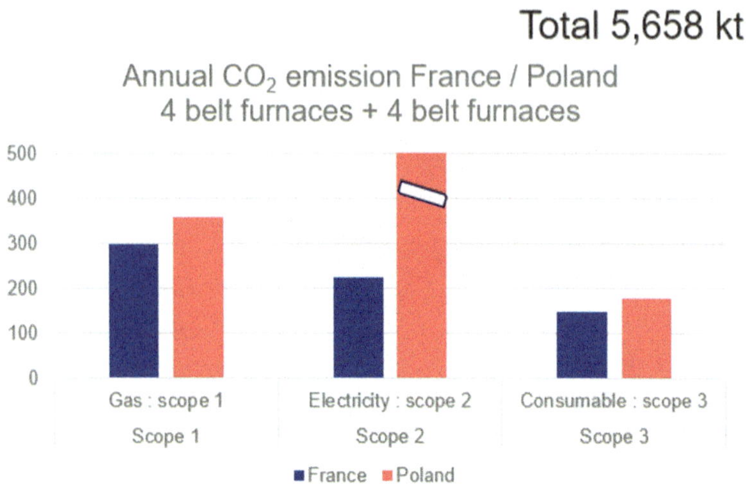

Fig. 7 Shares of scope 1, 2, and 3 regarding CO_2 emission for the internal heat-treatment facilities of Forvia Faurecia respectively in France and Poland—the scope 2 for Poland has been truncated for the graph readability (4420 t CO_2)

factor based on coal in Poland, scope 2 electricity in Poland rises more than 4000 t CO_2. The net zero CO_2 target by 2050 engages Faurecia Forvia group in a harsh quest for CO_2 reduction.

Energy and CO_2 emission are both tied, emphasizing the previously introduced need for better monitoring temperature and atmosphere regulation with a help of digitalization. Especially real-time monitoring atmosphere regulation is important to get control of an important source of energy loss through gas flow renewal. For a continuous process running 24/24 h, most of the time parameters are fixed along operations, even if parts sequences vary. New digitalized functionalities could be integrated to modulate the temperature regulations, and the enriching gas: the only means to absorb discontinuities of any kind, source of energy loss and CO_2 emission, in a continuous operative system.

2 How to Impact Cost, Energy Efficiency, CO_2 with Industry 4.0

2.1 Planification and Production Decision Making

A batch is a fixed number of parts with personalized process parameters (mostly cycle time and cadency). Interest of production planification is to enhance sequencing of batches on the right furnace, to minimize time lost when parameters are changed for another production following. Additionally, it is important to avoid situations where an equipment is running without any production. Before starting a new production reference associated with a different cycle time, furnace should be empty in order to avoid mixing and to respect cycle time difference between batches. Thus, he planification should minimize such situation by regrouping production with the same cycle times on the same equipment. Moreover, with a right planification, it is easier to predict production volume and make decision regarding either a furnace stop or a furnace stand-by at a temperature which minimizes cost and time. Take the decision to stop a furnace has consequences because furnace requires 3 days from ambient temperature to reach stabilized treatment temperature and right conditions for an operational re-start. Furnace running cost are correlated to temperature, energy required, and time. Decrease to intermediate temperatures enables to adjust delay of the stand-by and so minimize cost due to empty furnace on-work at higher temperature, as illustrated in Fig. 8. Thus, implementation of planification calculation brings immediate value.

Additionally, maintenance [18, 19] is sharply scheduled according to real temperature cycle duration recorded on each furnace. The system can learn from past production to give insights about volume and better scheduling on a real scale.

Since the Covid period, automotive industry has been confronted with rising fluctuations, variable production cycles and uncertanty on deliveries, increasing the request for an algorythm with predictive data acquired by learning [20, 21]. Production decision being complexified, this type of tool is significant to increase efficiency of equipment and staff. Moreover, the week-end production is not continuous, most

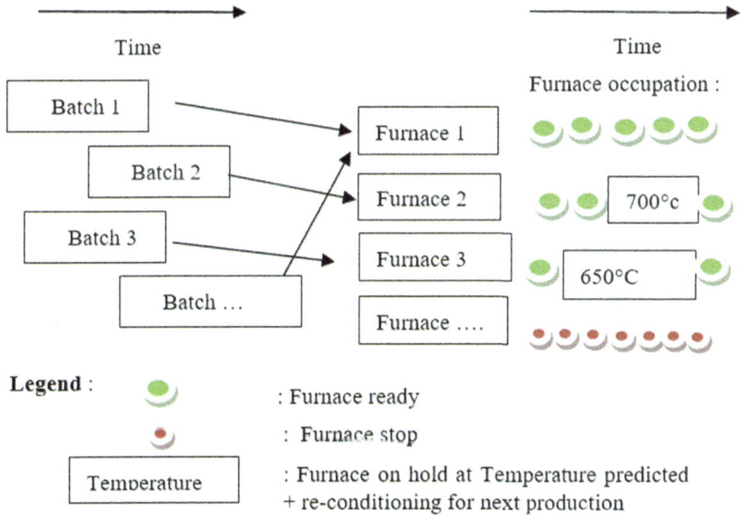

Fig. 8 Example of furnace planning optimizing cost and time according to batch parameters and due time delivery for assembly lines. The algorythm figures out the planification to minimize running cost and make decision which furnace to stop or put in stand-by at the lowest temperature predicted by cycle time to re-condition for next production

of the times, furnaces are still running at full power and energy lost is significant. The planification tool could help to predict week-end load and furnace heating power accordingly. This tool is an absolute priority.

On the other hand, planification being a human decision, a large safety margin is usually applied to absorb variability or unexpected peak due for instance to an incident in the part delivery for primary press workshop where parts flow. That means some planning maker decides for a safety margin of 20% (which represents almost one furnace), while another decides only for 15%, when in reality only 5% is needed. Digitalization allows to test all configurations at the same time to retain the minimum time and cost planification, while humain brain is not able to process all data in real-time.

By the time, it is difficult to evaluate precisely such tools but Faurecia has already implemented one to make decision about consumable purchase (like a gas tank refilling [22]) and substantial transport cost has been reduced (– 20%). A vizualization of the process steps running is given as a real example in Fig. 9, in that case the algorithm of planification shares the loads on the different equipments to minimize running cost but has not yet implemented the automatic sleep of each equipment and adjusted its parameters sleep at the optimum duration sleep.

2.2 *Reduction of Inter-Batch Time*

The previous paragraph targets the optimization of batch sequences on furnaces, now the focus is on the way to decrease the time between two batches [23].

In the case of two batches succeeding with the same cycle time, a stand-by is integrated to enable the production operator to change manually the crate to convert parts of the new batches. The standby time is roughly 5–10% of the production time. Here, digitalization is to correlate the last part of each batch with automatization of the rotative crate feeding system. This automatized step does not require any complexity or training. It has not been implemented yet because this function alone exhibits insufficient impact to the total production rate, but becomes necessary for the following automatized step to reduce time loss.

Two batches succeeding with different cycle times: in that case a full empty furnace should separate them to respect cycle time for each part of the batch. The loss of production is above one hour, critical for a jus- in-time automotive plant. The new possibility offered by digitalization solution with a calculator enables to connect part position and batch to an adjustable cycle time.

New cycle time $= f$ (cycle time 1, cycle time 2) tends to cycle time 2 when last part from batch 1 goes out the furnace. Additionally, a criterion on a minimum holding time in the maximum temperature hearth is to be conditioned to the new cycle time and figure out the new inter-batch time as shown in Fig. 10.

The modification on the time of the cycle time is assumed flexible regarding the 'tunnel process' mock-up inside the furnace, with multi-zones: front and bottom

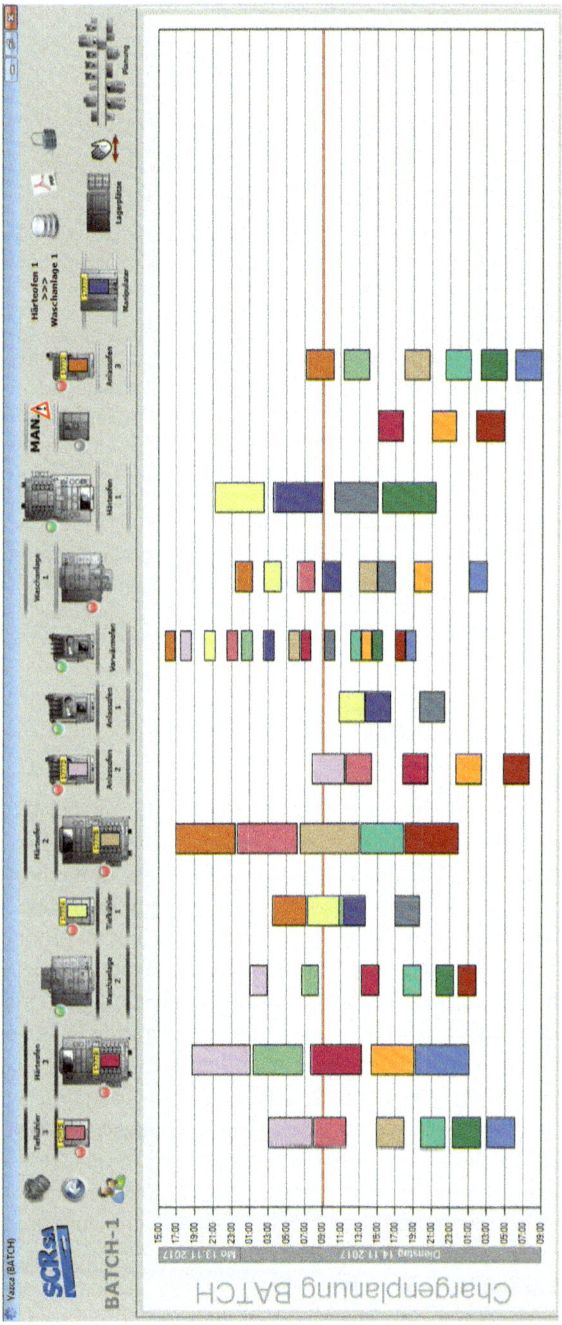

Fig. 9 Example of digitalized realization of charge planning: each batch is represented by a colored rectangle for each process equipment, and the tool is based on minimization of energy loss by minimizing total operating time (source: SCR Crevoiserat SA)

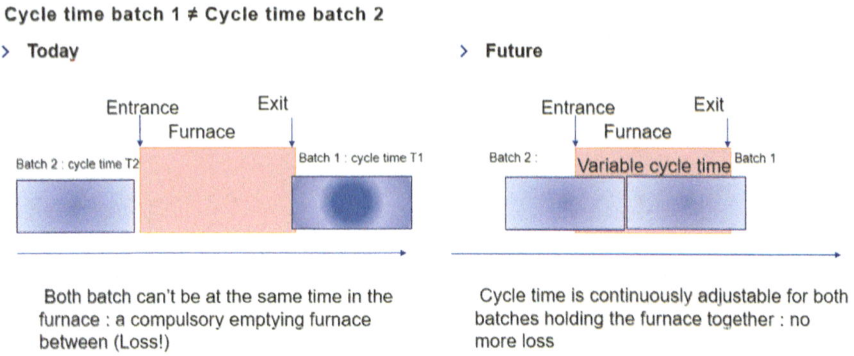

Fig. 10 Principle of a production multi-batches succeeding: as of today, on the left, and on the right, of the future target which is based on a continuous adjustment of cycle time. Each part, with its inner cycle time and its position into the load, is correlated to a variable cycle time which is computed according to its distance to a part which has a different inner cycle time. The variable cycle time enables a minimization of inter-batch time

ramp-up for heating and cooling respectively, should be considered as adjustable holding time. Only holding time in the stabilized temperature where thermochemical treatment occurs is fixed to respect rigoruous treatments specifications. The return in terms of cost is significant and calculation will be made in next chapter.

2.3 Process Parameters Adjustment

The power demand varies with the batch size and density, the gap between different batches, inducing regulation by cooling water which is a substantial loss of energy, as figured out in Chap. 1, by 15%. Furnace is the major block of the heat-treatment chain, but connected to a tempering furnace, washing equipment, etc. These are flexible and can be switched-off and switched-on when the flow of production is cut and no part are produced. Yet, that is not the case for the main furnace. Automatic driving of the switching is a source of saving. Note that switching the main furnace and planification of stand-by state at an intermediate temperature have been already introduced in the paragraph dedicated to load planification.

Gas atmosphere is a major source of energy loss and CO_2 footprint, even with no part running. Gas circulation drains heating energy in the exhausted fumes. In case of incident, or stand-by state, the system turns off automatically the gas flow, but there is a condition about duration of the incident, it must not overcome the time to recalibrate the atmosphere flow. The sum of stand-by states during production in a heat-treatment chain when operating, is between 10 and 20%, although their frequent occurrence, there is currently no action because it demands workforce attendance which is also expensive. The value brought by an 'intelligent' actuating is direct savings.

2.4 Quality Management

The management of incident is difficult for a machine operator as it is difficult to decide whether the production is to be discarded. Without any machine criteria all incidents are firstly discarded for re-inforced checking. The visualization on time of parts with over-treatment/insufficient treatment enables application of the right criteria for an automatic decision. In terms of work time, substantial gain is possible through quality checking. At the same time, a charge that burdens operators, and quality inspectors, is alleviated. On the other hand, optimizing heat-treatment programs [24] allows to improve part quality.

2.5 Machine Learning Continuous Improvement

Accessible parameters offer a great opportunity to compute correlation and optimization: energy consumption and gas consumption with multiple factors interacting: material composition, material mechanical properties, furnace parameters, parts, cadency, etc. One example of correlation and machine learning is to couple some gas inlet opening to stabilize atmosphere regulation and minimize gas consumption (see Fig. 11). These models are usually complex because they include more than 50 parameters which vary along one cycle time of roughly one hundred minutes, it represents a quite dense machine learning modus operandi.

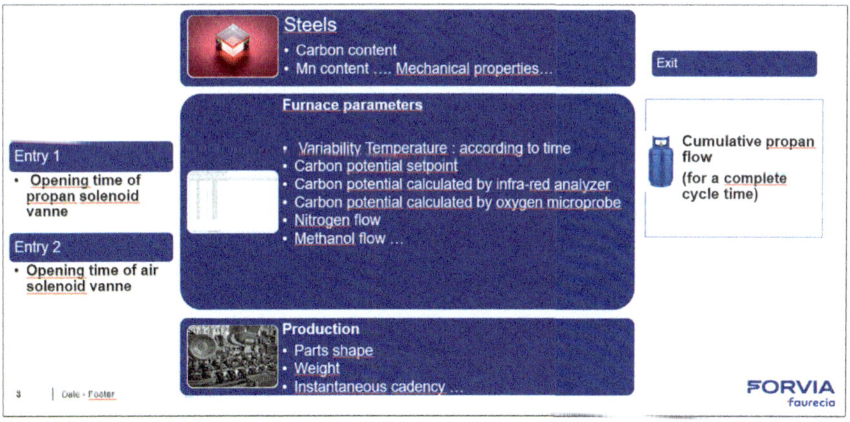

Fig. 11 Example of variables examined for a mutli-parameters modeling: steels, furnace parameters and production, for optimizing cumulative propane consumption

3 Calculation of the Impact of Full Digitalization on Energy, CO_2 Emission, and Cost

- Considering a typical interval time on a production journey, the value of each contributor can be estimated as shown in Fig. 12:
- Impact of inter-batch time separating each batch to adjust furnace parameters and crate for parts: it has been measured on site, being close to 15% of the whole opening time,
- Empty running estimated by the impact of week-end running at one third, by the fact equipment over-runs empty in a 'waiting state' to absorb production fluctuations and upstream incidents in the part feeding. Additionally, absence of automatic and predictive cutting during daily runs. The impact is close or even above inter-batch time impact,
- Process parameters optimizer assessed between 5 and 10% of global opening time,
- And quality checking of incident.

Major impact of digitalization deals with the empty running issue, which is figured out, on the one hand, by computed planification: minimizing and optimizing stand-by equipment, and on other hand, by management of automatic energy cutting and gas cutting according to online production advancement. Inter-batch time management is the second lever by enabling direct successive batch passing with a minimum gap in between. Cycle times are online adjusted according to part positioning in the cycle. Process optimizing by a better knowledge of multi-parameters interacting is another key, followed by quality management.

The room for improvement is huge when just considering the combination of inter-batch time and empty running of the process, close to 30% of the whole process occupation. In reality, half of the 30% can be considered as 'easily achievable'. Additionally, process parameters optimization is expected to improve by at least,

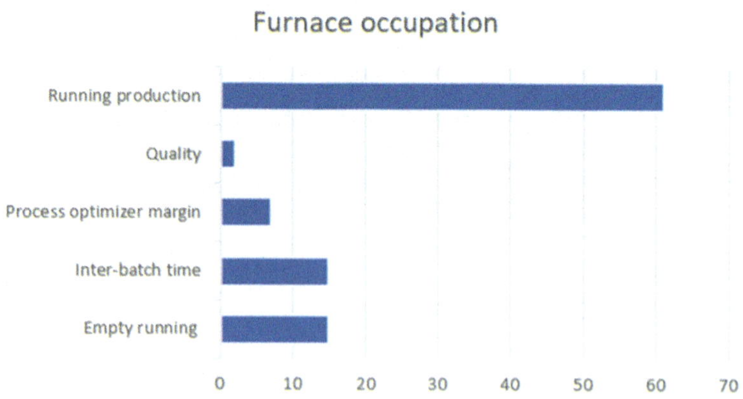

Fig. 12 Typical furnace occupation

5% the cost of process, meaning in total, a saving representing 20% of the furnace occupation is expected. The impact is direct in terms of cost.

Regarding the CO_2 emission, it has been shown how scope 1 to 3 are shared. Saving 20% of furnace occupation impacts the whole CO_2 emission rate by 20%. Moreover, the management of gas flow possible for a short stand-by period and a better regulation of the process, offers even larger positive impact, which has not yet been really valued.

4 Conclusion

A new step in digitalization strongly impacts the heat-treatment process productiveness. The connection among a large range of sensors would enable a just in time piloting of the equipment along the process chain according to part advancements, like never ever. The first major target is smoothing a discontinuous production planning on a continuous facility running 24/24 h: an algorithm optimizing charge planning with criteria of minimizing running cost and delivery on time is immediately beneficial, once each element of the process are connected and exchange datas. It allows for intelligent stop or stand-by at the adequate temperature for a quick re-start when necessary through automatic cutting or fitting of gas or temperature adjustment of the furnace in the stand-by state. The second improvement deals with the successive management of inter-batches: progressive cycle time adjustment enables reduction of distances between two batches. Finally, multi-parameters large models lead to fine-tuning optimization of the process. At least, a total 20% improvement on the whole process cost is expected and even higher regarding CO_2 emission.

References

1. Behrens, B.A., Bouguecha, A., Vuecetic, M., Peshekhodov, I., Kazhai, M., Ganin, S., Matveev, M.: Physical simulation of precipitation hardened ferrite-pearlite steels during hot deformation processing. Mater. Phys. Mech. **25**(016), 9–15 (2016)
2. Keul, C., Mosecker, L., Bleck, W., Rekersdrees, T., Stüber, A., Schliephale, M., Beyer, C., Raedt, H.W.: New developments in the material and process design of forged components in the automobile industry, Proceedings of the International Conference for Steels in Cars and Trucks, (2011)
3. Woimbee, V., Pouloux, P.: CO2 emission and carburizing treatment : comparison between a classic belt furnace and a low pressure carburizing process, European Conference on Heat Treatment, Toulouse 2024 A3TS June (2024)
4. Bisht, P.S.: Parametric energy efficiency: impact analysis for industrial process heating furnaces using MEASUR Graduate Theses, Dissertations Problem Rep., 12073 (2023)
5. Sobottka, T., Kamhuber, F., Heinzl, B.: Simulation-based multi-criteria optimization of parallel heat-treatment furnaces at a casting manufacturer. J. Manuf. Mater. Process. **4**(3), 94 (2020)
6. He, K., Wang, L.: A review of energy use and energy-efficient technologies for the iron and steel industry. Renew. Sustain. Energy Rev. **70**, 1022–1039 (2017)

7. Wang, J., Qiao, F., Zhao, F., Sutherland, J.W.: Batch scheduling for minimal energy consumption and tardiness under uncertainties: a heat treatment application. CIRP Ann. **65**, 17–20 (2016)
8. Oda, J., Akimoto, K., Tomoda, T., Nagashima, N., Wada, K., Sano, F.: International comparisons of energy efficiency in power, steel and cement industries. Energy Policy **44**, 118–129 (2012)
9. Mondikor, I., Sorli, M., et al.: Heat-treatment process energy efficiency design and optimization. Proced. Eng. **63**, 303–309 (2013)
10. Campos, I.B., Arguirrezebala, N.N., Valdes, L.D.: Energy efficiency and line productivity improvements for a continuous Heat-treatment process, Ind. Effi. (2018)
11. I.E.A.: Net zero roadmap: a global pathway to keep the 1.5 °C goal in reach, Reports September 2023 WWW.IAE.org
12. Woimbee, V.: Evaluation des emissions de gaz à effet de serre lors des operations de traitement thermique conventionnel. Traitement et Materiaux, Mars-avril, N°469, P23 (2021)
13. Woimbee, V.: Decarbonization plan for conventional heat-treatment facilities: neutrality targets in 2025, Conference Bodycote Lyon, 15th October (2021)
14. Woimbee, V.: Empreinte carbone et gestion de l'énergie dans les traitements thermiques, Conference, A3TS, Marseille, 24–25 novembre (2021)
15. Woimbee, V.: Empreinte carbone et gestion de l'énergie dans les traitements thermiques, Traitement et matériaux, Octobre-novembre, N°472, P16 (2021)
16. Woimbee, V.: Decarbonization plan for conventional heat-treatment facilities: CO_2 contributors assessment, neutrality targets, levers and action plan', IFHTSE, Int. Fed. Heat-treat. Surf. Eng. Salzburg, September pp. 5–8 (2022)
17. Woimbee, V.: Decarbonization plan for conventional heat-treatment facilities—Global vision for the company, Conference A3TS, Antwerpen, May 11 (2023)
18. Maierhofer, A., Trojahn, S., Ryll, F.: Concept for a robust and reliable manufacturing and logistics system that combine production planning and control with predictive maintenance. Proced. Comput. Sci. **232**, 3054–3062 (2024)
19. Braghirolli, L.F., Mendes, L.G., Engbers, H., Leohold, S., Triska, Y., Flores, M.R., Odebrecht de Souza, R., Freitag, M., Frazzon, E.M.: Improving production and maintenance planning with meta-learning-based failure prediction. J. Manufact. Syst. **75**, 42–55 (2024)
20. Shoorkand, H.D., Nourelfath, M., Hajji, A.: A hybrid deep learning approach to integrate predictive maintenance and production planning for multi-state systems. J. Manuf. Syst. **74**, 397–410 (2024)
21. Wu, X., Dai, J.T., Yan, J., Li, L.: Optimization of heat-treatment furnace based on automatic control system. J. Phys. Conf. Series **2044**, 012194 (2021)
22. Brunaud, B., Lainez-Aguirre, J.M., Pinto, J.M., Grossmann, I.E.: Inventory policies and safety stock optimization for supply chain planning, Aiche Journal, (2018)
23. Trovinger, S., Bohn, R.E.: Set-up time reduction for electronics assembly: combining simple SMED and IT based methods. Prod. Oper. Manag. **14**(2), 205–217 (2005)
24. Rink, M., Joritz, D., Ozkan, B.: Vârmebehandlingskonferensen, May 2022 pa Aronsborh Konferens i Balsta, pp. 4–5, (2022)

Energy Saving in Retail Based on the Development of IT-Services Architecture for Consolidation and Monitoring of Energy Efficiency Indicators

Olga Voronova, Tatyana Khnykina, and Artem Li

Abstract This research considers the theoretical foundations of a systematic approach to management of energy consumption in network retailing. It also evaluates the existing market offers in this area in order to develop a functional IT service architecture aimed at optimizing energy savings. The main energy monitoring tools in retail outlets were identified and specified, together with the key disadvantages associated with their implementation. In this paper, the authors describe the business process of energy management and present a traditional model of a relevant system. Based on the revision of successful examples of IT products in energy consumption, the main problems associated with their implementation in Russia were identified. As a result, the authors have developed models of stakeholders' requirements for IT services and their direct architectural model. The obtained results can be used as a method of IT system development at retail enterprises in order to ensure efficient energy management.

Keywords Energy management · Network retailing · FMCG · Enterprise architecture · Monitoring tools · IT services · Automation

1 Introduction

As retail trade volumes increase and the total retail space grows, the issue of energy-saving technologies is becoming more and more urgent. Such drivers as digitalization of retail, ESG agenda, and cost minimization are actively influencing the formation

O. Voronova (✉) · T. Khnykina
Peter the Great Saint-Petersburg Polytechnic University, Saint-Petersburg 195251, Russia
e-mail: iliina_ov@spbstu.ru

A. Li
G1 Software, Saint-Petersburg 199004, Russia

I. Ilin and M. Youzhong (eds.), *Digital Systems and Information Technologies in the Energy Sector*, Lecture Notes in Networks and Systems 1244,
https://doi.org/10.1007/978-3-031-80710-7_22

of innovative IT products aimed at automating the process of monitoring energy indicators in order to ensure their analysis and further decision-making on energy saving. In this regard, it is necessary to systematize the basics of the system approach to energy management, analyze the existing IT market offers in this area, and develop universal architectural models for IT services to support business processes of energy management in network retailing.

2 Materials and Methods

This research employs a combination of theoretical research methods (analysis, synthesis, classification, deduction, and induction) as well as practical research methods (description and modelling).

The methods of theoretical research provided an opportunity to review and evaluate regulatory legal documents, scientific research, and ready-made proposals presented on the IT market associated with functioning and optimization of energy consumption in retail.

The application of practical research methods allowed the authors to suggest the architectural models for further research and development in accordance with the TOGAF ADM method.

3 Results and Discussion

Energy consumption makes an integral part of any retail outlet in chain FMCG-retail. In the context of expanding trade and the subsequent growth of retail's influence on the economic, social, and environmental facets of society, energy consumption and energy-saving management become one of the key functional strategies [1–5]. The role of energy strategy in the structure of the company's general strategy is provided in Fig. 1.

With the development of R&D in engineering and energy efficiency [6, 7], energy assessment becomes a key source of data that shapes the energy strategy.

According to ISO 5001, company energy assessment is the data-based analysis of energy performance and consumption aimed at identification of sources of excessive energy consumption and further decision-making on the measures to improve energy efficiency and conservation [8].

Thus, two key technical activities (generalized functions) associated with energy consumption are represented by "Energy Management" and "Energy Data Management". In order to see the bigger picture, it is important to observe specifics on these functions in detail (Table 1).

Thereby, the "Energy management" business process can be represented as follows, using the ArchiMate notation [9] (Fig. 2).

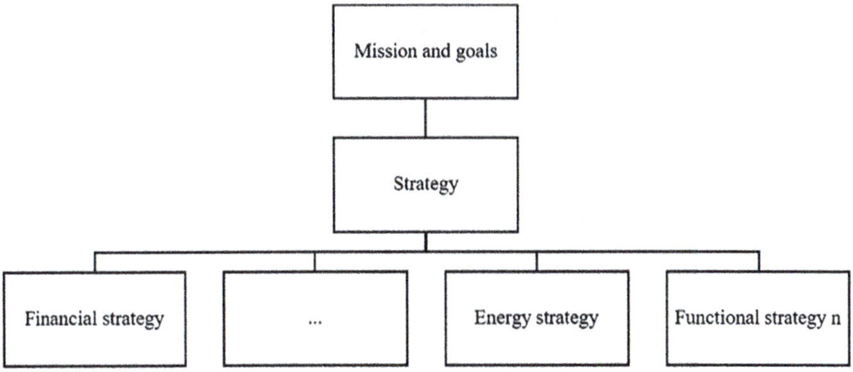

Fig. 1 Energy strategy in the structure of the company's general strategy (designed by the authors)

Table 1 Specifics of generalized functions "energy management" and "energy data management" in the functional area "energy consumption management and energy saving," retail-based case (designed by the authors)

General function assessment feature	Energy management	Management of energy consumption data
Role at the retail enterprise	Organization and coordination of energy-consuming equipment in order to maintain operability of the engineering complex at retail outlets	System planning, control and analysis of energy utilization at retail outlets in order to improve energy efficiency and reduce costs
Document management of implementation	Operational documents and regulations on energy use	Energy analysis regulations
Management document	Energy policy	
Outgoing data list	Indicators of energy intensity and operational mode of equipment at retail outlets	Assessment of energy efficiency of retail outlets, list of energy saving opportunities and risks
Functional responsibility areas	Chief engineer, operations director	Chief executive officer, development director
Upper level business process	Management of operation and equipment condition	Energy data management
Business process	Energy management	

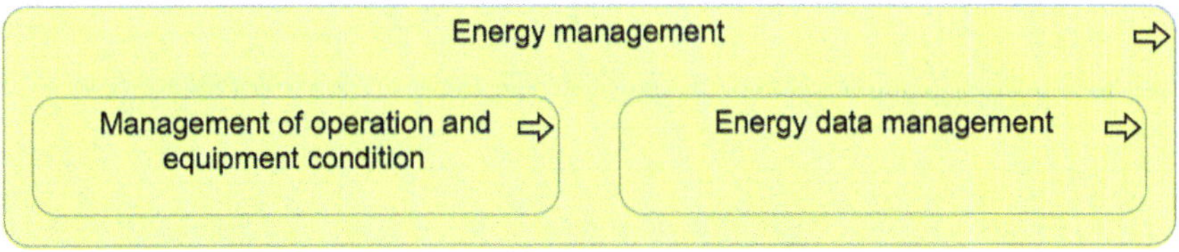

Fig. 2 Structure of "energy management" in network retail (designed by the authors)

According to the figure, the business process "Energy Management" in network retail includes two sub-processes aimed at providing retail outlets with energy resources and monitoring consumption while analyzing the obtained data and finding solutions to improve energy efficiency. In this regard, the system approach to energy management implies the formation of energy policy, goals, and objectives, as well as direct plans and processes. In accordance with ISO 5001 and a number of research works e.g., [10], the implementation algorithm of the energy management system can be presented as follows (Fig. 3).

The key metrics that are formed, collected, and analyzed in the associated processes are presented by the indicators of energy consumption (intensity) and efficiency. It is important to note that in accordance with ISO 5001, energy efficiency

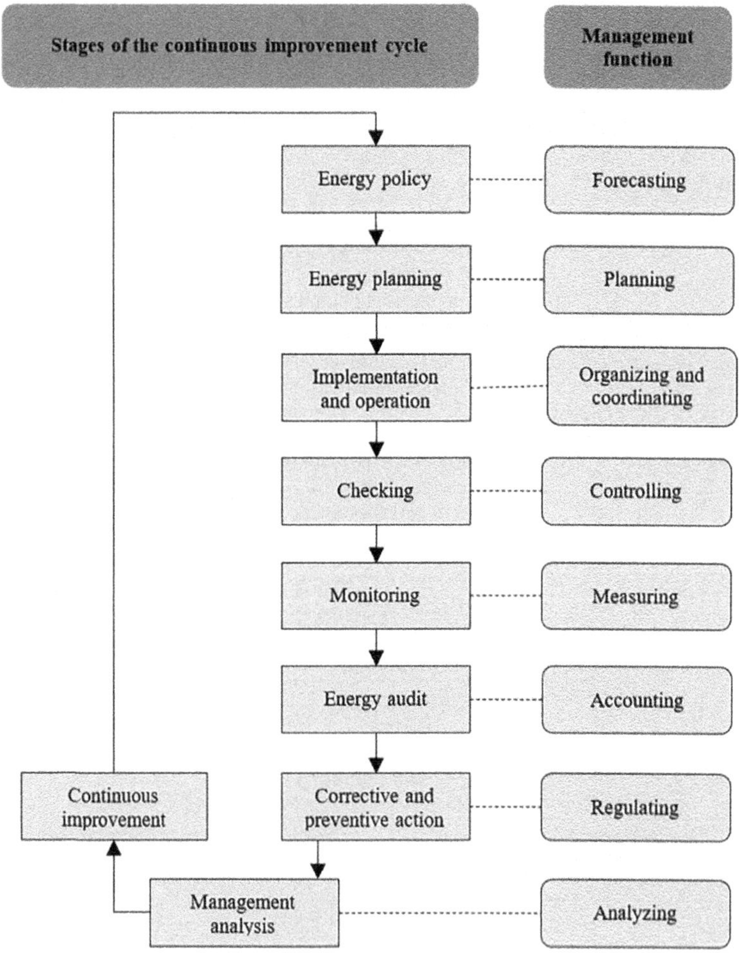

Fig. 3 Energy management system, functions and management methods [11]

indicators are considered the absolute or specific value of consumption or loss of energy resources required for production and technical processes. In this regard, it is necessary to structure the list of assessed metrics of the energy management system.

According to the UN "International Recommendations on Energy Statistics", the energy industry consists of 3 major sectors of energy use: industry, services, and transportation. Retail is an enlarged group within the services sector ("Commercial and Public Services"). In addition to retail, this sector includes other industries represented by both private and state-owned enterprises, as well as public–private partnerships. In total, the service industries include office real estate, warehouse real estate, retail (commercial real estate), state and municipal administration, health care, education, art, hospitality, food industry, and entertainment industry.

Each of the economic sectors that make up the services has different recommended units of measurement used to calculate energy efficiency indicators. For example, it is recommended to measure the activity of hotel enterprises in the number of rooms, nights, and floor space, while the educational institutions consider the number of students. In its turn, for retail trade, it is customary to measure the volume of premises and staff [12].

Apparently, today there is no unified approach to the development of indicators associated with energy consumption in retail. Using the standards of the International Energy Agency ("Energy Efficiency Indicators: Fundamentals of Statistics"), the authors developed a general block diagram that reflects the energy consumption indicators in retail (Fig. 4).

Figure 4 shows that due to the wide range of energy resources, disparate energy-consuming equipment, as well as differences in the measurement, each retail

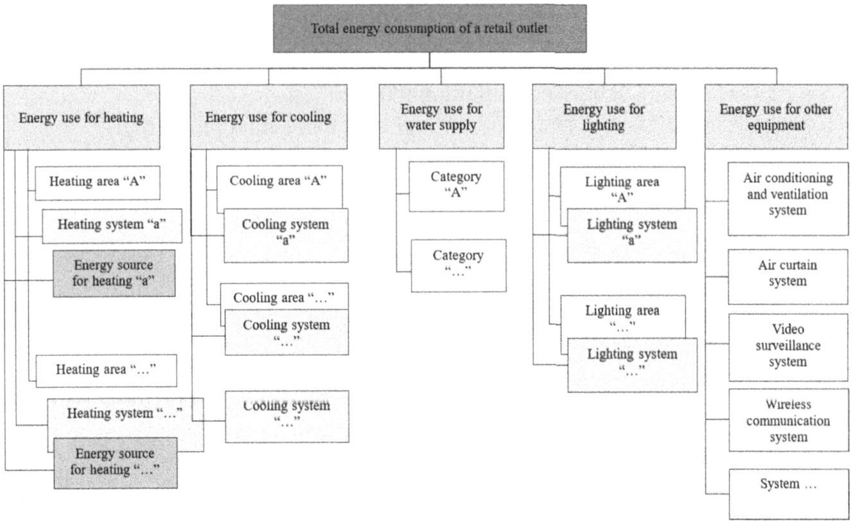

Fig. 4 General block diagram on energy data required for assessment of energy efficiency indicators at retail outlets (designed by the authors)

company needs up-to-date IT solutions that will ensure the implementation of an individual energy strategy. At the same time, with the overall digitalization of business, such products are supposed to be developed via the architectural approach [13].

4 Results and Discussion

The IT industry today provides a wide range of opportunities to track energy consumption levels in real time thanks to the automated systems involved in equipment management. At the same time, with the environmental agenda and the popularization of the search for new approaches to cost reduction, retailers around the world are engaged in projects aimed at increasing energy savings.

For large chain retailers, the difficulty in optimizing energy consumption systems lies in the scale of operations and the need to collect data from every outlet. However, in today's market, this problem is solved by the cloud technologies.

"Facilio" system makes one of such solutions. According to the developer [14], this IT system has rests on the following functions:

1. Tracking and consumption:

    1.1. Analyzing bills;
    1.2. Collecting data on energy consumption;
    1.3. Verification of bills;
    1.4. Budgeting energy consumption;
    1.5 Identifying best and worst performing retail outlets.

2. Benchmarking:

    2.1 Analyzing energy consumption index;
    2.2 Evaluation of consumption (retail outlet and equipment level);
    2.3 Measurement and verification.

3. Real-time monitoring and analytics:

    3.1. Indoor air quality and CO2 emissions;
    3.2. Heatmaps;
    3.3. Regression analysis;
    3.4. Load curves;
    3.5. Operational mode of equipment and retail outlets.

4. Cloud energy optimization based on the current strategies.

Practice shows that the application of this IT solution provides the rapid collection of energy data from all retail outlets in a single service, which allows operational managers to obtain detailed information on the energy intensity of their facilities and optimize equipment operation and energy savings. According to a large UK chain retailer, the software saved $5 million and improved energy efficiency by 12% in just one year. In turn, in the Middle East, the implementation of this cloud-based IT product reduced water consumption by 24% within a few months.

However, the network retail market of the Russian Federation faces a number of technical, social, and investment challenges associated with the implementation of IT solutions for monitoring data on energy consumption at retail outlets [14–19].

Due to the typically remote location of central offices in Russia, it is practically impossible to ensure the rapid data collection and exchange. What is more, the very issue of controlling the execution of the central office's energy strategy is exacerbated by a number of negative socio-cultural factors.

One of the successful examples of energy saving in Russian retail is X5's energy efficiency program. According to it, the company's retail outlets reduce the level of electricity consumption per 1 square meter of retail space through the system of automated control and metering of the volume and quality of electricity supplied. They also rely on a single centre for monitoring and management of these systems (based on the 2020–2024 results, 97% of retail outlets of the "Pyatyorochka" brand provide data on energy consumption in real time). In addition, the company has initiated active investments in the implementation of integrated systems of sensors, controllers, and software that ensure automatic control and minimize emergencies.

5 Conclusion

Following the ISO recommendations, a modern retailer should shape an energy management system that would be properly integrated into the overall management system. At the same time, due to the scale of retail activities, this system should be adaptable to each retail outlet and its engineering configuration. What is more, it is expected to allow the retailer to promptly monitor, plan, implement, and evaluate the actions associated with energy efficiency.

Using the ArchiMate notation, the authors have developed models of the requirements of top management and operational managers (stakeholders) to IT services in energy management (Figs. 5 and 6).

The data presented in the figures show that the functional areas of responsibility and requirements for IT services will differ significantly from one management level to another. This statement rests on the fact that different levels of stakeholders are responsible for specific business roles and processes. For instance, operational managers will administrate the business process "Management of operation and equipment condition," while top management will be responsible for "Energy Data Management." In this regard, the optimal set of IT services for "Energy Management" will consist of a number of services aimed at maintaining the business functions of a particular business process (Fig. 7).

The given model reflects the whole-scale combination of subprocesses associated with the "Energy Consumption Management" business processes. It also presents the technological services involved in their maintenance. In addition to their main purpose, some of the proposed services are aimed at direct support of managerial

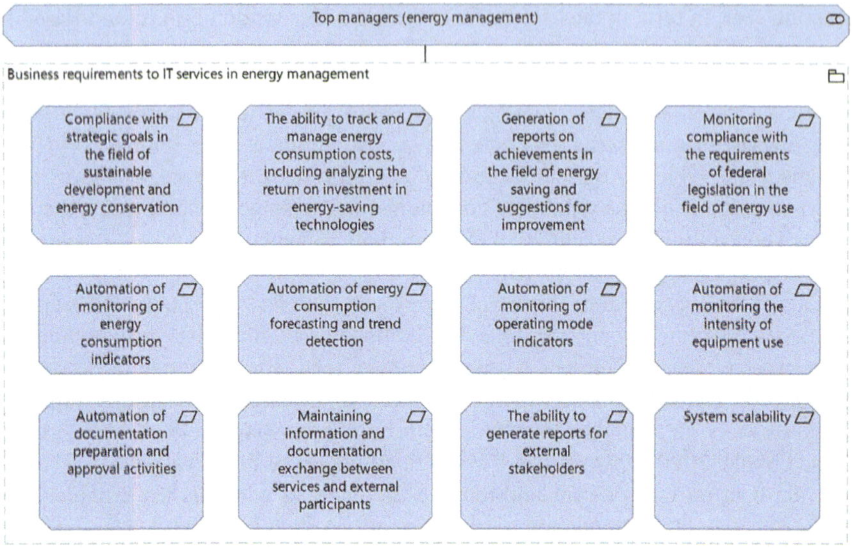

Fig. 5 Model of top managers' requirements to IT services in energy management (designed by the authors)

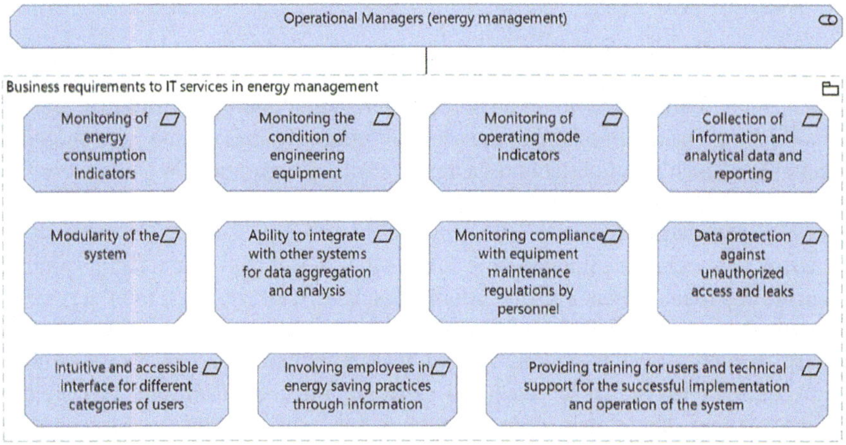

Fig. 6 Model of operational managers' requirements to IT services in energy management (designed by the authors)

decision-making in energy consumption and saving. Thus, for example, at the level of retail outlets (operational level), Remote Equipment Management Service and Short-Term Planning of Equipment Operation Service will allow operational managers to control microclimatic equipment zonally in accordance with the planogram of the retail floor and the operational mode of the enterprise. For top managers, the Energy

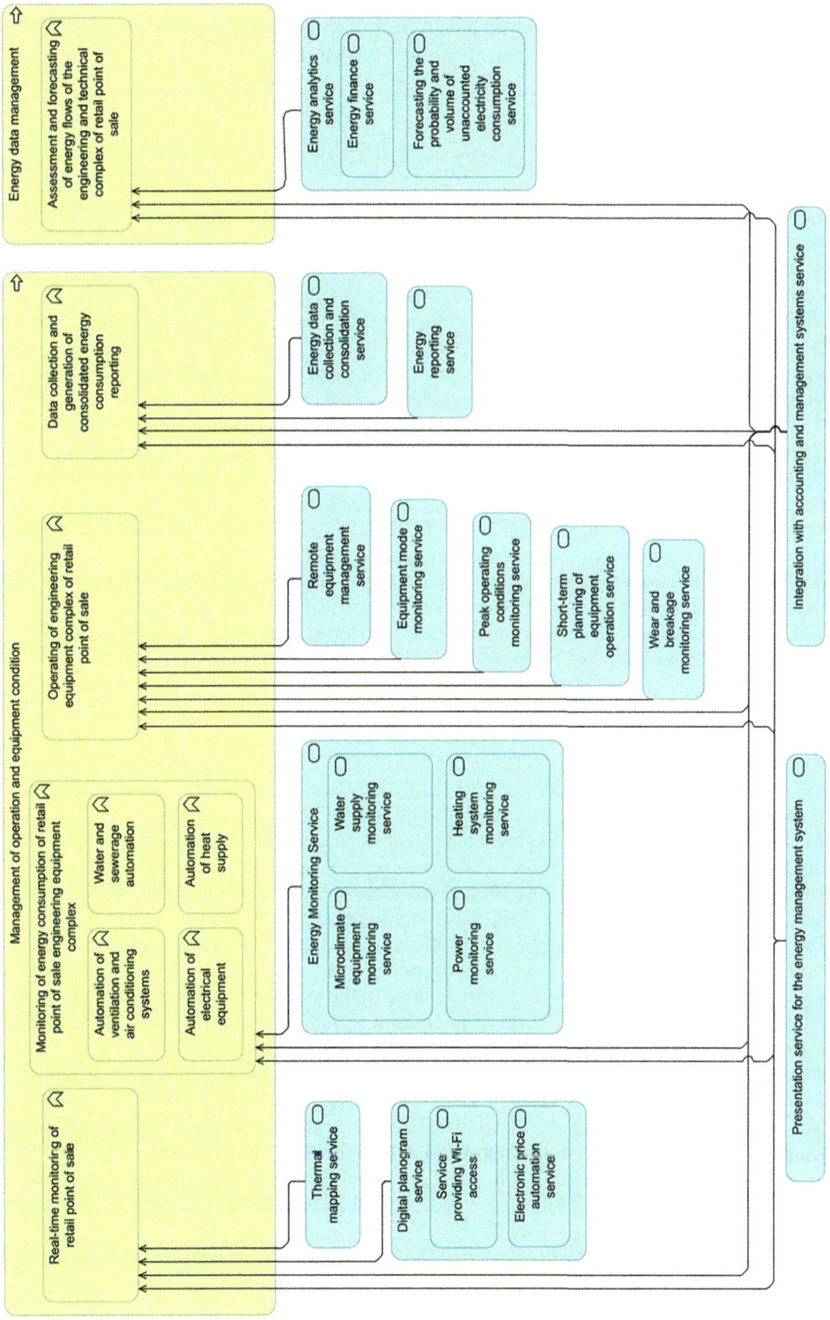

Fig. 7 Architectural model of IT services of the "energy consumption management" business process (designed by the authors)

Analytics Service will provide a comprehensive approach to energy management based on energy performance as recommended in ISO 50001. The designed model can be further implemented in the development of energy management information systems as part of a unified platform of network FMCG-retail.

References

1. Bakharev, V., Kapustina, I., Mityashin, G., Katrashova, Y.: Ecologization of retail trade: analysis of strategies. Siberian J. Life Sci. Agric. **12**(5), 79–96 (2020)
2. Politaeva, N.A., Illin, I.V., Oparina, A.M., Donetskova, A.S.: New energy approaches to the use of waste biosorbents of microalgae Chlorella kessleri / Povolzhskii Ekologicheskii Zhurnal, (3), pp. 322–335 (2022)
3. Ilin, I., Vasilenok, V., Marchenko, R. Use of renewable energy and tax burden on CO2 emissions in industrial enterprises. In: E3S Web of Conferences 110, 02102 (2019)
4. Ilin, I., Levina, A., Iliashenko, O.: Enterprise architecture analysis for energy efficiency of Saint-Petersburg underground. Adv. Intell. Syst. Comput. **692**, 1214–1223 (2018)
5. Poljanskihh, A., Levina, A., Dubgorn, A.: Investment in renewable energy: practical case in Estonia. In: MATEC Web of Conferences 193, 05065 (2018)
6. Kapustina, I., Pereverzeva, T., Stepanova, T., Rusu, I.: Convergence of institutes of retail traditional and digital economy. Mater. Sci. Eng. **497**, 012120 (2019)
7. Kalinina, O., Kapustina, I., Buniak, V., Golubnichaya, G.: Economic features of investment nature of energy-saving projects in Russia. EDP Sciences **110**, 02089 (2019)
8. GOST R ISO 50001–2023 Energy management systems. Requirements and Guidelines for Application, https://docs.cntd.ru/document/1200195836. Last accessed 2024/08/15
9. ArchiMate standard. The Open Group, https://pubs.opengroup.org/architecture/archimate-doc/ts_archimate/toc.html. Last accessed 2024/08/15
10. Sikorsky, A.: Energy management: standard update and directions of development. Energy Ind. Russ. **10**, 366 (2019)
11. UN International Recommendations for Energy Statistics, https://unstats.un.org/unsd/energystats/methodology/ires/2017. Last accessed 2024/08/15
12. International Energy Agency. Energy efficiency indicators: statistical basics, https://www.iea.org/. Last accessed 2024/08/15
13. Ilyin, I., Levina, A., Dubgorn, A.: Digital transformation as a factor in shaping enterprise architecture and IT architecture. Econ. Environ. Manag. **3**, 50–55 (2019)
14. Fasilio.: Decarbonization in building operations. A technology roadmap for smart building operations, https://facilio.com/blog/decarbonization-tech-roadmap/. Last accessed 2024/08/15
15. Ilyin, I., Voronova, O., Khareva, V.: Development of the system of requirements to the architecture of business services in terms of interaction with consumers within the chain trading companies. Glob. Challenges Digital Trans. Markets, pp. 261–276 (2022)
16. Niyazbekova, S., Barykin, S., Sergeev, S.: Sustainable energy efficient human-centered digital solutions for ESG megacities development. Front. Energy Res. **10**, 938768 (2022)
17. Voronova, O., Khareva, V.: Network retail FMCG-segment in the Russian Federation: current state and problems of development. Int. Sci. J. **2**, 7–16 (2019)
18. Voronova, O., Ilyin, I., Khareva, V.: Methodological foundations of forming a system of requirements for the architecture of services of network trading companies. Izvestiya St. Petersburg State Univ. Econ. **5**(125), 117–124 (2022)
19. X5. Energy Resources Reduction Program, https://esg.x5.ru/ru/programs/energy-resources/. Last accessed 2024/08/15

Improving Methods for Protecting Confidential Information when Processed in Energy Company Information Systems

Olga S. Chemeris, Sherzod Gulomov, Anatoly Tsaregorodtsev, Anna Kogteva, and Ekaterina Gubert

Abstract Protecting confidential information in energy companies is a critical task with serious national security implications. The article emphasises that inadequate protection can lead to serious consequences, including damage to national infrastructure and even military conflicts. The article analyses three levels of information protection: legal, organisational and technical. Particular attention is paid to the growing threat of insider intruders, which has become increasingly important in recent years. The research is aimed at using mathematical methods to estimate the probability of an attack on confidential information by an insider to identify priority areas for prevention and threat reduction. A multivariate regression-based model is proposed that takes into account many factors that influence the probability of an attack, such as the level of access to information, motivation, knowledge and skills of the intruder, as well as the probability of detection and punishment. To predict the probability of attack, we propose to use the logistic regression method, which allows predicting values in the range from 0 to 1. The model takes into account the interaction between the parameters and their mutual influence on each other, helping to develop a comprehensive approach to protecting confidential information and reduce the risk of information leakage in energy companies.

Keywords Protection of confidential information · Information leakage · Assessment of the probability of an attack

O. S. Chemeris (✉)
Peter the Great St. Petersburg Polytechnic University, St. Petersburg, Russia
e-mail: o.s.pogarskaya@gmail.com

S. Gulomov
Tashkent University of Information Technologies Named After Muhammad Al-Khorezmi, Tashkent, Republic of Uzbekistan

A. Tsaregorodtsev · A. Kogteva · E. Gubert
Federal State Budgetary Educational Institution of Higher Education "Financial University Under the Government of the Russian Federation", Moscow, Russia

I. Ilin and M. Youzhong (eds.), *Digital Systems and Information Technologies in the Energy Sector*, Lecture Notes in Networks and Systems 1244,
https://doi.org/10.1007/978-3-031-80710-7_23

1 Confidential Information in the Energy Industry

1.1 The Relevance of Protecting Confidential Information

Confidential information is a category of data that is of particular importance to the energy industry, as well as to the personal lives of employees [1]. It may include customer data, financial information, development plans, technological secrets and other types of trade secrets. This data is a valuable company asset and requires strict access control and protection from unauthorized use. The leak of such information can have serious consequences, which can negatively affect the operation of the entire energy system, lead to significant economic losses and even pose a threat to national security [2].

Protecting confidential information in an energy company is one of the key tasks that requires close attention and an effective approach.

Insufficient protection of confidential information can lead to a number of negative consequences—from damaging national infrastructure to creating conditions for military conflict.

Today, in the era of the digital economy and globalization, maintaining confidential information has become even more challenging. Modern technologies allow information to be quickly disseminated, which increases the risk of its leakage or misuse [3]. Therefore, improving mechanisms for protecting confidential information has become one of the key tasks for energy companies seeking to maintain their competitiveness and protect the interests of their employees.

The movement of confidential information in an energy company is a complex process that includes many stages and requires strict control. An approximate flow diagram of information is shown in Fig. 1 and begins with its receipt from external sources, such as clients, partners, suppliers and other market participants [4].

The information is then stored in secure databases and information management systems. Employees then use this information to perform their work tasks, such as

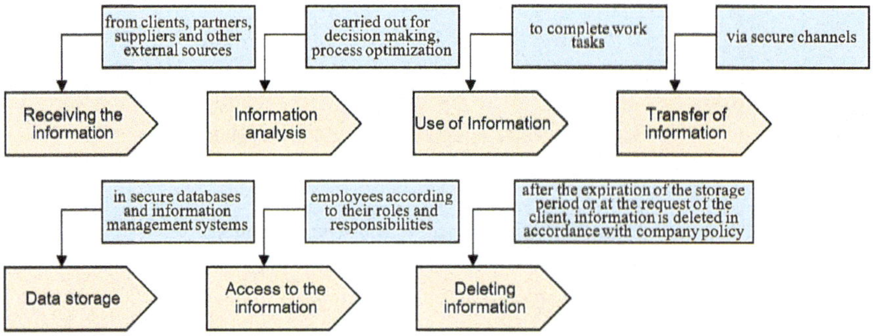

Fig. 1 Confidential information flow pattern

energy management, production planning, financial accounting and others. Information is analysed to make decisions, optimise processes and identify potential threats. Regular backups allow data to be restored in case of loss or damage. After the expiry of the retention period or at the customer's request, the information is deleted in accordance with the company's policy.

1.2 Review of Existing Measures to Protect Confidential Information

To ensure comprehensive protection of confidential information, three levels of protection are used: legal, organizational and technical [5]. In general, the above interacting and complementary levels are shown in the Venn diagram in Fig. 2.

The legal level of information protection in the Russian Federation is regulated by the Constitution of the Russian Federation [6], Federal laws [7–10] and Decrees of the Government of the Russian Federation [11, 12], which determine the requirements for the processing and protection of personal data, trade secrets and other confidential information.

Technical measures to protect confidential information are a set of tools and technologies for detecting and preventing unauthorized access to confidential information [13]:

- Use of encryption algorithms when transmitting and storing information;
- Installation of firewalls and firewalls to control access to the network and protect against external threats;

Fig. 2 The intersection of three levels of confidential information protection

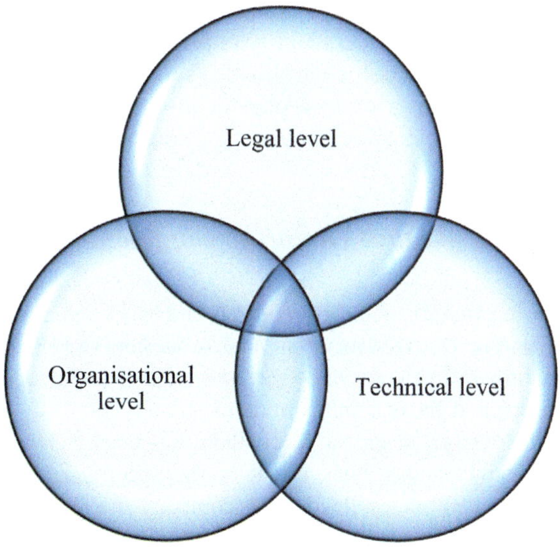

- Installation of antivirus software;
- Using intrusion detection systems to detect unauthorized attempts to access data and alert administrators to potential threats;
- Regularly creating backup copies of confidential information to allow quick recovery in case of data loss or damage;
- Implementation of two-factor authentication and the use of biometric data to enhance the security of access to confidential information;
- Restriction of physical access to confidential information storages.
- Regular software updates to close known vulnerabilities and prevent possible attacks.

Organizational measures to protect confidential information include a set of measures aimed at ensuring the safety and security of information within the energy organization. They include the following aspects [14]:

- Development and implementation of a security policy that defines the rules and procedures for the processing, storage and transmission of confidential information, including requirements for access authorization, use of passwords, data encryption, backup and other security measures;
- Creating clear procedures and processes for handling confidential information;
- Implementation of educational programs and trainings on information security;
- Regular monitoring and evaluation of the effectiveness of protection measures;
- Cooperation with law enforcement agencies, specialized information security companies and other external organizations to exchange experience and receive advice;
- Assessing the risks associated with the processing of confidential information and developing plans to manage these risks;
- The use of cryptographic methods to protect the confidentiality and integrity of data, including encryption and digital signatures;
- Ensuring physical security of premises where confidential information is stored, including access control, video surveillance and other measures;
- Developing a response plan for information security incidents, including notifying responsible parties, collecting evidence, and taking corrective action.

1.3 Risks and Threats to Confidential Information

In information security research and work, much attention has traditionally been paid to external adversaries, such as hackers and cybercriminals [15]. This is because external threats are often the most visible and dangerous, as they can be global in scope and cause significant damage.

However, in recent years, there has been growing awareness that insiders pose a distinct and serious threat to the security of confidential information [16] in energy companies. Insiders are company employees who have access to confi-dential information and can misuse it to commit crime or sabotage [17].

According to statistics [18], 79% of cases of confidential information leakage are associated with the actions of internal violators. Of these, the vast majority (70%) occur in cases where the leak was caused by intentional actions of em-ployees. This highlights the importance of insider threats and the need to strengthen measures to manage risks associated with human factors.

It is important to note that accidental violations can be no less dangerous than intentional ones [19]. Accidental errors by employees can occur for many rea-sons—careless attitude to their duties, neglect of safety rules, lack of necessary knowledge and skills to work with information systems, ignoring education and security training, from fatigue, inattention or haste, nevertheless, despite Without malicious intent, such violations can have negative consequences, including fi-nancial loss, threats to reputation and legal liability.

Internal offenders can be motivated by a variety of factors, such as financial gain, revenge, personal beliefs, or the desire to gain power. In addition, insiders often have knowledge of an energy company's internal processes and security systems, making them especially dangerous. They can use this knowledge to by-pass security systems and carry out attacks [20].

2 Mathematical Methods for Assessing the Likelihood of an Attack on Confidential Information

2.1 Model for Assessing the Likelihood of an Attack by an Insider

Let us present a model for assessing the likelihood of an attack by an insider, which is a generalization and synthesis of various studies [20–24] and works in the field of information security and risk assessment:

$$\mathrm{P(A)} = f\ (x_1,\ x_2,\ \ldots,\ x_n), \tag{1}$$

where P(A) is the probability of an attack by an insider, $x_1, x_2, \ldots, x_n$—are parameters that affect the probability of an attack.

To determine the values of parameters $x_1, x_2, \ldots, x_n$ various methods can be used, such as expert assessments, data analysis, statistical methods, etc.

Table 1 lists some factors that can be parameters $x_1, x_2, \ldots, x_n$.

The function $f(x_1, x_2, \ldots, x_n)$ can be defined in various ways depending on the specific conditions and goals of the simulation. For example, it could be a linear function, a quadratic function, or a higher order function.

It is important to note that this model is simplistic and does not take into account all possible factors affecting the probability of an attack. In addition, it does not take into account the interaction between the parameters and their mutual influence on each other. Therefore, we will use a multivariate regression method to estimate the

Table 1 Characteristics of an insider

x_1	Level of access to confidential information	Amount and types of data that an insider has access to
x_2	Significance of information for an insider	Connection with personal interests such as financial gain, political influence or personal revenge
x_3	Opportunity to benefit from the use of information	The possibility of selling information on the black market, using it for blackmail or other criminal activities
x_4	Motivation level	The emotional state of the offender, his attitude towards the company or his personal goals
x_5	Availability of knowledge and skills to carry out an attack	Education, work experience and knowledge in the field of information technology
x_6	Probability of detection and punishment	Effectiveness of monitoring and detection systems, severity of penalties for violations
x_7	Time and resources	The time required to prepare and carry out an attack, as well as the availability of resources such as hardware and software
x_8	Attack difficulty	The need to use sophisticated tools and techniques

probability of an attack more accurately, taking into account the interaction between the parameters and their mutual influence on each other.

Multivariate regression is a statistical technique that is used to model the relationship between several independent variables (predictors) and one dependent variable. In the context of assessing the likelihood of an insider attack, we can use multivariate regression to model the effect of various factors on the likelihood of an attack.

Let us have n observations, each of which is characterized by a set of parameters:

$$x_i = (x_{1i}, x_{2i}, \ldots, x_{ki}) \quad (2)$$

where x_{ji} is the value of the jth parameter for the ith observation.

Logistic regression can be used to predict the probability of an attack P(A) based on these parameters, which is suitable for probability modeling because it can predict values between 0 and 1.

A mathematical model based on the multivariate regression method can be presented as follows:

$$\mathrm{P(A)} = \frac{1}{1 + e^{(-\beta_0 - \beta_1 x_1 - \beta_2 x_2 - \ldots - \beta_k x_k)}} \quad (3)$$

where β_0, β_1, β_2, ..., β_k are regression coefficients that we must determine based on the available data.

The coefficients β_j can be estimated using least squares or other optimization methods. Once the coefficients are estimated, we can use the model to predict the probability of an attack for any new set of parameters.

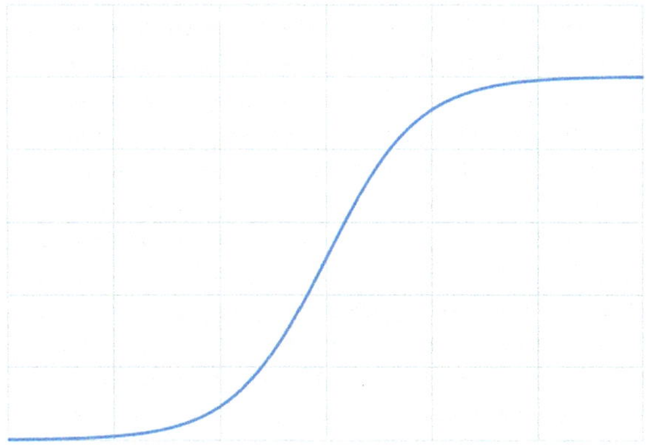

Fig. 3 Logistic (S-shaped) curve

In general, the graph will have and S- shape known as a logistic curve. On the horizontal axis the values of parameters x_1, x_2, ..., x_k, will be plotted, and on the vertical axis the corresponding probability values P(A) will be plotted (see Fig. 3).

Assuming that all parameters x_1, x_2, ..., x_k are positive and increase with increasing probability P(A), then the graph will start at point (0,0) and gradually rise upward, approaching but never reaching unity.

If some of the parameters are negative or have an inverse relationship with probability P(A), then the shape of the graph may change. For example, if the parameter x_1 decreases as the probability increases, then the graph will have a shape resembling an inverted letter U (see Fig. 4).

Thus, the specific type of the graph will depend on the values of coefficients β_0, β_1, β_2, ..., β_k and on how parameters x_1, x_2, ..., x_k affect the probability P(A).

Methodologies for assessing the probability of an attack by an internal intruder using mathematical methods will help to identify priority areas for the development of measures to prevent and reduce the level of potential threats.

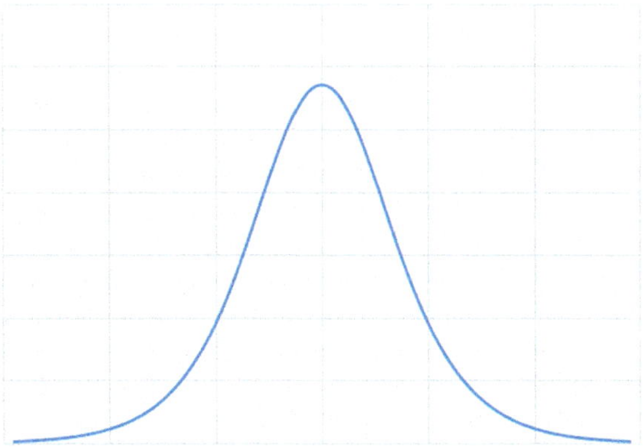

Fig. 4 Inverse parameter dependence

2.2 *Methods of Prevention and Reduction of Potential Threats*

The problem of confidential information leaks in energy companies remains topical. However, there are effective tools and strategies to minimise this threat [25]: a clear security policy, strict procedures for gaining access to confidential information, regular training on information security issues and raising employee awareness of risks and security rules.

Technical solutions include access control, monitoring and auditing systems [26, 27]. Implementation of modern leak detection and personnel action control systems, such as DLP (Data Loss Prevention) solutions, allows to detect and prevent unauthorised distribution of confidential information. Access Control Systems help to control and restrict access to information resources, preventing unauthorised access. Advanced Machine Learning Analytics automatically detect anomalies in user behaviour and identify potential threats. Regular training programmes and information security training for employees will help reduce the risk of errors and violations by staff.

The combination of these measures and strategies can significantly reduce the risks of confidential information leakage in energy companies, among which internal employees continue to be a major threat.

References

1. Chamzo, O.D.: Confidential information and methods of protecting it. Bull. Master's Degree Program. **2–1**(137), 58–59 (2023)
2. Senapati, K.K., Kumar, A., Sinha, K.: Impact of information leakage and conserving digital privacy. Malware analysis and intrusion detection in cyber-physical systems. IGI Global, pp. 166–188 (2023)
3. Wang, Z., et al.: Attrleaks on the edge: exploiting information leakage from privacy-preserving co-inference. Chin. J. Electron. **32**(1), 1–12 (2023)
4. Dexkonov, T.A.: Edo (electronic document management). Galaxy Int. Interdiscip. Res. J. **10**(4), 328–331 (2022)
5. Somasundaram S.P.: Enhancing organizational data protection: advanced security measures for database systems. Int. J. Res. Comput. Appl. Inform. Technol. (IJRCAIT) 6(1) (2023)
6. Constitution of the Russian Federation, http://publication.pravo.gov.ru/Document/View / 0001202007040001. Last accessed 2024/08/02
7. Federal Law of the Russian Federation No. 149 of July 27, 2006 "On information, in-formation technologies and information protection"
8. Federal Law No. 152-FZ of July 27, 2006 (as amended on July 29, 2017) "On Personal Data"
9. Federal Law of July 27, 2006 No. 149-FZ (as amended on November 25, 2017) "On information, information technologies and information protection"
10. Federal Law of July 26, 2017 No. 187-FZ "On the security of critical information infrastructure of the Russian Federation."
11. Decree of the Government of the Russian Federation of November 17, 2007. No. 781 "On approval of the Regulations on ensuring the security of personal data during their processing in personal data information systems"

12. Decree of the Government of the Russian Federation of November 1, 2012 No. 1119 "On approval of requirements for the protection of personal data during their processing in personal data information systems"
13. Saltzer, J.H., Schroeder, M.D.: The protection of information in computer systems. Proceed. IEEE **63**(9), 1278–1308 (1975)
14. Da Veiga, A., et al.: Defining organisational information security culture—perspectives from academia and industry. Comput. Secur. **92**, 101713 (2020)
15. Toro-Alvarez, M.M.: Hacking. Handbook on crime and technology. Edward Elgar Publishing. pp. 334–357 (2023)
16. Chu, A.M.Y., So, M.K.P.: Organizational information security management for sustainable information systems: an unethical employee information security behavior perspective. Sustainability **12**(8), 3163 (2020)
17. Saxena, N., et al.: Impact and key challenges of insider threats on organizations and critical businesses. Electronics **9**(9), 1460 (2020)
18. Official website of the InfoWatch group of companies, http://www.infowatch.ru. Last accessed 2024/06/28
19. Official website of the Solar Group, https://rt-solar.ru. Last accessed 2024/06/29
20. Burns, A.J., et al.: Going beyond deterrence: a middle-range theory of motives and controls for insider computer abuse. Inf. Syst. Res. **34**(1), 342–362 (2023)
21. Shedden, P. et al.: Towards a knowledge perspective in information security risk assessments–an illustrative case study (2009)
22. Shameli-Sendi, A., Aghababaei-Barzegar, R., Cheriet, M.: Taxonomy of information security risk assessment (ISRA). Comput. Secur. **57**, 14–30 (2016)
23. Landoll, D.: The security risk assessment handbook: a complete guide for performing security risk assessments. CRC press (2021)
24. McIlwraith, A.: Information security and employee behaviour: how to reduce risk through employee education, training and awareness. Routledge (2021)
25. Paananen, H., Lapke, M., Siponen, M.: State of the art in information security policy development. Comput. Secur. **88**, 101608 (2020)
26. Gubert, E.S., Kozminykh, S.I.: Protection of confidential document flow at the facilities of the fuel and energy complex Modern science: current problems of theory and practice. Series Nat. Tech. Sci. **11**, 60–69 (2023)
27. Ilin, I., Levina, A., Iliashenko, O.: Enterprise architecture analysis for energy efficiency of Saint-Petersburg underground. Adv. Intell. Syst. Comput. **692**, 1214–1223 (2018)

Application of Blockchain Technology in the Power Sector

Olga S. Chemeris, Sergei Kitanin, Victoria Tinyakova, Anton Pushkin, and Sofia Kalyazina

Abstract The integration of blockchain technology into the energy sector represents a paradigm shift with the potential to redefine market interactions, promote renewable adoption, and reshape industry infrastructure. However, there are significant scientific and practical challenges to be addressed to realize its full transformative potential. This article explores the multifaceted dimensions of blockchain application in energy sector, considering the potential for decentralization and transparency enhancement that could redefine industry dynamics and market roles. It critically analyzes blockchain's scalability and performance limitations that challenge its widespread implementation as a universal platform for energy infrastructures. Furthermore, the discussion extends to the security and regulatory concerns that blockchain introduces, questioning the feasibility of complete autonomy from centralized systems and the compatibility with existing legislative frameworks. The paper also investigates how blockchain could disrupt traditional energy company operations, hypothesize the emergence of new business models based on principles of decentralization and collaborative consumption, and scrutinize the power balance shift and responsibility reallocation between central power plants and local energy sources. Lastly, the implications of blockchain on the acceleration towards renewable energy sources are assessed to determine whether blockchain could act as a sufficient catalyst in this transition. In conclusion, the discourse on blockchain's impact on the energy sector uncovers contrasting points and potential research directions, necessitating a critical evaluation of current scientific hypotheses and practical

O. S. Chemeris (✉) · S. Kalyazina
Peter the Great St. Petersburg Polytechnic University, Saint Petersburg, Russia
e-mail: o.s.pogarskaya@gmail.com

S. Kitanin
Moscow University for Industry and Finance "Synergy", Moscow, Russia

V. Tinyakova
State University of Management, Moscow, Russia

A. Pushkin
EPAM Systems, Burgas, Bulgaria

I. Ilin and M. Youzhong (eds.), *Digital Systems and Information Technologies in the Energy Sector*, Lecture Notes in Networks and Systems 1244,
https://doi.org/10.1007/978-3-031-80710-7_24

implementations to unravel the technology's comprehensive impact on the energy industry and society at large.

Keywords Blockchain · Energy sector · Decentralization · Renewable energy · Energy infrastructure · Business models · Market dynamics

1 Introduction

The current environment is conducive to innovative development in key economic sectors. The focus of policy strategies on digital transformation of the economy and ecosystem, along with increased attention to environmental initiatives and the transition to green energy, creates the preconditions for increasing the technological sophistication of industries.

According to VTB Capital's analysis, one of the economically beneficial ways to achieve a 50–60% reduction in CO2 by 2050 is the process of decarbonization in power generation (Fig. 1). At the same time, a significant part of carbon dioxide reduction is expected in the areas of power generation and forestry [1].

Decarbonization requires changes in the way energy resources are used and distributed, with energy efficiency, decentralization and the integration of digital technologies in the management of energy infrastructure. These trends contribute to the reformatting of energy system architecture and the cooperation of market players on a new basis, which in turn opens the way for the application of innovative solutions, including blockchain technology, which can act as a mechanism for finding new promising business models.

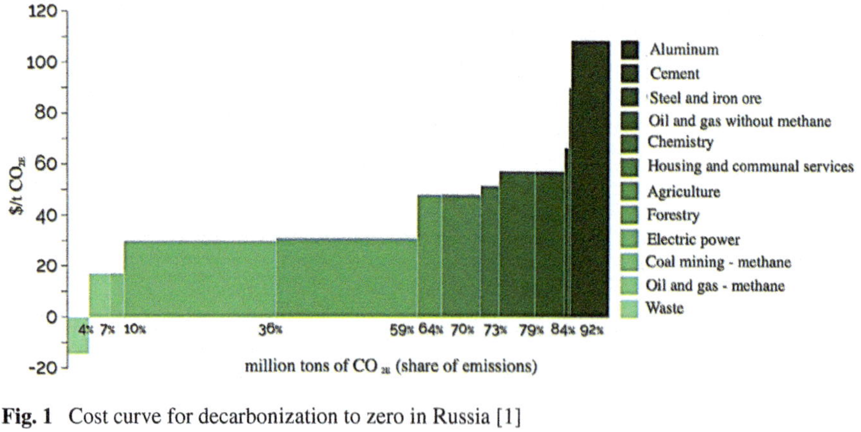

Fig. 1 Cost curve for decarbonization to zero in Russia [1]

2 Materials and Methods

The text on the application of blockchain in the energy industry was prepared using a review of academic and professional literature, reports from expert energy agencies and analytical companies, and a comparative analysis of various blockchain projects. The study was based on the analysis of public data up to 2024, including case studies and environmental assessments, with a focus on the accuracy, update and objectivity of the information provided.

3 Results

The existing system of centralized power supply has significant drawbacks:

1. Annual electricity losses reach 9.1 percent per year [2].
2. The cost of services is constantly increasing.
3. The population's debts and the total costs of ensuring the operation of the unified energy system are growing.
4. Remote regions are not electrified, and the quality of supplied energy does not meet the required indicators.

The main causes of these problems may be:

1. Defective or outdated meters.
2. Mistakes in taking readings.
3. Incorrect filling in of receipts.
4. Illegal connection.
5. Technical problems.

To solve the problems, blockchain technology, also known as distributed registry technology, may be a promising technology to provide the necessary level of information exchange, monitoring and management in the power system.

Blockchain technology is suitable for use in a variety of aspects of the electricity industry [3]:

1. Control systems monitoring. The use of blockchain will provide continuous monitoring of the activity of decentralized grid elements and real-time analytics of electricity distribution and consumption.
2. Integration with smart grids and data exchange. Blockchain can be used to create a network linking smart infrastructure devices, facilitating information exchange, enhancing data protection, and improving process management and coordination. This system will include smart meters, sensors, grid surveillance devices, control systems and controllers to rationalize consumption in smart homes and building monitoring systems.
3. Energy grid regulation. Blockchain can help establish control over energy networks by facilitating data collection, storage and analysis. This will achieve

managerial flexibility in trading and resource optimization, preventing unnecessary upgrades and potentially affecting service costs and tariffs.

4. Increased transparency in compliance verification processes. Blockchain promotes data privacy and ensures reliable maintenance of customer identities, building trust in verification processes.
5. Automated billing. By combining blockchain, smart contracts, and meters into one, a billing system for electricity consumption can be developed. Energy supply companies will be able to benefit through microtransactions, prepayment, and energy pre-orders, which will make demand more elastic and may help reduce payment arrears.

One of the vectors of energy development is the transition to distributed generation and building the Internet of Energy (IoE).

The use of the Internet of Energy will lead to increased efficiency of energy infrastructure by improving infrastructure reliability and increasing data reliability through the use of smart meters and other IoT technologies.

According to the analysis by MINDSMITN and the Center for Energy at the Moscow School of Management SKOLKOVO of the blockchain landscape of power projects and investors, government investors are focusing on the development of trading platforms, payment systems and scalable blockchain solutions, which is the application of blockchain as software. The specifics of government support vary from country to country. For example, the European Union emphasizes the development of innovative and systemic solutions, including green energy. In Australia, the priorities are divided between renewable energy and ensuring the reliability of electricity supply. In Asia–Pacific, policies are focused on opening markets and incorporating renewable energy, which is driving demand for blockchain projects. In particular, the rapid growth of blockchain projects in the US, which has enabled the country to emerge as a leader, is the result of an active government strategy to invest in digital initiatives and, in particular, blockchain solutions for the energy sector. The government's contribution in the US is noticeably reflected in the use of blockchain to build reliable energy systems, which is an element of enhancing the country's national security [4].

According to MINDSMITN and the Center for Energy at the Moscow School of Management SKOLKOVO, blockchain projects in the energy sector account for 6.4% of all projects among all investors (Fig. 2), and in particular, 7.7% of projects among government investors are data management projects (Fig. 3). These projects involve recording information about energy production and utilization, system transactions and ownership and management of property on the basis of blockchain technology. Smart metering systems also fall into this category. This type of projects, for example, includes the Spanish Pylon Network, which organized a database of energy data (on electricity consumption and production) that allows energy service companies to form new innovative services [4].

Validation and data integrity functions can best be realized on the basis of a distributed registry. For example, the Erachain Blockchain Platform automates billing

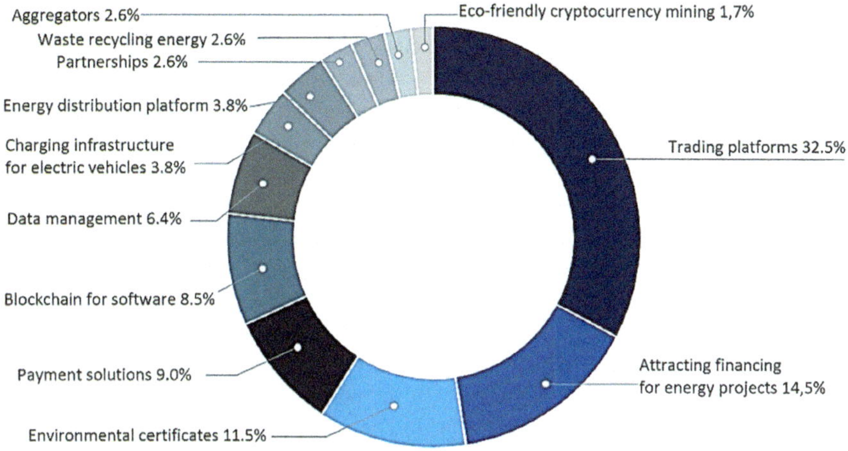

Fig. 2 Typology of blockchain projects in the energy sector [4]

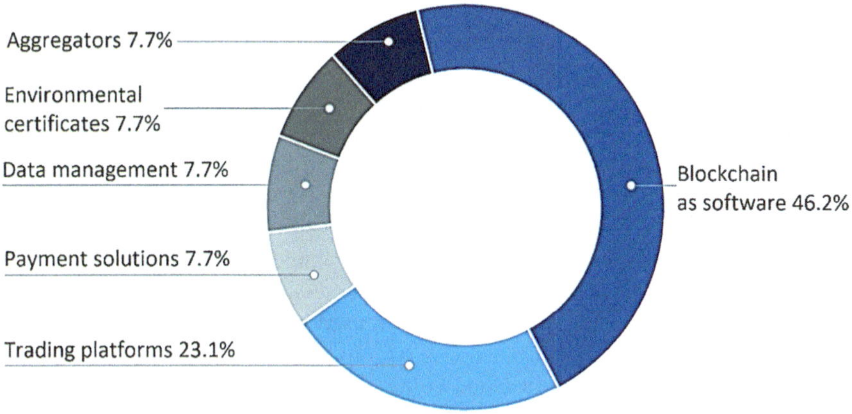

Fig. 3 Types of blockchain projects that have attracted funding from public investors [4]

along the entire energy supply chain from generators to end users and provides extensive protection against unauthorized access to data at all levels.

The system allows creating records not only by the amount of energy resources, but also for a certain period of time. The platform can be integrated into the infrastructure of a city, a region or an entire country. Erachain allows you to create verified profiles of participants in the Internet of Energy ecosystem. This makes it possible to control all changes in meter software and changes in configuration parameters on specific metering devices.

The construction of distributed generation facilities creates a new market model where consumers can also be electricity producers, and all participants freely exchange energy resources and services. In a microgrid system, blockchain technology enables direct contractual relationships between consumers and producers of electricity. The key advantage of conducting transactions via blockchain is that all electricity supplied to power grids can be unambiguously allocated to the accounts of specific consumers within a very short time—up to a few minutes.

In this case, the calculation for all electricity produced and consumed can be done very accurately at variable prices. Simplifying the mutual settlement process will reduce the amount of balancing energy that market participants are billed for.

The use of blockchain technology will provide consumers with greater transaction transparency. Consumers will be able to track where the electricity they buy was generated.

As part of the transition to a more complex intelligent model of distributed energy, the Erachain platform was integrated with the program of electricity metering and remote data collection "YaEnergetik" [5]. The implementation of YaEnergetik in Pyaterochka stores to automate metering for 110 stores in Moscow ensured 100% generation of reports on hourly consumption and managed to completely get rid of energy supply company penalties, resulting in savings of 3 million rubles per month [6].

As one of the realizations of reducing electricity consumption by the end user it is worth considering Server Boilers with heating from the processors of the internal computing unit performing cloud computing of the AIZ company [7]. The demand for complex computational operations is continuously increasing, which leads to the creation of large-scale data centers equipped with significant server capacities. These processors, while performing computations, consume a significant amount of energy and generate heat, which in turn leads to additional energy costs for cooling. About 35–45% of a data center's budget is spent on cooling systems, and up to 30% of the cost is spent on the electricity used by the servers. Thus, energy is first used to process data and converted into heat, and then used again to keep the equipment at an optimal temperature.

With the growing interest in cloud computing, many educational and research organizations are spending money on renting computing power. AIZ experts have developed an innovative system of decentralized data centers in the form of a series of electric boilers embedded with servers, enabling highly complex computing over a network. The server boiler, connected to the internet, provides computing services and simultaneously heats rooms.

A person can use this computing power or rent it out to perform complex calculations such as scientific research (Fig. 4). The boiler is connected to the Internet through a standard RJ45 connector, allowing automatic download and retrieval of calculation results, and also the boiler can become part of a remote computing network of any organization [8].

Like any solution, blockchain technology has its limitations and challenges [9–11]. The integration of blockchain technology into the energy sector implies the creation of a decentralized network that can significantly improve the interaction

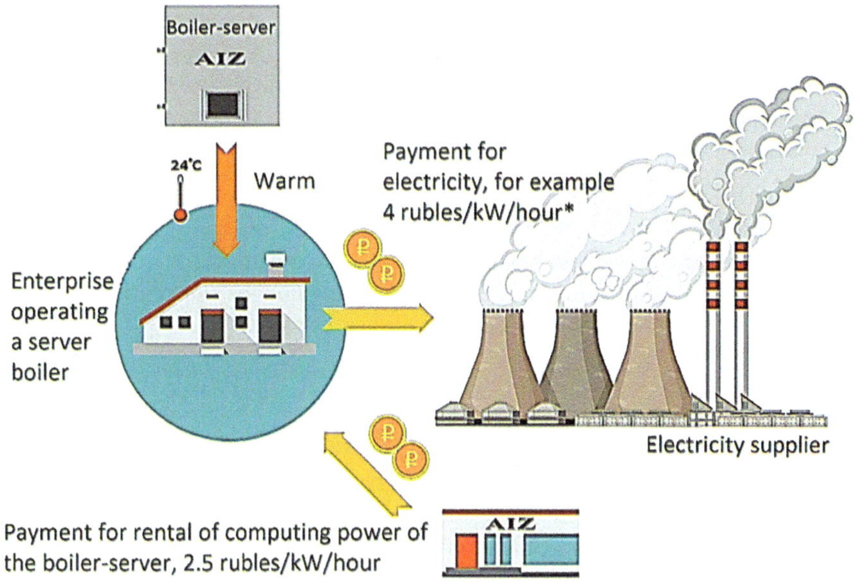

Fig. 4 Option of payment to person for calculations made on the boiler-server [8]

between energy producers, suppliers and consumers, thereby increasing the reliability and transparency of grid operations. However, the practical application of blockchain faces a number of challenges:

1. A necessary step for blockchain technology adoption is the modernization of existing infrastructure, which will require significant capital investment in upgrading hardware, software and staff training. This may also include the development of new energy metering devices capable of transmitting data to the blockchain network in real time.
2. Regulatory alignment and regulatory complexities. Since the energy sector is highly regulated in most countries, any innovation must meet existing legal requirements or wait for new regulations to be adopted. This requires active stakeholder dialog with regulators to ensure legal clarity and compliance.
3. Technical and scaling challenges also present a significant barrier. Blockchain infrastructure must be designed so that it can handle large volumes of transactions without delays and additional performance costs. Difficulties can also arise due to a lack of skilled labor or high network load with large numbers of transactions.
4. Cybersecurity and data protection issues require the development of strong encryption algorithms and solutions that will prevent unauthorized access to sensitive information and eliminate the possibility of manipulation within the network. Preserving customer privacy is a critical aspect as energy consumption data can be used for unscrupulous purposes.

5. The need for comprehensive cost-benefit analysis studies before embarking on a large-scale deployment of blockchain technology. A cost-benefit analysis, as well as an up-to-date risk assessment report, will help avoid economically unprofitable investments and enable organizations to make informed decisions based on complete and accurate data.

Within Russia, the movement towards energy efficiency, decentralization and integration of digital technologies in the management of energy infrastructure is progressing smoothly [12] and already has real examples working in the economy. The general trend both within large companies and within the end user is to embrace new opportunities for cost optimization in the power sector.

4 Discussion

The application of blockchain in the energy sector opens new horizons for industry transformation, but presents a number of scientific and practical challenges. The main scientific interest is the ability of blockchain to provide decentralization [12–14] and increase transparency, which can be a determining factor in developing hypotheses about the future interaction of energy market participants. However, there are zonal limitations on the scalability and performance of blockchain networks [15], which casts doubt on the possibility of their widespread use as a universal platform for energy infrastructure.

In the area of security and regulation, blockchain offers new mechanisms to create a genuine and resilient digital environment. This could dramatically change current approaches to data protection and network management, but raises doubts about the possibility of full autonomy from centralized systems, especially in the context of legal regulations.

The transformational potential of blockchain also has the potential to redefine the role and functions of traditional energy companies, stimulating hypotheses about the emergence of new business models based on decentralization and collaborative consumption. This may prompt critical analysis of how the balance of capacity and responsibility between centralized generating stations and local energy sources will change.

Finally, the impact of blockchain on the transition to renewable energy remains an open question. While the technology provides tools to facilitate decentralized energy trading and consumption management, it is unclear whether this will be a sufficient incentive to accelerate the integration of green technologies.

In conclusion, the discussion of blockchain's impact on the energy sector reveals a number of controversial points and potential areas for further research. It is important to continue to critically evaluate this technology in terms of current scientific hypotheses and feasibility in order to uncover all aspects of its impact on the energy industry and more broadly on society and the environment.

5 Conclusion

In conclusion, the opportunities that blockchain presents for the energy sector represent significant prospects for decentralization, increased reliability and transparency in the industry. However, realizing these opportunities requires overcoming significant challenges, such as the need for significant upfront investment, regulatory harmonization, technical complexities, cybersecurity, and cost–benefit and risk assessment studies. Nevertheless, the drive for energy efficiency and digital inclusion is already reflected in real projects in Russia and elsewhere.

Utilizing blockchain together with the Internet of Things, big data and machine learning in the context of the Internet of Energy will achieve the following system-wide effects:

1. Establishment of civilized relations between all participants of the process of generation, transmission, sales and consumption of energy resources.
2. Control over the reliability of the amount of energy resources submitted for payment by the supplier.
3. Saving budgetary and consumer funds to pay for resources.
4. Increasing the flexibility and efficiency of the country's energy system.

Progress on blockchain adoption in the Russian energy sector is driven by the opportunity to build an ecosystem and facilitate interaction between key players. The main role here will be played by establishing communication between energy companies, government agencies and developers of technological solutions. The success of blockchain initiatives depends on factors such as the quality of communication between market participants, consideration of general trends and unique regional conditions.

Building on these developments, we can expect to see the gradual adoption of blockchain technology as an important component of the future of energy. By overcoming initial barriers and innovating, companies and governments will be able to create a more flexible and interconnected energy system that better meets the needs of all market participants, from producers to end users. Successful and sustainable blockchain-based energy development will require not only technological advances and funding, but also the active engagement of all stakeholders—businesses, citizens and government agencies.

References

1. Scenarios of decarbonization in Russia, https://sk.skolkovo.ru/storage/file_storage/9b713c16-cfb4-4da3-aeaa-66061e1d14f5/SKOLKOVO_EneC_RU_Decarbonisation_Scenarios_in_Russia.pdf. Last accessed 2024/07/06
2. Pinto, R., et al.: The rise and stall of world electricity efficiency: 1900–2017, results and insights for the renewables transition. Energy **269**, 126775 (2023)

3. Belkin, P.A., Posmakov, N.P., Rostovsky, N.S.: Application of blockchain technology in the electric power industry as a cohesive digital technology in the transition to decentralized generation. Mod. Sci. Intensive Technol. **3**, 19–24 (2020)
4. Blockchain in the power sector: the landscape of projects and investors, https://energy.skolkovo.ru/downloads/documents/SEneC/Research/SKOLKOVO_EneC_RU_MINDSMITH_blockchain_investment_landscape2019.pdf. Last accessed: 2024/07/06
5. Completed Projects—Blockchain as a Service Provider Erachain, https://erachain.ru/assets/docs/Erachain_Company(ru).pdf. Last accessed: 2024/07/06
6. IEnergetik ASCME implementation results at our clients, https://yaenergetik.ru. Last accessed: 2024/07/06
7. Server boiler AIZ-KT18–2 with heating from processors, https://aiz.com/product/aiz-kt18-2. Last accessed: 2024/07/06
8. AIZ—How it works, https://aiz.com/how-it-works. Last accessed: 2024/07/06
9. Zielińska, A.: Application possibilities of blockchain technology in the energy sector. E3S Web Conf. **154**, 07003 (2020)
10. Chemeris, O.S., Borremans, A.D., Tick, J.: Analysis of Economic Consequences of digital solutions in logistics on the example of Russian railways holding. In: Ilin, I., Petrova, M.M., Kudryavtseva, T. (eds.) Digital Transformation on Manufacturing, Infrastructure & Service. DTMIS 2022. Lecture Notes in Networks and Systems 684. Springer, Cham (2023). https://doi.org/10.1007/978-3-031-32719-3_74
11. Mrowiec, D., Sołtysik, M.: Examples of possible blockchain technology applications in the power engineering sector. Zeszyty Naukowe Instytutu Gospodarki Surowcami Mineralnymi Polskiej Akademii Nauk, 133–144 (2018)
12. Maydanova, S., Ilin, I.: Strategic approach to global company digital transformation. In: Proceedings of the 33rd International Business Information Management Association Conference, IBIMA 2019: Education Excellence and Innovation Management through Vision, pp. 8818–8833 (2019)
13. Poljanskihh, A., Levina, A., Dubgorn, A.: Investment in renewable energy: practical case in Estonia. MATEC Web Conf. **193**, 05065 (2018)
14. Sokolov, B., Ushakov, V., Zaharov, V.: Optimal planning and scheduling of information processes during interaction among mobile objects. Int. J. Prod. Res. **62**(16), 5905–5924 (2024). https://doi.org/10.1080/00207543.2024.2302388
15. Andoni, M., et al.: Blockchain technology in the energy sector: a systematic review of challenges and opportunities. Renew. Sustain. Energy Rev. **100**, 143–174 (2019)
16. Di Silvestre, M.L., Gallo, P., Guerrero, J.M., Musca, R., Sanseverino, E.R., Sciumè, G., Zizzo, G.: Blockchain for power systems: current trends and future applications. Renew. Sustain. Energy Rev. **119**, 109585 (2020)
17. Nour, M., Chaves-Ávila, J.P., Sánchez-Miralles, Á.: Review of blockchain potential applications in the electricity sector and challenges for large scale adoption. IEEE Access **10**, 47384–47418 (2022)
18. Zhu, S., Song, M., Lim, M.K., Wang, J., Zhao, J.: The development of energy blockchain and its implications for China's energy sector. Resour. Policy **66**, 101595 (2020)
19. Wu, J., Tran, N.K.: Application of blockchain technology in sustainable energy systems: an overview. Sustainability **10**(9), 3067 (2018)
20. Zhang, T.Y., Feng, T.T., Cui, M.L.: Smart contract design and process optimization of carbon trading based on blockchain: the case of China's electric power sector. J. Clean. Prod. **397**, 136509 (2023)
21. Mika, B., Goudz, A.: Blockchain-technology in the energy industry: Blockchain as a driver of the energy revolution? With focus on the situation in Germany. Energy Syst. **12**(2), 285–355 (2021)
22. Basova, A.: Blockchain as a platform to digitally transform the electricity sector. IOP Conf. Series Earth Environ. Sci. **990**(1), 012056 (2022)
23. Henly, C., Hartnett, S., Endemann, B., Tejblum, B., Cohen, D.S.: Energizing the future with blockchain. Energy LJ **39**, 197 (2018)

Energy Saving Management in Russian Agriculture Using Digital Technologies

Elena Kuznetsova, Anna Kogteva, Nataliya Shevtsova, Oksana Balabneva, and Erik Akimjanov

Abstract The article defines the legislative basis for the process of digitalization of Russian agriculture, analyzes the level of digitalization of Russian agricultural organizations, identifies the main directions of digital technologies actively used in the agricultural sector and the level of their implementation in business entities. The article reveals priority software and hardware digitalization tools that are promising for managing energy saving in Russian agriculture, taking into account the specifics of the industry, the material and technical state of agricultural enterprises, the state of the market for Russian and foreign digital technologies, their accessibility, the level of development of Russian science and the innovation sphere. Particular attention is paid to the development of a precision farming system, which is the basis for the rational use of energy capacities and fuels and lubricants in crop production, and the prospects for managing energy saving in greenhouses and the livestock industry are highlighted. The issues of saving water resources and the use of alternative energy sources are touched upon. The main problems of digitalization of agriculture are also identified, which impede the intensification of the process of introducing digital technologies into the agricultural sector, including as part of energy saving management.

Keywords Digitalization · Agriculture · Energy saving management

E. Kuznetsova · N. Shevtsova
Voronezh State Agrarian University Named After the Emperor Peter the Great, Voronezh, Russia

A. Kogteva (✉)
Finance University under the Government of the Russian Federation, Moscow, Russia
e-mail: ankogteva@fa.ru

O. Balabneva
Peter the Great St. Petersburg Polytechnic University, Saint Petersburg, Russia

E. Akimjanov
Kostanay Academy of the Ministry of Internal Affairs of the Republic of Kazakhstan Named After Shrakbek Kabylbaev, Kostanay, Republic of Kazakhstan

I. Ilin and M. Youzhong (eds.), *Digital Systems and Information Technologies in the Energy Sector*, Lecture Notes in Networks and Systems 1244,
https://doi.org/10.1007/978-3-031-80710-7_25

1 General Provisions and the State of Digitalization of Agriculture in Russia

1.1 Regulatory Framework and Priority Directions of Digitalization of Agriculture in Russia

New realities of the agri-food market, of the socio-economic and political transformations, of increased global and local competition required the intensification of digitalization processes in all sectors of the economy, including in the agricultural sector.

The legal framework for the implementation of digitalization of the agricultural sector in Russia has become a package of regulations adopted in the last decade at various levels and the impact on the agriculture and related industries [1–15].

According to Russian legislation, digital means agriculture based on modern methods of producing agricultural products and food using digital technologies that ensure increased labor productivity and reduced production costs [15].

The following digital end-to-end technologies have been identified as the main strategic directions in the field of digital transformation of the agricultural sector for the next decade:

- Digital twins;
- Artificial intelligence technologies;
- Internet of things;
- Unmanned aerial vehicles and agricultural machinery and robotics;
- Earth remote sensing;
- Satellite communication and positioning systems;
- Big data processing;
- Sensors and beacons with a satellite data transmission channel, etc. [12]

1.2 Analysis of the Level of Digitalization of the Russian Agricultural Sector

Since the beginning of the intensification of the digitalization process in Russian agriculture, certain changes and results have already been obtained, but so far this economic branch lags behind other sectors in many indicators. According to the research by the Higher School of Economics, in 2022, fixed broadband Internet access with data transfer speeds above 100 Mbit/s is provided in only 7.3% of agricultural organizations. In the structure of internal costs, the expenses of agricultural enterprises for the creation, distribution and use of digital technologies and related products and services amounted to 7.9 billion rubles, or 0.3% of total costs, which in absolute terms is almost 5 times lower than expenditure in construction, in wholesale trade—more than 23 times, in the financial sector—63 times. Also, the cost of

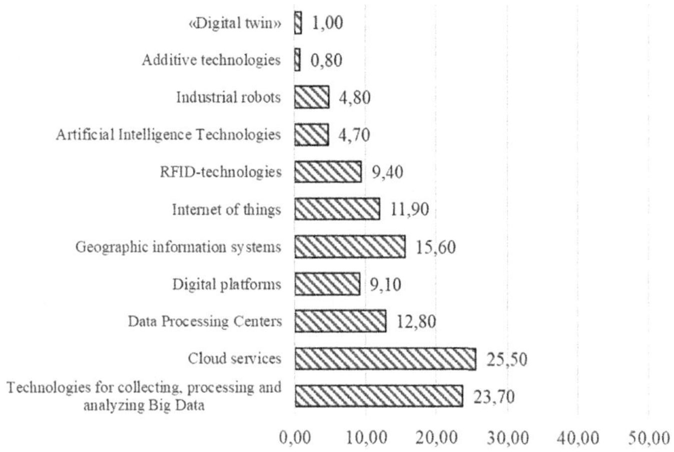

Fig. 1 Use of digital technologies in Russian agricultural organizations (% of the total), 2022 [16]

paying third-party organizations for the implementation of digital technologies for agricultural enterprises amounted to 1.9 billion rubles, which is lower than almost all other values observed in economic sectors [16].

Among agricultural organizations, the most popular were cloud technologies and big data technologies, which were implemented in 25.5% and 23.7% of sector organizations, respectively (Fig. 1). Geographic information systems, data centers and the Internet of Things are also relevant, used in 15.6%, 12.8% and 11.4% of agricultural enterprises. RFID technologies and systems, robotics and artificial intelligence systems are gaining momentum. [16]

Among the software tools, the most in demand are programs for automating financial settlements, electronic document management systems, reference and legal systems, and information security systems (Fig. 2). Just over 20% of enterprises use various management systems for purchasing, sales, warehouses, access to databases and training programs. Less actively introduces are the systems for managing relationships with clients, personnel management, integrated resource management, enterprise management and individual technological processes.

Activation of digitalization of the agricultural sector has become an important condition for ensuring the growth of energy saving of production and management processes in the agricultural sector and the rational use of energy resources. The latter in the agricultural sector include fuels and lubricants, thermal energy, gas and electricity, and in recent decades—alternative energy sources.

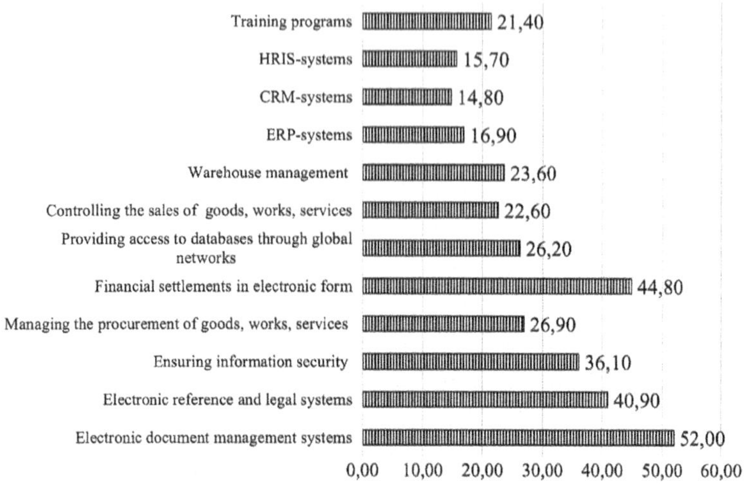

Fig. 2 Use of software in Russian agricultural organizations (% of the total), 2022 [16]

2 Directions for the use of Digital Technologies in the Agricultural Energy Saving Management System

From 2000 to 2022, the amount of capacity used by agricultural enterprises in absolute and relative size (per 100 hectares of crop area) decreased significantly: by 62.1% and 39.8%, respectively (Fig. 3). These trends are associated with the improvement of technologies for the production of agricultural products, scientific and technical progress in mechanical engineering for the agro-industrial complex, renewal of the machine and transport fleet in recent decades, partial modernization of energy supply enterprises, etc.

One of the modern factors of energy saving in the agricultural sector in recent years has been the digitalization of the industry, which has predetermined the following priority areas of energy saving.

1. Minimizing the consumption of fuels and lubricants, heat and electricity in technological operations in agricultural sectors.

First of all, we should note the introduction of precision farming technology, including the integrated use of the following components:

- Remote sensing of the Earth, which guarantees the collection of information about an object, territory or phenomenon through all types of non-contact surveys using various measuring platforms: aircraft and spacecraft, ships and submarines, ground stations. These technologies make it possible to map agricultural lands and adjacent territories, create reference plans and attribute databases, monitor the condition of crops based on the calculation of vegetation indices, assess the

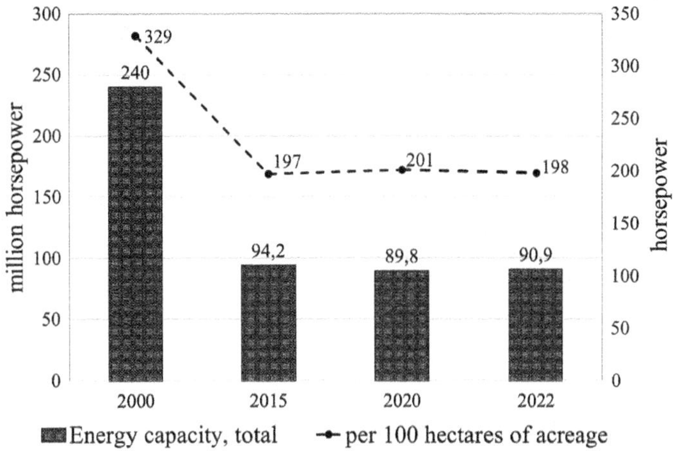

Fig. 3 Use of energy capacity in Russian agricultural organizations [24]

productivity of agricultural crops and forecast their yield, and map soil characteristics. Land cover, land degradation areas, planning and monitoring of agrotechnical and other activities;

- Satellite communication and positioning systems: GPS (USA), (GLONAS (Russia), GALILEO, COMPASS BEIDOU (China), Navigational Satellite System (IRNSS) (India);
- Geographic information systems, which are the basis of support systems for making technological decisions in precision agriculture and are designed for storing, analyzing, processing and changing spatial data. These systems form a digital model of the area on which agrotechnical operations are carried out, information on remote sensing, information on the properties and characteristics of soils, maps of crops by year, history of field cultivation, etc. Among Russian developments, the most popular are GeoGraph GIS, Panorama, etc.;
- Parallel and automatic driving systems that help automate the process of controlling agricultural machines. In this case, global satellite navigation systems are used to determine the location of equipment on the field and automatically control its movement along a predetermined route. Well-known Russian developments are such as "GloNASH" and "KURSACH AT5" from the Agrosturman Group of Companies, automatic steering systems from Getteile NAVI;
- Yield mapping systems such as Trimble, AFS (CaseIH), Ag Leader, GreenStar (John Deere), GrainTrak, (Micro-Trak), Agromap (Claas), RSM Yield map from Rostselmash;
- Differential fertilization systems, etc. [17].

Thus, a precision farming system RSM Agrotronic from Rostselmash (Rostov-on-Don) has been developed, which allows you to obtain remote control over technological processes, optimize operating modes of equipment, plan and effectively manage a fleet of equipment in real time.

An important digital addition is the use of artificial intelligence systems in conjunction with the Internet of Things. The industry actively uses sensors and equipment for managing agricultural and automobile transport to save energy: GPS navigators used in unmanned vehicles; tire pressure sensors; fuel control sensors; door (body) opening sensors; axle load sensors; tilt sensors; engine hour meter; tachographs; starter lock relay; GPS lock-seals; accelerometers, etc. [18].

Drones play an important role in crop production, which are used to apply fertilizer and control pests; for spraying; for assessing the condition of agricultural crops, for irrigation; for landing; for soil and field analysis, etc. To ensure the integrated use of precision farming, special sensors are used that transmit information about the state of crops (dendrometers; fruit size sensors; moisture sensors of fruits and leaves, etc.). Also the sensors are important that measure and transmit information about soil characteristics (soil moisture sensors, soil temperature sensors, oxygen content sensors, water volume fraction sensors, etc.), climate sensors (light sensors; radiation sensors; sensors humidity; temperature sensors; atmospheric pressure sensors, etc.; high-precision weather stations) [18].

These technologies together make it possible to reduce the consumption of fuels and lubricants for production processes, optimize the movement of agricultural machinery and equipment, control the execution of technological operations, and prevent fuel theft.

It should be specially noted that it is possible to reduce the consumption of electricity and heat in greenhouses, provided by the use of microclimate control systems based on sensors of light, humidity, CO2, air temperature, soil moisture and controlled automatically, which is especially important for fifth and sixth generation greenhouses. Active use of climate control devices can minimize energy consumption. Some foreign research has also developed various digital twin platforms as decision support tools for precise control and monitoring of microclimate parameters, including the thermal environment [19, 20]. Container farms are also becoming a trend, allowing energy modeling to take into account crop growth and energy saving potential in different climatic conditions [21].

In the livestock sector, energy saving in modern conditions is achieved by using microclimate control systems in enclosed spaces (poultry farms, pig farms) also using sensors and the Internet of things, big data analysis and the use of artificial intelligence systems, which also allow automating processes feeding, watering, manure removal.

An alternative to using controlled mechanization means is the introduction of robotics of various directions. For the plant growing industry in the global practice, robots are used to automate work in greenhouses and vegetable growing in protected soil: robots for planting flowers; lettuce thinning robots; hydroponic system for growing and harvesting strawberries; field robot for grafting fruits and vegetables, autonomous watering of plants; assistant robots for greenhouses, etc. To automate work in gardening, a robot sprayer has been developed; robot spreader; fertigator robot; robot mower; robot mulcher; robotic platform for organizing intra-garden logistics of equipment, agrochemicals, products and waste; pruning robot; citrus harvesting robot; fruit counting robot, etc. [22].

As a part of the automation of agricultural field work, lightweight autonomous robots are used for weeding; robots and drones for spot spraying; spreader with nitrogen-spectral sensors; unmanned tractors, incl. electric tractors, etc. Automation of feed procurement can be carried out by a robotic SUV for harvesting, mowing, removal and research; unmanned-controlled forage harvesting equipment. Among Russian developments, the most famous is an autonomous cableless mini-tractor controlled by artificial intelligence, developed by the Cognitive Pilot company. Robots are also actively used to automate additional work: mini-loader robots; mobile laboratories for soil chemical analysis; system for taking soil samples, etc.

In livestock farming, robots are used to automate animal feeding (mixer robot; feed preparation robot; dispensing robot; leveling feed poured on the floor; bot for—automation of milking of animals: milking robots) and automation of additional work (bot for cleaning the barn, removing manure, washing and disinfecting the floor), as well as robotic shepherds with functions for assessing the condition of pastures, monitoring the condition of animals, etc. These types of robotics have been developed and are actively used in European countries and the USA [22].

The Smart4Agro company (Yaroslavl) implements services for managing a dairy farm, monitoring storage facilities, monitoring, analyzing and managing greenhouse farms, monitoring ripening and storage indicators in silo trenches, etc. The Technoserv group of companies (Moscow) has developed IoT technology for beef cattle farming, production asset management systems based on IoT and GLONASS, etc.

The variety of AI applications has increased significantly in the last decade, especially in forecasting applications and optimization applications, respectively, with research literature mainly focusing on bioenergy (55%), energy use analysis (17%), process value chain (6%), energy efficient irrigation (6%), greenhouse energy (6%), livestock management (2%), farm capacity and technology (4%) and risk management (4%) [23].

2. Reducing water consumption for production needs. This direction is relevant not only in Russia, but also on a global scale and is caused by global warming, a decrease in the volume of groundwater and a general decrease in water quality due to the intensification of agricultural production, the increase in the application of mineral fertilizers, and plant protection products. According to Rosvodresurs of the Russian Federation, in 2022, the volume of fresh water used by agricultural enterprises for irrigation amounted to 6,199 million m3, the volume of fresh water for agricultural water supply—257 million m3, which is 4.4% lower than the previous year. And 15.7% respectively [24]. However, the trend of water consumption in agriculture is unstable in dynamics, which is explained both by the influence of natural and climatic factors, and by the level of organization and budget of water supply and water use, material and technical support both from farms and from the state.

In order to reduce water consumption while maintaining the efficiency of agricultural production, a systematic scientific substantiation of the need for water resources and their rational use is required. For these purposes, traditional directions are used, such as the formation of scientifically based crop rotation, taking into account the specific moisture needs of individual crops, the selection of drought-resistant crops, planting cover crops and forest plantations, etc. However, basic methods must be complemented and enhanced by new digital technologies, such as the use of air humidity sensors, incl. complete with "smart" weather stations, soil moisture, and smart drip, sprinkler and underground irrigation systems built on their basis, aeroponics and hydroponics systems (for greenhouses with a high level of automation). Smart irrigation installations help reduce water consumption, financial costs for irrigation, and conserve water resources. In world practice, artificial intelligence systems are being actively implemented to optimize water use. For example, a device has been developed to optimize the functioning of agricultural pumping stations, implemented using a neural network to predict the need for irrigation water depending on meteorological conditions and a programmable controller to control the electric pumps of the pumping station [25].

Increasingly, scientists are paying attention to alternative water sources, for example, in the United States, the reuse of filtered disinfected municipal wastewater, desalinated brackish water, purified produced water and seawater is being actively implemented. Economic feasibility can be increased through the introduction of autonomous, easy-to-operate, renewable energy-powered, decentralized desalination systems [26].

3. Use of energy-efficient machine and tractor fleet. Due to the economic and political situation in the world, the limited import of modern agricultural machinery, machinery and equipment has somewhat reduced the possibility of updating the machine and tractor fleet: by 2023, 1 tractor will account for 369 hectares of arable land, 1 grain harvester—454 hectares crops, for 1 corn harvester—2418 ha, potato harvester—72 ha, beet harvester—490 ha. Since 2000, the load on these agricultural machines has increased by 73.3%, 30.0%, 19 (!) times, 2.3 times and 6.9 times, respectively (Table 1).

At the beginning of 2023, 196.7 thousand tractors, 64.1 thousand seeders, 52.3 thousand grain harvesters were registered in agricultural enterprises, 10.7 thousand

Table 1 Provision of agricultural organizations with tractors and harvesters in the Russian Federation [24]

Indicator	2000	2005	2010	2015	2020	2023
Arable land load per tractor, ha	135	181	236	308	349	369
Crop land of the relevant crops per a machinery type unit, ha:						
Harvester						
Grain harvester	198	253	327	422	451	454
Corn harvester	120	215	817	2008	2974	2418
Potato harvester	22	31	62	67	66	72
Beet harvester excluding tops removing machines	62	93	278	396	431	490

units of forage harvesters, 1.8 thousand beet harvesters. Their number since 2015 has decreased by 18.7%, 31.5%, 14.8%, 23.5% and 18.1%, respectively.

Total production in Russia in 2022 amounted to 10.6 thousand tractors, 13.6 thousand seeders, 4.5 thousand grain harvesters, 43.5 thousand plows, 52.9 thousand cultivators, 2,406 fertilizer spreaders and etc., which is higher than the level of the previous several years.

In the livestock industry, the situation is less satisfactory: while 4.0% of milking installations were written off in 2022, the percentage of their renewal was only 2.5%. In 2022, agricultural enterprises had 19.6 thousand milking units, which is 21.9% lower than in 2015. The production of milking machines in 2022 amounted to 2.8 thousand units, which is 30.0% lower than the previous year [24].

In general, the degree of depreciation of fixed assets in agriculture in 2022 amounted to 41.3%, including machinery and equipment—52.4%. The share of completely worn-out fixed assets in the agricultural sector in 2022 was 1/10 [24].

4. Production and use of alternative types of energy. These include solar energy, wind, geothermal, hydropower and biomass processing. High rates of price growth in recent years for electricity both in the world and in Russia, and the increasingly complex environmental situation require increased use of non-traditional energy sources in agriculture. Switching to alternative energy sources for energy management in agriculture holds great promise for reducing greenhouse gas emissions, improving energy efficiency and promoting sustainability in food production. For the foreseeable future, solar energy is the only source of completely renewable energy. Sustainable solar energy can be produced through a variety of means, including wind turbines, hydroelectric power plants, solar thermal collectors and photovoltaic cells [21, 27]. However, the peculiarities of climate diversity in the Russian Federation predetermine the focus on biomass processing as priority sources [28–31].

The most common problems in intensifying the digitalization process both for the agro-industrial complex and for its energy complex are:

- Low level of awareness about accessible and effective digital solutions, features of their implementation, configuration and further operation;
- Lack of a unified information base or platform with a list of software and hardware available in modern realities with a description of the functionality, price range, developer or integrator data, capable of concentrating data on advanced digital technologies in the agricultural sector, including in the energy complex;
- The presence of uniform interaction interfaces and communication protocols when implementing IT solutions and their subsequent integration with existing complexes and specialized equipment and infrastructure;
- Shortage of highly qualified, highly specialized personnel. In agricultural organizations specialists in information and communication technologies make up only 0.2% of the total number of employees, other specialists who intensively use ICT—only 3.4%; [32].

- Insufficient government support for the state with low financial accessibility of digital solutions for medium and small businesses [33–35].

Thus, the digitalization of agriculture makes it possible to increase labor productivity through the use of high-speed and productive equipment, automation of most technological operations, uninterrupted and flexible operation of equipment, operational monitoring of the condition of production facilities and precise regulation of the functioning of technical equipment, optimization of modes and operating time, managing the consumption of working capital, reducing the risk of unexpected breakdowns. Accordingly, energy supply management using digital technologies makes it possible to reduce energy costs through operational controlling and timely adoption of management decisions using cloud technologies, big data and other digital technologies, and reduce the risk of cases of misuse of energy resources.

References

1. On the development of agriculture: Federal Law of the Russian Federation No. 264-FZ: adopted by the State Duma on December 22, 2006; approved by the Federation Council on December 27. 2006. Collections of legislation of the Russian Federation (2007)
2. Doctrine of food security of the Russian Federation: approved by the Decree of the President of the Russian Federation dated January 21 (2020)
3. Strategy for scientific and technological development of the Russian Federation: approved by Decree of the President of the Russian Federation dated December 1, No. 642. (2016)
4. On measures to implement state scientific and technical policy in the interests of agricultural development: Decree of the President of the Russian Federation dated July 21, No. 350. (2016)
5. On national goals and strategic objectives of the development of the Russian Federation for the period until 2024: Decree of the President of the Russian Federation of May 7, No. 204. (2018)
6. On national development goals for the period until 2030: Decree of the President of the Russian Federation of July 21, No. 474. (2020)
7. On the strategy for the development of the information society in the Russian Federation for 2017–2030: Decree of the President of the Russian Federation dated May 9, No. 203. (2017)
8. Passport of the national project National program, "Digital Economy of the Russian Federation": approved by the minutes of the meeting of the Presidium of the Council under the President of the Russian Federation for Strategic Development and National Projects dated June 4, No. 7, (2019) https://spa.msu.ru/wp-content/uploads/4-1.pdf. Last accessed 2024/07/15
9. Federal scientific and technical program for the development of agriculture for 2017–2030: approved by Decree of the Government of the Russian Federation of August 25, No. 996. (2017)
10. Concept of technological development for the period until 2030: approved by Decree of the Government of the Russian Federation dated May 20, No. 1315-r, (2023) https://www.garant.ru/products/ipo/prime/doc/406831204/. Last accessed 2024/07/15
11. Strategy for the development of the agro-industrial and fishery complexes of the Russian Federation for the period until 2030: approved by Order of the Government of the Russian Federation of April 12, N 993-r, (2020) https://www.garant.ru/products/ipo/prime/doc/405172287/, Last accessed 2024/07/15
12. On approval of the strategic direction in the field of digital transformation of the agro-industrial and fishery sectors of the Russian Federation for the period until 2030: Order of the Government

of the Russian Federation of December 29, No. 3971-r, (2021) https://www.garant.ru/products/ipo/prime/doc/403236609/. Last accessed 2024/07/15
13. Forecast of scientific and technological development of the agro-industrial complex of the Russian Federation until 2030: approved by Order of the Minister of Agriculture of the Russian Federation No. 3 dated January 12, (2017), https://www.garant.ru/products/ipo/prime/doc/71499570/. Last accessed 2024/07/15
14. Departmental digital transformation program of the Ministry of Agriculture of the Russian Federation for 2021–2023: approved by the presidium of the Government Commission on Digital Development, the use of information technologies to improve the quality of life and business conditions, protocol dated 10/07/2021 No. 34, https://base.garant.ru/402990172/. Last accessed 2024/07/15
15. Concept of "Scientific and technological development of digital agriculture "Digital Agriculture". Ministry of Agriculture of the Russian Federation, (2019), https://www.mcxac.ru/upload/iblock/97d/97d2448548e047b0952c3b9a1b10edde.pdf. Last accessed 2024/07/15
16. Abashkin, V.L., Abdrakhmanova, G.I., Vishnevsky, K.O., Gohberg, L.M. et al.: The digital economy: 2024: a brief statistical collection. 124 p., ISIEZ VSHE, Moscow (2024)
17. Ulezko, A.V., Kulev, S.A., Goryukhina, E.Y., Kusmagambetov, S.M., Kuznetsova, E.D.: Information technologies in professional activities. Textbook for students studying in the specialty 38.05.01 Economic security. 359 p. Federal State Budgetary Educational Institution of Higher Education VSAU, Voronezh (2022)
18. Kuznetsova, E.D.: Internet of things in the management of an agricultural enterprise in Russia: directions, problems and prospects. In: Theory and Practice of Innovative Technologies in the Agro-Industrial Complex: Materials of the National Scientific-Practical Conf., pp. 367–371 (2023)
19. Rahman, H., Shah, U.M., Riaz, S.M., Kifayat, K., Moqurrab, S.A., Yoo, J.: Digital twin framework for smart greenhouse management using next-gen mobile networks and machine learning. Futur. Gener. Comput. Syst. **156**, 285–300 (2024). https://doi.org/10.1016/j.future.2024.03.023
20. Slob, N., Hurst, W., Zedde, R., Tekinerdogan, B.: Virtual reality-based digital twins for greenhouses: a focus on human interaction. Comput. Electron. Agric. **208**, 107815 (2023). https://doi.org/10.1016/j.compag.2023.107815
21. Ilyin, I., Laskin, M., Logacheva, I., Sarygulov, A., Tick, A.: Land valuation for agriculture and green energy projects: how to overcome the contradiction using mathematics. Math **10**, 4376 (2022). https://doi.org/10.3390/math10224376
22. Kuznetsova, E.D.: Problems and prospects of robotization of agriculture in Russia. In: Production and Processing of Agricultural Products: Materials of the VIII International Scientific and Practical Conference, pp. 452–459 (2022)
23. Kabir, M., Ekici, S.: Energy-agriculture nexus: exploring the future of artificial intelligence applications. Energy Nexus **13**, 100263 (2024). https://doi.org/10.1016/j.nexus.2023.100263
24. Agriculture in Russia. 2023: Statistical collection/Rosstat. 103 p. (2023)
25. Manoliu, M., Toculescu, R., Dragoicea, M., Ralea, M.: Neural networks for energy saving in agriculture pumping stations. IFAC Proceed. **34**(11), 231–233 (2001). https://doi.org/10.1016/S1474-6670(17)34138-1
26. Nimanthi, E.M., Edirisooriya, T., Wang, H., Banerjee, S., Longley, K., Wright, W., Mizuno, W., Xu, P.: Economic feasibility of developing alternative water supplies for agricultural irrigation. Curr. Opin. Chem. Eng. **43**, 100987 (2024). https://doi.org/10.1016/j.coche.2023.100987
27. Burg, V., Rolli, C., Schnorf, V., Scharfy, D., Anspach, V., Bowman, G.: Agricultural biogas plants as a hub to foster circular economy and bioenergy: an assessment using substance and energy flow analysis. Resour. Conserv. Recycl. **190**, 106770 (2023). https://doi.org/10.1016/j.resconrec.2022.106770
28. Kusuma, H.S., Az-Zahra, K.D., Saputri, R.W., Utomo, M.D.P., Jaya, D.E.C.H., Amenaghawon, A.N., Darmokoesoemo, H.: Unlocking the potential of agricultural waste as biochar for sustainable biodiesel production: a comprehensive review. Bioresour. Technol. Rep. **26**, 101848 (2024). https://doi.org/10.1016/j.biteb.2024.101848

29. Saleem, M.: Possibility of utilizing agriculture biomass as a renewable and sustainable future energy source. Heliyon **8**(2), e08905 (2022). https://doi.org/10.1016/j.heliyon.2022.e08905
30. Rahman, M., Khan, I., Field, D.L., Techato, K., Alameh, K.: Powering agriculture: present status, future potential, and challenges of renewable energy applications. Renew. Energy **188**, 731–749 (2022). https://doi.org/10.1016/j.renene.2022.02.065/
31. Deshmukh, S.M., Shekh, M.I., Samal, K., Geed, S.R., Ganguly, A.: Utilization of biomass for energy conservation in agricultural usage. Bioresour. Technol. Rep. **22**, 101480 (2023). https://doi.org/10.1016/j.biteb.2023.101480
32. Abashkin, V.L., Abdrakhmanova, G.I., Vishnevsky, K.O., Gohberg, L.M. et al.: The digital economy: 2024: a statistical collection. 276 p., ISIEZ VSHE, Moscow (2024)
33. Ahamed, S.H., Sultan, M., Monfet, D., Rahman, S., Zhang, Y., Zahid, A., Bilal, M., Ahsan, T.M.A., Yasmine, A.: A critical review on efficient thermal environment controls in indoor vertical farming. J. Clean. Prod. **425**, 138923 (2023). https://doi.org/10.1016/j.jclepro.2023.138923
34. Nasimi, S., Fakhroleslam, M., Zarei, G., Sadrameli, S.M.: Passive energy-efficiency optimization in greenhouses using phase change materials; a comprehensive review. J. Energy Storage **90**(15), 111762 (2024). https://doi.org/10.1016/j.est.2024.111762
35. Majeed, Y., Khan, M.U., Waseem, M., Zahid, U., Mahmood, F., Majeed, F., Sultan, M., Raza, A.: Renewable energy as an alternative source for energy management in agriculture. Energy Rep. **10**, 344–359 (2023). https://doi.org/10.1016/j.egyr.2023.06.032

Energy Sector of the Economy and Digital Opportunities of Modernity

Oksana Mashevskaya, Evgenia Karlovskaya, Oksana Vaganova, Daria Gordya, and Nina Trifonova

Abstract The article considers the digital advantages that have arisen for the energy industry in the current conditions, as well as objectives that the authors highlighted as a necessary condition for the digital transformation of the energy industry. In the course of study, the authors give examples of digitalization in the industry from international practice and from the practice of countries of the Commonwealth of Independent States (CIS). The objective of the paper is to study the features of digital transformation in the energy industry, using international practice and national experience of transformation. Methods of analysis and synthesis, methods of comparison and aggregation, as well as private research methods, such as graphical and tabular methods, have been used to conduct the study. The theoretical review and considered applied aspects of the study of the energy industry will make it possible to take into account the economic factors of the implementation of digital transformation in the industry as a whole, and in particular, in the energy system of the Russian Federation and Belarus.

Keywords Energy sector · Energy industry · Energy consumption · Digital transformation · Digitalization of energy sector · Smart grids

O. Mashevskaya
Belarusian State University, Minsk, Republic of Belarus

E. Karlovskaya · O. Vaganova (✉) · D. Gordya
Belgorod State National Research University, Belgorod, Russia
e-mail: vaganova@bsuedu.ru

N. Trifonova
Peter the Great St. Petersburg Polytechnic University, Politekhnicheskaya St., 29, Saint-Petersburg 195251, Russia

I. Ilin and M. Youzhong (eds.), *Digital Systems and Information Technologies in the Energy Sector*, Lecture Notes in Networks and Systems 1244,
https://doi.org/10.1007/978-3-031-80710-7_26

1 Introduction

Throughout the world, the importance of digitalization and digital technology is undeniable and is increasing every year. An entire generation has become fully immersed in the digital environment and has become an active participant in the digital transformation process, which involves the widespread introduction of digital technology to change business models, improve efficiency and productivity of customer service. These trends (e.g. intelligent sensors, as well as smart industrial production and cloud technology) are forcing a rethinking of the basic foundations of existence. The digitalization of energy management systems and automation underpin the emergence of the digital economy and the ongoing digital transformation makes it possible to create a new energy landscape, completely transform the concept of the energy industry and to produce an innovative revolution in the field of interaction with users.

There is an urgent need worldwide to change the way resources are used, especially in the energy sector. This situation occurs due, firstly, to the threat of an increase in the population of the Earth and the associated growth of the resource consumption and, secondly, to the fact that access to modern energy services has not yet been provided to low-income populations. The response to these issues, we believe, is digitalization and the introduction of digital technology into the energy sector. Digital technology will become the basis for the integration of renewable energy sources into electric power systems, will increase the reliability of electrical grids, will ensure the delivery of energy at the lowest possible price, promoting a socially just and equitable transition to digital energy.

2 Literature Review

Many scientists are concerned about the amount of energy consumed in the world. And, therefore, this problem is given due attention. Regional energy production and consumption is particularly worth noting. This issue has been studied by such scientists as E. K. Stephen, R. N. Singh, H. Mohammadi [1], Kapatina, Belov [2], Jan, Liu, Zhang, Mielnik, etc.

The problems of the energy factor in European countries have been dealt with by Russian and foreign authors, such as Grachev, Kuzheleva et al. [3, 4, 5, 6, 7 and 8].

A small number of works are presented on analysis and development of modern energy policy of Russia in the countries of South-Eastern Europe, e.g. by Stanojevic P., Jeftic [9], et al.

Among Belarusian scientists who dealt with this issue, it is worth noting Zorina, Prusov [10, 11]. Particular studies by Baynev [12], Tsedrik and joint works of Russian and Belarusian scientists Danilova, Novikova, Krishtanosov [13, 14] are of scientific interest. They have paid attention to assessing the digitalization of the energy sector and the electric power complex in their scientific papers.

3 Results and Discussion

3.1 *Digitalization and the Energy Sector: Essence, Advantages and Challenges*

Global energy sector is undergoing a paradigm shift towards digitalization, driven by recent technological developments and the need to increase efficiency and productivity, sustainability and competitiveness [15]. In this connection, the concept of digital energy has emerged, which is conditioned by the re-collection and development of a set of production and economic relations in the industry on the basis of digital approaches and means. Therefore, the main goal in digital energy is to advance the new business model, the potential of which is being realized through communication exchange, inter-machine interaction and digital modelling. Examples of such business models are demand aggregators, virtual power plants, virtual distributed energy storage, energy hedging, etc.

The digitalization process offers a number of opportunities for the energy sector, primarily:

1. Improving the performance of the energy sector through the introduction of innovative digital measurement and monitoring systems.
2. Better forecasting of system performance and signaling of the time of preventive maintenance of equipment and networks.
3. Meeting the needs of energy consumers adequately [16].
4. Reducing electricity consumption and energy supply costs.
5. Digitizing business processes to optimize project staff deployment, reducing the number of operational errors and accidents.
6. Creating value at an early stage of value creation and optimizing investment.

It should be remembered that, according to US analysts, technology-intensive manufacturing has a multiplicative effect of 16, i.e. one job in high-tech production creates 16 additional jobs, while classical production has a multiplier effect of 4.6; for agriculture it is 1.5 and for retail it is 0.8 [17].

Emerging digital environment has proclaimed the principle "change yourself or be changed" [18]. Such principle is an incentive for healthy competition and has confirmed the non-viability of a number of firms that failed to switch to digital technology in time and subsequently left the Fortune 500 list.

As for the tasks in the existing digital conditions in the energy sector, according to Belarusian scientists Tatiana Zorina and Stanislav Prusov [19], they should be:

1. To expand regional digital space and information security:
   a. developing programs and projects aimed at the sustainable digital development of the country's energy system, both regionally and internationally;
   b. networking of digital platforms and technology (e.g., at the level of countries of the Commonwealth of Independent States (CIS)), including industrial Internet, Big Data, wireless communications, virtual and augmented reality

technology, neurotechnology, artificial intelligence, etc., as well as analysis of trends in the industry and developing recommendations for global digital development of the energy industry;

   c. guaranteeing the protection of interests for individuals, businesses and the State within the existing legal framework;
   d. creating competitive conditions for the leading position of the country in the field of export of services and information security, taking into account national interests;
   e. maintaining the integrity, authentication, confidentiality and accessibility of transmitted information and its processing;
   f. using predominantly national software and equipment, taking into account international obligations.

2. To develop the information infrastructure of the industry:

   a. strengthening communication networks that will fully meet the needs of the country's economy for the collection and transmission of big data, taking into account the stated technical requirements for digital technology;
   b. creating national information processing centers that provide accessible, safe and cost-effective data storage and processing services to citizens, the State and the businesses, enabling these services to be exported;
   c. introducing digital data platforms;
   d. establishing a platform or other digital infrastructure facility that would allow information and data to be collected, processed, stored and provided to consumers at minimal cost;
   e. improving the quality of digital services provided through the use of the best digital technology available to the industry.

3. For the education, research and training system:

   a. creating conditions for training specialists for the energy industry, able to work in the emerging digital environment;
   b. improving and transforming national education system by means of cooperation in the development and implementation of professional standards and educational programs for higher education, taking into account the emerging digital skills in the research industry;
   c. developing digital skills and competencies in the digital energy sector;
   d. stimulating the labor market that places digital demands on the workforce and specialists in the energy industry;
   e. creating a motivational system for mastering basic digital competencies required for the energy industry;
   f. developing a KPI system for employees interested in the early implementation of digitalization in the industry and ready to improve their own qualifications;
   g. developing human capital through the enhancement of digital literacy and digital skills, beginning at schools.

3.2 Digitalization and the Energy Sector: Essence, Advantages and Challenges

Let us turn to the world experience of digital transformation. In this regard, it is worth noting the experience of the UK, where the key task was to integrate renewable energy into the energy system of the country, which, as the British scientists themselves note, "...will allow the country to reach net zero" [20]. An intelligent platform, Kaluza, was created in the UK, which made it possible to charge and connect to smart devices (such as electric vehicles, battery heaters and home batteries, manage their charging, etc.), use alternative energy sources and reduce the demand for electricity in the grid (Fig. 1).

The advantage of the platform is that owners of mobile devices, using web applications, mark the time of travel in an electric vehicle and the platform prepares the car for operation. The use of artificial intelligence and machine learning creates possibilities to collect information about customers who consume energy in real time from working smart devices, electricity markets, retailers, meteorological and system operators to predict demand and further operation of the electricity system. Moreover, it acts as an additional source of power for connected devices to support the network. Kaluza collects minute-by-minute data and monitors the electricity consumption of smart devices during periods of low system demand, and during periods of high generation it connects energy from renewable sources.

Such positive experience demonstrates all the advantages of digital transformation and the efficiency of the digital technology market in the energy sector, which is estimated by Navigant Research at approximately $56 billion and should grow to $64 billion by the end of 2025 (Fig. 2).

Schneider Electric, a French power engineering corporation, is another great example in global practice. Schneider Electric collaborates with Esri, a leading geographic information systems (GIS) software provider. Schneider Electric's partnership with Esri is crucial to using the best of the real and digital world for success, especially in the context of aligning classical software with the new features of the Esri Utility Network. Seamless business process integration and focus on maximizing the benefits of both ArcFM Editor XI and ArcGIS Pro demonstrate a commitment to providing integrated solutions for network management. In addition, Schneider Electric's software certification development process with each release of ArcGIS Pro and ArcGIS Enterprise network management ensures the compatibility and reliability of the products and services offered.

Schneider Electric strives to achieve parity between its classic solutions and new developments, guaranteeing customers a smooth transition to digitalization, beneficial for all. The company is constantly working to improve Esri's current ArcGIS network management capabilities and meet evolving industry standards and customer requirements (Fig. 3). This collaboration allows customers to benefit, and the company at the same time creates the image of a reliable partner focused on innovation and digitalization.

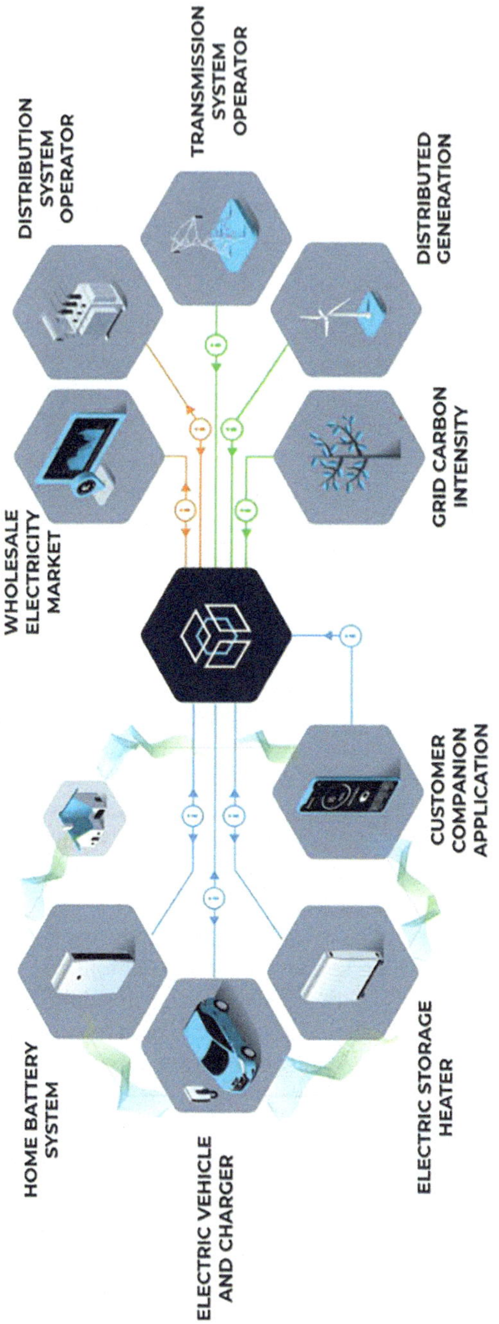

Fig. 1 Kaluza's intelligent device charging [20]

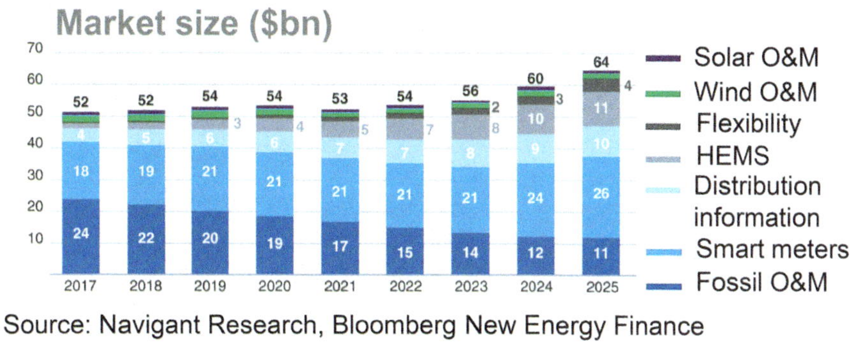

Fig. 2 International digital technology market

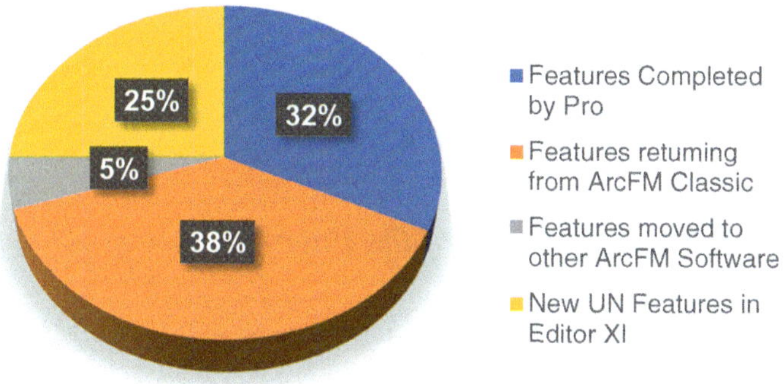

Fig. 3 Digital solutions for network management [21]

Schneider Electric in Mumbai (India) and the Indian energy company Tata Power have developed and implemented a self-healing electrical power supply network for 350,000 homes, which in a matter of minutes determines where power is lost in the network and restores it itself. Network self-repair occurs through automatic fault localization and recovery without operator intervention. Built with EcoStruxure ADMS, the solution connects smart ring network units, circuit breakers, block systems and digital substations. Such integration has enabled decentralized management of a complex and large electrical network, reducing malfunctions to seconds, increasing customer satisfaction and minimizing revenue losses.

The company stimulates productivity improvements of its partners and allows them to maintain a leading position in the rapidly evolving energy industry. Due to a rich heritage of delivering exceptional software, a comprehensive support ecosystem, a focus on security and digital transformation, Schneider Electric is a partner for businesses seeking to achieve unparalleled productivity and efficiency in digital environments.

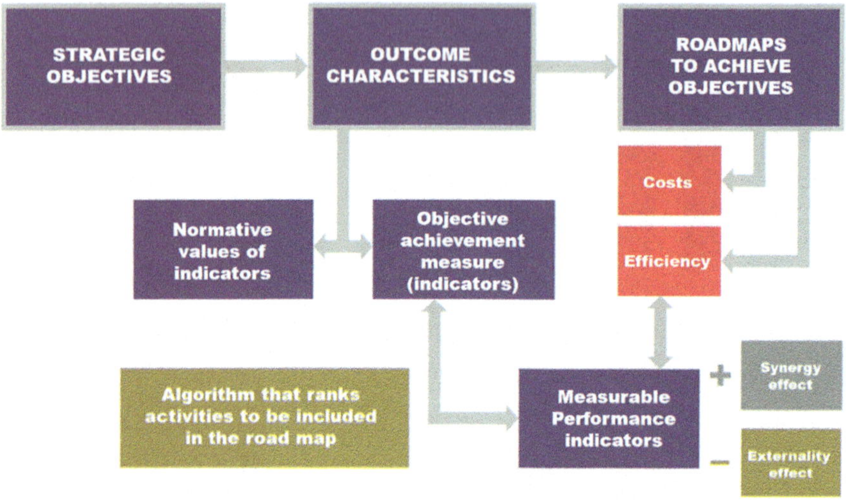

Fig. 4 Algorithm for ranking activities within the framework of the digital transformation of the energy system of Belarus [23]

3.3 Digital Transformation of the Energy Sector in Belarus

The energy sector in Belarus is undergoing the same digital changes. In 2023, the Program for the modernization of eclectic energy accounting facilities until 2023 was completed. A total of $175 million was allocated from the national budget to finance the Program. The economic impact in 2024 reached $191.2 million [22].

The digital transition itself was presented, for example, by Prusov [23] as an algorithm for assessing the level of digital transformation of Belarus (Fig. 4).

There are 16 quantitative indicators for measuring the digital transformation in Belarus, some of which we will consider in Table 1.

In addition to quantitative indicators, if it is necessary to effectively manage the energy system in real time, planning and interconnections in order to ensure and improve the operation of electricity and capacity markets using digital technology, we recommend using the Common Information Model (CIM) [25] which is presented in Fig. 5. This will ensure the smooth and effective exchange of information between current entities and participants in the electricity industry.

This model of information interaction between electric power enterprises represents the possibility of effective integration of heterogeneous automated systems and provides a unified method of management of energy facilities, regardless of their purpose, quality of equipment and different manufacturers [26]. In practice, the model can be applied in CIS countries, but we must take into account that the level of digitalization of electricity in all countries is different (Table 2). This can also create certain difficulties in integration and make some of the countries uncompetitive outside the general information model.

Table 1 Indicators for measuring the digital transformation of the Belarusian energy system

No	Indicator	Calculation formula	Variables	Values
1	Share of 35 kV and above digital substations	$As_{e1} = k/n$	k—number of 35 kV and above digital substations; n—total number of 35 kV and above digital substations	Entities that have a website, users of cloud services and wireless technologies to access the global Internet computer network, specific weight, %
2	Degree of automation of 0.4–10 kV distribution networks	$As_{e2} = \frac{\sum_{i=1}^{4} k_i}{4}$	k_1—automated dispatching management system and automated enterprise (organization) resource planning and management system are implemented; k_2—automatic switching control in case of damage to the electrical network is implemented; k_3—intelligent decision support system is implemented, including automatic digital data collection and subsequent dynamic real-time analytics. The end result is a digital asset that is used to make quick and optimal decisions; k_4—enterprise ecosystem created in which all processes for managing and monitoring electrical networks are formed and implemented automatically	Entities that have access to global information and local computer networks, specific weight, %

(continued)

Table 1 (continued)

No	Indicator	Calculation formula	Variables	Values
3	Share of energy industry facilities integrated into an automated system for monitoring and accounting of electrical energy of interstate intersystem flows and generation	$\mathrm{As}_{e3} = k/n$	k—number of energy industry facilities integrated into an automated system for monitoring and accounting of electrical energy of interstate intersystem flows and generation; n—total number of energy industry facilities	Entities that use special software, specific weight, %
4	Share of energy industry facilities integrated into a regional automated system for monitoring and accounting of electrical energy	$\mathrm{As}_{e4} = k/n$	k—number of energy industry facilities integrated into a regional automated system for monitoring and accounting of electrical energy; n—total number of energy industry facilities	Number of economic entities that use information and communication technology, units
5	Share of industrial and equivalent consumers with connected capacity of 250 kV·A and above	$\mathrm{As}_{e5} = k/n$	k—number of industrial and equivalent consumers with connected capacity of 250 kV·A and above, integrated into ASCAPC; n—total number of industrial and equivalent consumers with connected capacity of 250 kV·A and above	Availability (quantity) of machinery and equipment in the industry, created on the basis of advanced production technologies, by type of economic activity, both units and specific weight, %
6	Share of household consumers integrated into an automated system for monitoring and accounting of electrical energy	$As_{e6} = k/n$	k—number of household consumers integrated into an automated system for monitoring and accounting of electrical energy; n—total number of household electricity consumers	Distribution of organizations in the industry by type of Internet connection (e.g. fixed broadband, wireless, wireless broadband access)

(continued)

Table 1 (continued)

No	Indicator	Calculation formula	Variables	Values
7	Share of energy sources equipped with an automated process control system	$A_{pc} = \frac{k}{n}$	*k*—number of energy sources equipped with an automated process control system; n—total number of energy sources	Share of organizations in the industry by maximum data transfer speed via WWW (Mbit/s) in the total number of organizations that have access to the Internet, %
8	Share of energy industry facilities equipped with an automated power quality control system	$A_{pqc2} = \frac{k}{n}$	*k*—energy industry facilities equipped with an automated power quality control system; *n*—total number of energy industry facilities	Number of employees using ICT in the industry at the end of the year (both as a percentage of all employed in the economy and in the number of people)

(continued)

Table 1 (continued)

No	Indicator	Calculation formula	Variables	Values
9	Degree of automation of decision making	$A_{DM} = \frac{\sum_{i=1}^{4} k_i}{3}$	k_i—expert assessment answer to each of 4 questions with answers on the following scale (0–3 points): 0—not automated; 0,5—electronic regulatory and reference information is used; 0,5—passive Decision Support System (DDS) is implemented to collect and structure the information needed for decision making; 1—active DSS is implemented to analyze the available information and generate possible solutions; 1—cooperative DSS is implemented to modify, replenish or improve possible solutions offered by the system and further send them to the system for verification	Distribution of organizations in the industry by purpose of using the Internet (for communication with consumers) (% of the total number of organizations that have access to the Internet)
10	Assessing the availability of a system to assess the quality of products and services online	$A_{DM} = \begin{cases} \frac{0-\text{no}}{1-\text{yes}} \end{cases}$	Binary value 0 or 1 is used when answering the question: "Is your organization (enterprise) hosted on a common online platform for assessing product quality, service quality?" (these platforms can include resources: services.bel, Google My Business, etc.)	Data on entities in the industry engaged in electronic sales and purchases of goods (works, services), % of the total number of energy sector organizations surveyed

Source Compiled by authors based on [19, 24]

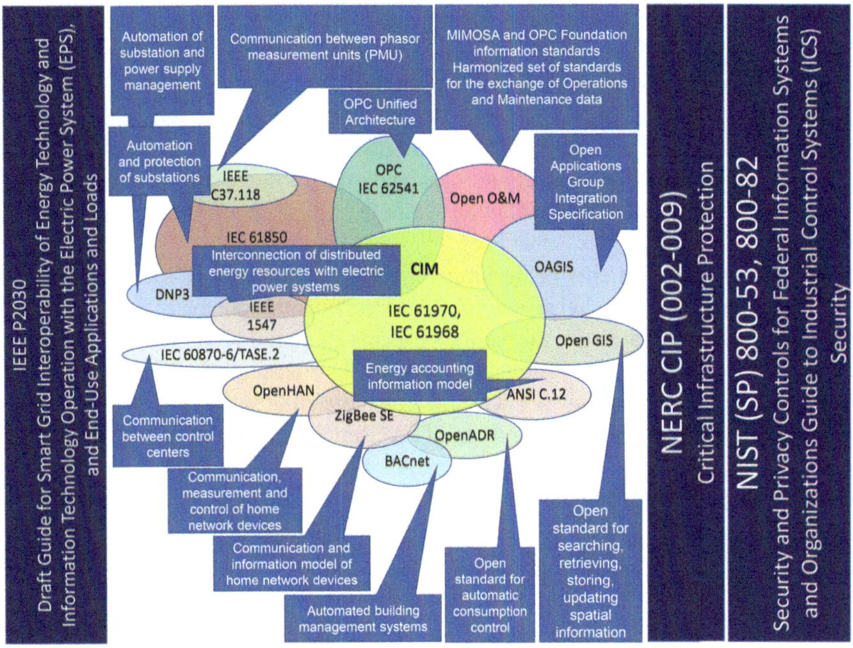

Fig. 5 Common Information Model [25]

Table 2 Technologies in use (and/or under implementation) in the Commonwealth of Independent States (CIS)

CIS Country	ASCAPC	SCADA	Digital sub-stations	CIM	Smart Grid	Demand Response	Digitals twins
Republic of Azerbaijan							
Republic of Armenia							
Republic of Belarus							
Republic of Kazakhstan							
Republic of Kyrgyzstan							
Republic of Moldova	No information provided						
Russian Federation							
Republic of Tajikistan							
Turkmenistan	No information provided						
Republic of Uzbekistan							

Pilot stage	Massive adoption	In progress/Is to be introduced	Is not to be introduced

Note ASCAPC is an automated system for commercial accounting of power consumption. *Source* [25]

In order to assess the ongoing digitalization processes in Belarus, comparative analysis among the States with which the Republic has the closest economic, cooperative and political relations is important. The EAEU is an example of such economic union [27]. A number of States, while conducting the digital transformation of the industry [28, 29], are guided by existing national standards, and at the same time take into account the specifics of international standards (Table 3), adapt them to national conditions. This could be seen from the example of the Republic of Belarus, Russian Federation and the Kyrgyz Republic.

In general, the elements of the transformation of the energy sector considered will allow a more comprehensive approach to the quantifiable measurement of digitization, taking into account the various aspects of functioning and the prevailing conditions. In addition, the implementation of the measures presented above will improve the quality of energy services, increase consumer satisfaction and contribute to the sustainable energy development of the Republic of Belarus. In the future, it is worth paying attention to studies of economic consequences of digital transformation related to both cybercrime and the system for managing the state of energy equipment.

Table 3 Energy standards in CIS countries

CIS Country		Application of international standards						Application of national standards								
	ISO/IEC	ISO	NIST SP	IEC	IEC	IEC	IEC	GOST R	GOST R	GOST R	GOST R	GOST R	Technical Code of Common Practice (TKP)	Enterprise Standard (STP)	Enterprise Standard (STP)	Organization Standard (STO)
	27000	27001	800-53	60870	61850	61968	61970	599472021	58651	571142016	599502021	599482021	609-2017	33240.20.1 17-18	33240.01.1 08-22	5694700729.240.10. 299-2020
Republic of Armenia																
Republic of Belarus																
Republic of Kazakhstan																
Kyrgyz Republic																
Russian Federation																

Source [25]

4 Conclusion

To summarize, it is worth noting that the digitalization of energy represents an important stage in the development of the industry, which is becoming increasingly relevant in the modern world. The application of digital technology in the energy sector makes it possible to improve the efficiency of energy production, distribution and consumption and to make management system more flexible and transparent.

One of the areas of digitalization in the energy sector is the introduction of smart energy management systems. This helps to optimize energy consumption in real time, taking into account supply and demand, as well as individual consumer preferences. Due to the implementation of digital technology, namely the use of sensors, the Internet of Things, Big Data technologies, management decisions have become much easier and more sustainable.

Digitalization also affected the development of data and energy transmission networks. The introduction of smart networks contributes to more reliable and efficient use and distribution of energy, load management and quick response to emergencies, and sometimes even blocking them, which will certainly improve the overall energy efficiency of the entire system.

The digitalization of energy also opens up new opportunities for renewable energy. Digital technology has made it possible to integrate solar and wind power more efficiently into a common network, store and distribute the generated energy on demand (UK example). International experience shows that digital transformation requires the integration of all economic entities, including state and government programs, large energy companies, who strive to introduce digital technology into current operations, and "reasonable needs" of users of such services.

Overall, digitalization of energy contributes to the creation of a more sustainable, flexible and environmentally friendly energy system that can effectively meet society's energy needs, reduce resource consumption (as the world's population grows) and reduce the negative impact on the environment.

References

1. Mohammadi, H.: Electricity prices and fuel costs: long-run relations and short-run dynamics. Energy Econ. **31**(3), 503–509 (2009)
2. Belov, V.I.: Jenergopotreblenie kak faktor jekonomicheskogo rosta regional'nyh social'no-jekonomicheskih sistem Energy consumption as a factor of economic growth of regional socio-economic systems. Res. Res. Econ. Res. **9**(3), 52–60 (2023). https://doi.org/10.18413/2409-1634-2023-9-3-0-5. (in Russian)
3. Kuzheleva, K.S., Grachev, B.A.: Jenergeticheskaja politika ES v oblasti VIJe, jenergojeffektivnosti i vnedrenija novyh resursosberegajushhih tehnologij [EU Energy Policy for RES, Energy Efficiency and Implementation of New Resourse Saving Technologies]. Jenergeticheskaja politika **1**, 8–15 (2018). (in Russian)
4. Kirichenko, A.B., Komlev, S.L., Loginov, L.O.: Negativnye pobochnye jeffekty politiki dekarbonizacii dlja jenergobezopasnosti Evropy: vremja rasstavanija s illjuzijami negative side

effects of decarbonization policy for European energy security: time to part with illusions. Gazovaja Promyshlennost **S4**(762), 20–27 (2017). (in Russian)
5. Ponomareva, E.G., Popadyeva, T.I.: Bosnija i Gercegovina: jetnopoliticheskij faktor problemnoj gosudarstvennosti Bosnia and Herzegovina ethnopolitical factor of problematic statehood. Obozrevatel **11**(346), 63–76 (2018). (in Russian)
6. Entin, M.L., Entina, E.G., Tnelm N.I.: V poiskah partnerskih otnoshenij-VIII. Rossija i Evropejskij Sojuz v 2018—pervoj polovine 2019 godov [In Search of Partnerships-VIII. Russia and the European Union in 2018—the First Half of 2019]. Zebra-E, Moscow (2019) (in Russian)
7. Todorova, R.V.: Vozobnovljaemye istochniki jenergii kak dvizhushhaja sila jenergeticheskoj politiki Respubliki Bolgarija Renewable energy sources as the driving force of the energy policy of the Republic of Bulgaria. Voprosy Politol. **3**(31), 118–129 (2018). (in Russian)
8. Kulikova, N.V., Orlik, I.I., Feit, N.V.: Rossija i Central'no-Vostochnaja Evropa: vzaimootnoshenija v XXI veke [Russia and Central-Eastern Europe: relationships in the 21st century]. Institute of Economics of the Russian Academy of Sciences, Moscow (2012). (in Russian)
9. Stanojevic, P., Jeftic, Z.: Russian—Serbian cooperation in the sector of oil and gas: past, present and possible future. In book: Russia and Serbia in the Contemporary World: Bilateral Relations, Challenges and Opportunities, (2020) https://doi.org/10.18485/iipe_ru_sr.2020.ch6
10. Zorina, T.G., Prusov, S.G.: Sovershenstvovanie metodologii ocenki cifrovoj transformacii ob##edinennoj jenergeticheskoj sistemy Respubliki Belarus'. Problemy i perspektivy [improving the methodology for assessing the digital transformation of the united energy system of the republic of Belarus. problems and prospects]. Bull. Kazan State Energy Univ. **13**(3), 99–113 (2021). (in Russian)
11. Zorina, T.G., Prusov, S.G.: Metodicheskie podhody k sovershenstvovaniju ocenki jeffektivnosti cifrovoj transformacii jenergetiki Respubliki Belarus' [methodical approaches to improving the assessment of the efficiency of the digital energy transformation of the Republic of Belarus]. Bull. Brest State Tech. Univ. **2**, 176–181 (2022). (in Russian)
12. Baynev, V.F.: Jenergija kak faktor proizvodstva i dvizhushhaja sila industrializacii [Energy as a Factor in Production and the Driving Force of Industrialization]. In: Economic Science Today Proceedings, vol. 10, pp. 26–40 (2019) (in Russian)
13. Danilova, O.V., Novikova, I.V., Krishtanosov, V.B.: Problemy cifrovizacii kljuchevyh sektorov jekonomiki v Rossijskoj Federacii i Respublike Belarus': smart grid v jelektrojenergetike [problems of digitalization of key sectors of the economy in the Russian Federation and the Republic of Belarus: smart grid in the electric power industry]. Proceed. BSTU **250**(2), 5–14 (2021). (in Russian)
14. Munenge, S., Moskovkin, V.M., Vaganova, O.V.: Prostranstvenno-vremennoj analiz processov cifrovizacii rossijskih regionov [Spatio-temporal analysis of digitalization protsesses in Russian regions]. Res. Res. Econ. Res. **8**(3), 48–62 (2021). https://doi.org/10.18413/2409-1634-2022-8-3-0-4. (in Russian)
15. Danilova, O.V.: Cifrovye tehnologii i perspektivy razvitija jelektrosetevogo kompleksa Rossii [digital technologies and prospects for the development of the electric grid complex of Russia]. Bull. Tver State Univ. **46**(2), 95–104 (2019). (in Russian)
16. Digitalization & Energy, https://www.unep.org/topics/energy/digitalization/digitalization-energy#:~:text=The%20digitalization%20process%20unlocks%20multiple,reduction%20in%20cyber%20security%20concerns. Last accessed 2024/06/10
17. Cifrovaja jenergetika [Digital Energy], digitalsubstation.com. Last accessed 2024/06/10 (in Russian)
18. Tricoire, J.P.: Disrupt or be disrupted: innovation for the sake of customers, https://blog.schneider-electric.com/energy-management-energy-efficiency/2018/04/05/disrupt-or-be-disrupted-innovation. Last accessed 2024/06/10
19. Zorina, T., Prusov, S.: Cifrovaja transformacija jelektrojenergetiki Belarusi [Digital transformation of the Belarus' electric power industry within the bounds of common approaches to the CIS fuel and energy sector digitalization]. Nauka i Innov. **2**(228), 59–65 (2022). (in Russian)
20. Digitalising our Energy System for Net Zero, https://www.gov.uk/government/publications/digitalising-our-energy-system-for-net-zero-strategy-and-action-plan/energy-digitalisation-taskforce-report-joint-response-by-beis-ofgem-and-innovate-uk. Last accessed 2024/06/10

21. Elevating productivity with EcoStruxure ArcFM Editor XI: Embracing tradition and innovation, https://blog.se.com/digital-transformation/2024/07/12/elevating-productivity-ecostruxure-arcfm-editor-xi-embracing-tradition-innovation. Last accessed 2024/06/10
22. Programma modernizacii sredstv ucheta jelektricheskoj jenergii do 2023 g. [Program for the modernization of electrical energy metering up to 2023], https://www.energosbyt.by/ru/novosti/modernizatsiya pribornogo ucheta. Last accessed 2024/06/10 (in Russian)
23. Prusov, S.G.: Ocenka jeffektivnosti investicionnyh proektov v jelektrojenergetike [assessing the effectiveness of investment projects in the electric power industry]. Misanta, Minsk, (2016)
24. Tsedrik, A.V.: Cifrovizacija jenergeticheskogo kompleksa Respubliki Belarus': analiz sostojanija, perspektivnye napravlenija, al'ternativnoe predlozhenie po ocenke jeffektivnosti Digitalization of the Energy Complex of the Republic of Belarus: Analysis of the State Promising Directions, Alternative Proposal for Assessing Efficiency//. Cifrovaja transformacija **30**(1), 16–27 (2024). https://doi.org/10.35596/1729-7648-2024-30-1-16-27. (in Russian)
25. Cifrovizacija «okno vozmozhnostej» dlja povyshenija jeffektivnosti i upravljaemosti jenergosistem gosudarstv-uchastnikov SNG [Digitalization "window of opportunity" to improve the efficiency and controllability of energy systems of the CIS member states, http://energo-cis.ru/rumain51211/. Last accessed 2024/06/10 (in Russian)
26. Common Information Model (CIM), https://www.techtarget.com/searchstorage/definition/Common-Information-Model#:~:text=The%20Common%20Information%20Model%20(CIM)%20is%20a%20. Last accessed 2024/06/12
27. Vladyka, M.V., Chistnikova, I.V., Ermachenko, P.M.: Mezhregional'naja integracija v prostranstve makroregiona kak faktor razvitija nacional'noj jekonomiki [Interregional integration in the space of the macroregion as a factor of the development of the national economy]. Res. Res. Econ. Res. **6**(4), 33–43 (2020). https://doi.org/10.18413/2409-1634-2020-6-4-0-4. (in Russian)
28. Faustov, A., Chemeris, O., Halada, K.: Using digital solutions in railway logistics: the experience of medium-sized businesses. In: Ilin, I., Jahn, C., Tick, A. (eds.) Digital Technologies in Logistics and Infrastructure. ICDT 2021. Lecture Notes on Data Engineering and Communications Technologies, Springer, Cham. Vol. 157. (2023) https://doi.org/10.1007/978-3-031-24434-6_17
29. Poljanskihh, A., Levina, A., Dubgorn, A.: Investment in renewable energy: practical case in Estonia. MATEC Web Conf. **193**, 05065 (2018)

Mathematical Models for Digital Solutions

Imprecise SHAP as a Tool for Explaining the Class Probability Distributions Under Limited Training Data

Lev Utkin, Andrei Konstantinov, Kirill Vishniakov, and Igor Ilin

Abstract Applications of artificial intelligence in predictive analytics of the energy complex meets with risks of incorrect predictions caused by data errors, lack of data, and overtraining of models. Risks can be minimized if to use meta-models which could interpret and explain the provided predictions. One of the most popular methods of the machine learning prediction explanation (XAI) is the SHapley Additive exPlanations method (SHAP). An imprecise SHAP as a modification of the original SHAP is proposed for cases when the class probability distributions are imprecise and represented by sets of distributions. The first idea behind the imprecise SHAP is a new approach for computing the marginal contribution of a feature, which fulfils the important efficiency property of Shapley values. The second idea is an attempt to consider a general approach to calculating and reducing interval-valued Shapley values, which is similar to the idea of reachable probability intervals in the imprecise probability theory. A simple special implementation of the general approach in the form of linear optimization problems is proposed. It is based on using the Kolmogorov–Smirnov distance and imprecise contamination models. Numerical examples with synthetic and real data illustrate the imprecise SHAP.

Keywords Machine learning · XAI · SHAP · Imprecise probabilities · Contamination model · Kolmogorov–Smirnov distance

1 Introduction

One of the key applications of artificial intelligence in the energy complex is predictive analytics for reliability and preventive maintenance, where elements of energy systems are constantly monitored and analyzed to identify potential faults and the anomalous behavior. Developing intelligent models for processing heterogeneous

L. Utkin (✉) · A. Konstantinov · K. Vishniakov · I. Ilin
Peter the Great St. Petersburg Polytechnic University, St. Petersburg, Russia
e-mail: utkin_lv@spbstu.ru

I. Ilin and M. Youzhong (eds.), *Digital Systems and Information Technologies in the Energy Sector*, Lecture Notes in Networks and Systems 1244,
https://doi.org/10.1007/978-3-031-80710-7_27

information coming from sensors that monitor the state of the energy systems to ensure their continuity of functioning is an extremely urgent task that can be solved using machine learning methods.

However, risks associated with artificial intelligence also need to be considered and addressed before the technology is scaled across the entire sector. These include the impact of data errors, lack of data, and overtraining of models which are viewed as black boxes because users do not know how the models act and predict. Risks can be minimized if to use meta-models which could interpret and explain the provided predictions. The explanation means that we select features of an analyzed example which are responsible for the prediction of the black-box model or significantly impact on the corresponding prediction. By considering the explanation of a single example, we say about the so-called local explanation methods which aim to explain the black-box model locally around the example. Another explanation methods try to explain predictions considering the whole dataset or its part. The need of explaining the black-box models by using the local or global explanations in many applications motivated developing a large number of explanation methods which are described in detail in several survey papers [1–3].

Among all explanation methods, we select two methods: the Local Interpretable Model-Agnostic Explanation (LIME) [4] and SHapley Additive exPlanations (SHAP) [5]. The basic idea behind LIME is to build an approximating linear model at an explained example by generating many synthetic examples around the example which are the training set for constructing the linear regression. Its coefficients can be regarded as quantitative representation of impacts of the corresponding features on the prediction. SHAP is inspired by game-theoretic Shapley values which can be interpreted as expected marginal contributions over all possible subsets of features to the black-box model prediction. Despite two important shortcomings of SHAP, including its computational complexity and some ambiguity of approaches for removing features from consideration [6], SHAP is widely used in practice and can be viewed as the most promising and theoretically justified explanation method which fulfils several nice properties. Nevertheless, another difficulty of using SHAP is how to deal with predictions in the form of probability distributions which arise in multi-class classification and in machine learning survival analysis. The problem is that SHAP in each iteration calculates a difference between two predictions defined by different subsets of features in the explained example. Therefore, the question is how to define the difference between the probability distributions. One of the simplest ways in classification is to consider a class with the largest probability. However, this way may lead to incorrect results when the probabilities are comparable. Moreover, the same approach cannot be applied to explanation of the survival model predictions which are in the form of survival functions. The well-known Kullback–Leibler (KL) divergence as well as other distances between distributions can be applied to explanation. However, their application to SHAP violates important properties of the Shapley values. Therefore, one of the problems for solving is to define a meaningful distance between two predictions represented in the form of the class probability distributions, which fulfils the Shapley value properties.

The class probability distributions in classification may be imprecise due to a limited number of training data. Therefore, we should consider the imprecision of probability distributions due to the lack of sufficient training data. The imprecision can be referred to the epistemic uncertainty [7] which represents our ignorance about a model caused by the lack of observation data. Epistemic uncertainty can be resolved by observing more data. There are several approaches to formalize the imprecision. One of them is to consider a set of probability distributions instead of the single one. Sets of distributions are produced by imprecise statistical models, for example, by the imprecise ε-contamination model [8]. It is obvious that imprecise statistical models produce interval-valued probabilities of events or classes, and the intervals can be regarded as measures of the data insufficiency.

Imprecision of the class probability distribution leads to imprecision of Shapley values when SHAP is used for explaining the classification model predictions, i.e., Shapley values become interval-valued. This motivates us to construct a new framework of the imprecise SHAP or imprecise Shapley values, which provide a real accounting of the prediction uncertainty. Having the interval-valued Shapley values, we can compare intervals to select the most important features. It is important to point out that properties of the efficiency and the linearity of Shapley values play a crucial role in constructing the framework of imprecise Shapley values.

In summary, the following contributions are made in this paper:

1. A new approach for computing the marginal contribution of the i-th feature in SHAP for interpretation of the class probability distributions as predictions of a black-box machine learning model is proposed. It is based on considering a distance between two probability distributions. Shapley values using the proposed approach fulfil the efficiency property which is very important in correct explanation by means of SHAP.
2. An imprecise SHAP is proposed for cases when the class probability distributions are imprecise, i.e., they are represented by convex sets of distributions. Outcomes of the imprecise SHAP are interval-valued Shapley values.
3. Implementation of the imprecise SHAP by using the Kolmogorov-Smirnov distance (KSD) is proposed. Complex optimization problems for computing interval-valued Shapley values are reduced to finite sets of simple linear programming problems.
4. The imprecise SHAP is illustrated by means of numerical experiments with synthetic and real data.

2 Related Work

Local interpretation methods. An increasing importance of machine learning models leads to development of new explanation methods taking into account various peculiarities of applied problems. As a result, many models of the local interpretation have been proposed. Success and simplicity of LIME resulted in development of several its modifications, for example, Anchor LIME [9], LIME-Aleph [10], GraphLIME [11], SurvLIME [12], etc. A comprehensive analysis of LIME, including the study of its applicability to different data types is provided in [13]. To relax the linearity condition for the local interpretation models and to adequately approximate a black-box model, several methods using Generalized Additive Models were proposed in [14–16].

Another explanation method, SHAP [5], takes a game-theoretic approach for optimizing a regression loss function based on Shapley values. General questions of the computational efficiency of SHAP were investigated in [17]. An approach to applying SHAP to ensemble model was presented in [18]. The so-called off- and on-manifold Shapley values for high-dimensional multi-type data were introduced in [19]. Application of SHAP to explanation of recurrent neural networks was studied in [20]. Antwarg et al. [21] studied how to explain anomalies detected by autoencoders using SHAP. SHAP is also applied to problems of explaining individual predictions when features are dependent [22]. A lot of interpretation methods, their analysis, and critical review can be found also in survey papers [23, 24].

Imprecise probabilities in classification and regression. One of the first ideas of applying imprecise probability theory to classification decision trees was presented in [25], where probabilities of classes at decision tree leaves are estimated by using an imprecise model. Following this work, several papers devoted to applications of imprecise probabilities to decision trees and random forests were presented [26, 27]. Imprecise probabilities have also been used in classification problems in [28–30]. A focus of interest in the present paper is not imprecise probabilities in classification models, but imprecision of Shapley values as a result of the model prediction imprecision when SHAP is used to explain the prediction.

3 Shapley Values and Model Explainability

According to the concept of the Shapley values in explanation, the total gain of a game is distributed among players such that desirable properties, including efficiency, symmetry, and linearity, are fulfilled. In the framework of the machine learning, the gain is viewed as the machine learning model prediction or the model output, and a player is a feature of inputs. Hence, contributions of features to the prediction can be estimated by Shapley values ϕ_i. The i-th feature importance is defined

$$\phi_i(f) = \phi_i = \sum_{S \subseteq N \setminus \{i\}} B(S, N) \left[f(S \cup \{i\}) - f(S) \right], \tag{1}$$

where $f(S)$ is the black-box model prediction under condition that a subset S of features is used as the input; N is the set of all features; $B(S, N)$ is defined as

$$B(S, N) = |S|!(|N| - |S| - 1)!/|N|!. \tag{2}$$

Shapley values have the following well-known properties:

Efficiency. The total gain is distributed as $\sum_{k=0}^{m} \phi_k = f(\mathbf{x}) - f(\emptyset)$.
Symmetry. If two players with numbers i and j make equal contributions, i.e., $f(S \cup \{i\}) = f(S \cup \{j\})$ for all subsets S which contain neither i nor j, then $\phi_i = \phi_j$.
Dummy. If a player makes zero contribution, i.e., $f(S \cup \{j\}) = f(S)$ for a player j and all $S \subseteq N \backslash \{j\}$, then $\phi_j = 0$.
Linearity. A linear combination of multiple games $f_1, \ldots, f_n$, represented as $f(S) = \sum_{k=1}^{n} c_k f_k(S)$, has gains derived from f: $\phi_i(f) = \sum_{k=1}^{m} c_k \phi_i(f_k)$ for every i.

Suppose that there is a dataset $\{(\mathbf{x}_1, y_1), \ldots, (\mathbf{x}_n, y_n)\}$ where $\mathbf{x}_i \in \mathcal{X} \subset \mathbb{R}^m$ is a vector consisting of m features, y_i is the output for the vector $\mathbf{x}_i$ such that $y_i \in \mathbb{R}$ in the regression and $y_i \in \{1, 2, \ldots, C\}$ in the classification with C classes. If the task is to explain prediction $f(\mathbf{x}^*)$ from the model at a local feature vector $\mathbf{x}^*$, then the prediction $f(\mathbf{x}^*)$ can be represented by using Shapley values as [5]: $f(\mathbf{x}^*) = \phi_0 + \sum_{j=1}^{m} \phi_j^*$, where $\phi_0 = \mathbb{E}[f(\mathbf{x})]$, ϕ_j^* is the value ϕ_j for the prediction $\mathbf{x} = \mathbf{x}^*$. This implies that Shapley values explain the difference between prediction $f(\mathbf{x}^*)$ and a global average prediction.

One of the crucial questions for implementing SHAP is how to remove features from subset $N \backslash S$, i.e., how to fill input features to get predictions $f(S)$ of the black-box model. Various ways for removing features are presented in [6].

4 SHAP and Multi-classification Problems

Consider a classification model f whose predictions for the subset S are probabilities $p_1, \ldots, p_C$ of C classes under conditions $p_1 + \ldots + p_C = 1$ and $p_k \geq 0$ for all k. The probabilities can be viewed as a distribution $P(S) = (p_1, \ldots, p_C)$ over a C-class variable. Suppose that the prediction of the model f for the subset S is a distribution denoted as $Q(S) = (q_1, \ldots, q_C)$, where $q_1 + \ldots + q_C = 1$ and $q_k \geq 0$ for all k. It is justified in [31] that the Kullback–Leibler divergence $KL(P\|Q)$ is a way to measure the deviation of predictions from P and Q for the global explanation. This is an interesting idea to consider the KL divergence instead of the difference $f(S \cup \{i\}) - f(S)$. Lower the KL divergence value, the better we have matched the true distribution with our approximation, and the smaller we have impact of the i-th feature. Then we can write

$$\phi_i(f) = \phi_i = \sum_{S \subseteq N \backslash \{i\}} B(S, N) KL(P(S) \| Q(S)). \tag{3}$$

The KL divergence is a non-symmetric measure, i.e., $f(S \cup \{i\}) - f(S) > 0$ and it does not mean that $f(S) - f(S \cup \{i\}) < 0$. Even if we use a symmetric divergence, the symmetry property of the divergence measure and the symmetry property of the Shapley values are different concepts.

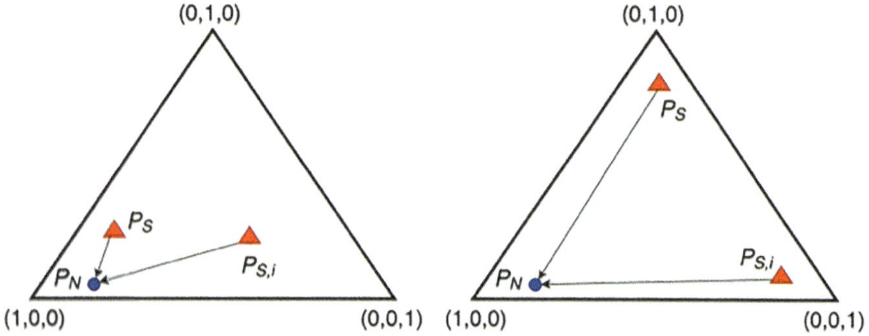

Fig. 1 Two cases of the relationship between probability distributions P_N, P_S, $P_{S,i}$

Suppose that a classification model provides a predicted class probability distribution $P_N = f(N)$ for inputs with features from the set N. Without loss of generality, we assume that P_N is close to the vector $(1, 0, \dots, 0)$. To use SHAP, we consider two types of probability distributions: $P_{S\cup i} = f(S \cup \{i\})$ and $P_S = f(S)$, $S \subseteq N \backslash \{i\}$. Define functions of predictions P_S or $P_{S,i}$ for all S and i as distances between P_N and P_S or P_N and $P_{S\cup i}$, denoted as $D(P_S, P_N)$ or $D\big(P_{S,i}, P_N\big)$, respectively. Introduce the i-th feature marginal contribution as $D(P_S, P_N) - D\big(P_{S,i}, P_N\big)$. This is an unusual definition of the marginal contribution can be explained by means of Fig. 1 where two cases of the predicted distributions of examples having three features are illustrated in the unit simplex with vertices $(1, 0, 0)$, $(0, 1, 0)$, $(0, 0, 1)$. $P_{S\cup i}$ and P_S are depicted by small triangles, P_N is depicted by the small circle.

At first glance, it seems that the i-th feature contribution should be measured by the distance between P_S and $P_{S,i}$. If they coincide, then the distance is zero, and there is no contribution of the considered feature into the prediction. We could use $KL\big(P_{S,i} \| P_S\big)$ to characterize the feature marginal contribution. However, if to look at the right unit simplex in Fig. 1, then one can see that the distance between P_S and $P_{S,i}$ is large and the marginal contribution of the i-th feature should be large. However, after adding the i-th feature, we remain at the same distance from P_N. We do not improve our knowledge about the true class or about P_N after adding the i-th feature to S and computing the corresponding prediction. Note that we do not search for the contribution of the i-th feature into the prediction P_S. We aim to estimate how the i-th feature contributes into P_N. Points P_S and $P_{S,i}$ in the right simplex are approximately at the same distance from P_N, therefore, the difference $D(P_S, P_N) - D\big(P_{S,i}, P_N\big)$ is close to 0 though points P_S and $P_{S,i}$ are far from each other. The left unit simplex in Fig. 1 shows another case when $P_{S,i}$ negatively contributes into our knowledge about the class distribution P_N because it makes the probability distribution to be more uncertain in comparison with P_S. This implies that the difference $D(P_S, P_N) - D\big(P_{S,i}, P_N\big)$ should be negative in this case.

Assuming that $D(P_S, P_N)$ is the KL divergence, this distance can be viewed as an information gain achieved if P_S would be used instead of P_N. A similar information gain is achieved if $P_{S,i}$ would be used instead of P_N. The difference $D(P_S, P_N) -$

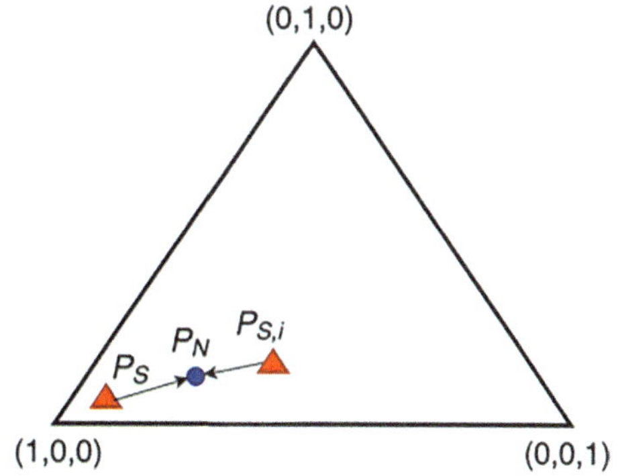

Fig. 2 A specific case of the probability distribution location

$D\left(P_{S,i}, P_N\right)$ can be viewed as the i-th feature contribution to change of the information gain. The problem arises when we consider a case illustrated in Fig. 2, where points P_S and $P_{S,i}$ are at the same distance from P_N. However, this case can also be explained. We do not explain a final decision about a class of the example. The distribution P_0 is explained here. But it is correctly explained because $P_{S,i}$ does not contribute into P_N in comparison with P_S. The distribution P_S explains the decision $(1, 0, 0)$ better than $P_{S,i}$, but not than P_N. We meet a situation when interpretation strongly depends on its aim. If its aim is to interpret P_N without considering a final decision about a class, then we should use the above approach. The i-th feature contribution in this case illustrated in Fig. 2 is 0.

Another case is when we explain the *predicted class* by having predictions in the form of probability distributions. The class can be represented by the distribution $P_0 = (1, 0, \ldots, 0)$ depicted in Fig. 3 by a small square. In this case, we compare P_S and $P_{S,i}$ with P_0, but not with P_N because we analyze how the i-th feature contributes into change of the predicted class or how the feature contributes into our knowledge about the true class. This is an important difference from the case when we interpreted P_N. It can be seen from Fig. 3 that $D(P_S, P_0)$ and $D\left(P_{S,i}, P_0\right)$ instead of $D(P_S, P_N)$ and $D\left(P_{S,i}, P_N\right)$ are computed. When interpretation of the class is performed, the contribution of the i-th feature for the case shown in Fig. 3 is no longer zero. Therefore, we will distinguish the above two cases: interpretation of the *class probability distribution* and interpretation of the *predicted class*.

It seems that $D(P_S, P_0)$ could be viewed as a measure of uncertainty of the prediction P_S. However, if we assume that $P_0 = (1, 0, 0)$ and $P_S = (0, 1, 0)$, then the distance between distributions is largest, but P_S is certain. To consider the uncertainty measure, $D(P_S, P_u)$ should be studied in place of $D(P_S, P_0)$, where P_u is the uniform distribution $(1/C, \ldots, 1/C)$. In this case, $D(P_S, P_u)$ is a certainty measure of P_S, because the most uncertain case $P_S = P_u$ leads to $D(P_S, P_u) = 0$. In sum, the Shapley values are computed now as follows:

$$\phi_i = \sum_{S \subseteq N \setminus \{i\}} B(S, N)\left[D(P_S, P_N) - D\left(P_{S,i}, P_N\right)\right]. \tag{4}$$

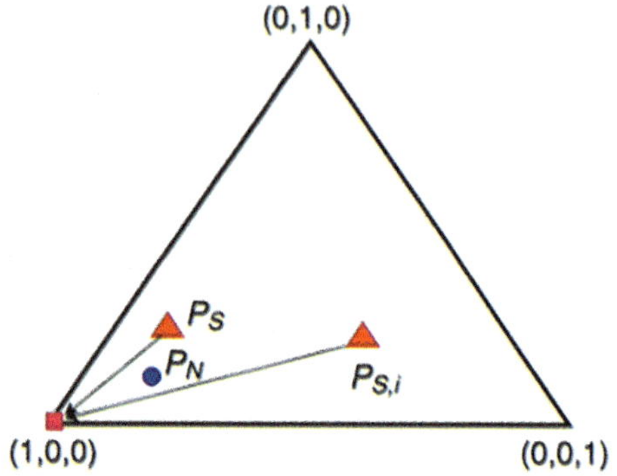

Fig. 3 Interpretation of the predicted class

The introduced distances fulfil basic properties of Shapley values as directly follows from the representation of the prediction $f(S)$ as $-D(P_S, P_N)$ or $-D(P_S, P_0)$. Since the condition $f(\emptyset) = 0$ may be not valid (the prediction by some predefined feature values), then the efficiency property for the case of the interpretation of P_S is

$$\sum_{k=1}^{m} \phi_k = f(\mathbf{x}) - f(\emptyset) = -D(P_N, P_N) + D(P_\emptyset, P_N) = D(P_\emptyset, P_N). \tag{5}$$

Here $P_\emptyset$ is the model prediction by replacing all features with predefined values, for example, mean values. Other properties of the Shapley values, including the symmetry, dummy and linearity properties, remain without changes.

5 Imprecise Extension of SHAP

We assume that a prediction of the black-box model is imprecise due to a small amount of training data, due to our ignorance about a model caused by the lack of observation data. In order to take into account the imprecision, we propose to replace precise probability distributions of classes with sets of distributions denoted as $\mathcal{P}$, which can be constructed in accordance with one of the well-known imprecise statistical models [8]. The imprecision of predictions changes the definition of distances between distributions because precise distributions are replaced with sets $\mathcal{P}$. This implies that distances between sets $\mathcal{P}_1$ and $\mathcal{P}_2$ of distributions should be defined instead of distances between single distributions. We have now a set of distances between all pairs of points such that one point in each pair belongs to $\mathcal{P}_1$, and another point in the pair belongs to $\mathcal{P}_2$. As a results, the set of distances produces an interval with lower and upper bounds corresponding to the smallest and the largest distances, respectively. The same can be said about pairs of distributions which produce an interval of Shapley values ϕ_i denoted as $\left[\phi_i^L, \phi_i^U\right]$. This implies that (4) can be rewritten as

$$\phi_i^L = \sum_{S \subseteq N \setminus \{i\}} B(S,N) \min_{P \in \mathcal{P}(P_S), |R \in \mathcal{P}(P_{S,i}), |Q \in \mathcal{P}(P_N)} \big[D(P,Q) - D(R,Q)\big], \quad (6)$$

$$\phi_i^U = \sum_{S \subseteq N \setminus \{i\}} B(S,N) \max_{P \in \mathcal{P}(P_S), |R \in \mathcal{P}(P_{S,i}), |Q \in \mathcal{P}(P_N)} \big[D(P,Q) - D(R,Q)\big]. \quad (7)$$

where P, R, Q are distributions from subsets $\mathcal{P}(P_S)$, $\mathcal{P}\big(P_{S,i}\big)$, $\mathcal{P}(P_N)$, respectively.

We assume convexity of $\mathcal{P}(P_S)$, $\mathcal{P}\big(P_{S,i}\big)$, $\mathcal{P}(P_N)$. The next question is how to calculate intervals for Shapley values. A simple way is to separately compute the minimum and the maximum of distances $D(P,Q)$ and $D(R,Q)$. However, obtained intervals of Shapley values may be too wide because we consider extreme cases assuming that the distribution $Q \in \mathcal{P}(P_N)$ may be different in $D(P,Q)$ and $D(R,Q)$. To obtain tighter intervals, we assume that there exists a single class probability distribution Q from $\mathcal{P}(P_N)$, which is unknown, but it provides the smallest or the largest value of the difference $D(P,Q) - D(R,Q)$ over all distributions from $\mathcal{P}(P_N)$. Of course, we relax this condition in other terms of the sum over $S \subseteq N \setminus \{i\}$ because it is extremely difficult to solve the obtained optimization problem under this condition. The same can be said about distributions P and R. As a result, we also get wide intervals for Shapley values, but they will be reduced taking into account intervals for all Shapley values and the efficiency property of the values.

Solutions of optimization problems in (6) and (7) depend on definitions of D. Let us return to the efficiency property (5) which is briefly written as $\sum_{k=0}^{m} \phi_k = D(P_\emptyset, P_N)$. Suppose that we have computed ϕ_i^L and ϕ_i^U. Let us denote the lower and upper bounds for $D(P_\emptyset, P_N)$ as D^L and D^U, respectively. Then there hold

$$D^L = \min_{T \in \mathcal{P}(P_\emptyset), |Q \in \mathcal{P}(P_N)} D(T,Q), \; D^U = \max_{T \in \mathcal{P}(P_\emptyset), |Q \in \mathcal{P}(P_N)} D(T,Q). \quad (8)$$

Hence, the efficiency property of Shapley values can be rewritten by considering interval-valued Shapley values as

$$D^L \leq \sum_{k=1}^{m} \phi_k \leq D^U. \quad (9)$$

It follows from (9) that we cannot write a precise version of the efficiency property because we do not know precise distributions $P_\emptyset$ and P_N. Therefore, we use bounds for the total gain, but they help us to reduce intervals of ϕ_i, which can be obtained by solving the following linear optimization problems:

$$\tilde{\phi}_k^L = \min \phi_k, \; \tilde{\phi}_k^U = \max \phi_k, \quad (10)$$

subject to (9) and $\phi_k^L \leq \phi_k \leq \phi_k^U, \; k = 1, \dots, m$.

The above optimization problems for all $k = 1, \dots, m$ allow us to get tighter bounds for Shapley values. It turns out that problems (10) can be explicitly solved.

To avoid introducing new notations, we assume that all Shapley values are positive. It can be done by subtracting value $\min_{i=1,\dots,m}\phi_i^L$ from all variables ϕ_i.

Proposition 1. *Assume that* $\min_{i=1,\dots,m}\phi_i^L \geq 0$ *and* $\sum_{i=1}^m \phi_i^L \leq D^L$, $\sum_{i=1}^m \phi_i^U \geq D^U$. *Then problems* (10) *have the following solutions for all* $k \in \{1,\dots,m\}$: $\tilde{\phi}_k^U = \min\left(\phi_k^U, D^U - \sum_{i=1,i\neq k}^m \phi_i^L\right)$, $\tilde{\phi}_k^L = \max\left(\phi_k^L, D^L - \sum_{i=1,i\neq k}^m \phi_i^U\right)$.

It is interesting to note that the above problem statement is like the definitions of reachable probability intervals [11] in the framework of the imprecise probability theory [8]. It can be regarded as an extension of the probabilistic definitions which consider sets of probability distributions as convex subsets of the unit simplex. Moreover, vectors of Shapley values produce a convex set. In contrast to the imprecise probabilities, imprecise Shapley values are not a part of the unit simplex. Transferring definitions of reachable probability intervals to the above problem, conditions of Proposition 1 imply that intervals $\left[\phi_i^L, \phi_i^U\right]$ are proper. This implies that the corresponding set of Shapley values is not empty. Moreover, it can be simply proved that bounds $\tilde{\phi}_k^L$ and $\tilde{\phi}_k^U$ are reachable. Proposition 1 can be extended to a more general case when we would like to find bounds for a linear function $g(\phi_1,\dots,\phi_m) = \sum_{i=1}^m a_i\phi_i = \langle a, \phi\rangle$ of imprecise Shapley values. Here $a = (a_1,\dots,a_m)$, $\phi = (\phi_1,\dots,\phi_m)$. Intervals of $g(\phi|a)$ are computed by solving the following two linear programming problems (minimization and maximization):

$$g^L(\phi|a)\left(g^U(\phi|a)\right) = \min(\max)a, \phi, \tag{11}$$

subject to (9) and $\phi_k^L \leq \phi_k \leq \phi_k^U$, $k = 1,\dots,m$.

The dual optimization problems are

$$g^L(\phi|a) = \max\left(D^L v_0 - D^U w_0 + \sum_{i=1}^m \phi_i^L v_i - \sum_{i=1}^m \phi_i^U w_i\right), \tag{12}$$

subject to $v_0 - w_0 + \sum_{i=1}^m (v_i - w_i) \leq a_k$, $k = 1,\dots,m$, and

$$g^U(\phi|a) = \min\left(D^U v_0 - D^L w_0 + \sum_{i=1}^m \phi_i^U v_i - \sum_{i=1}^m \phi_i^L w_i\right), \tag{13}$$

subject to $v_0 - w_0 + \sum_{i=1}^m (v_i - w_i) \geq a_k$, $k = 1,\dots,m$.

The next question is how to define models producing sets $\mathcal{P}$ and distances D to simply compute bounds (6, 7).

6 Imprecise Contamination Model

There are several imprecise models of probability distributions. One of the interesting models is the imprecise $\boldsymbol{\varepsilon}$-contamination model [8]. It produces a set $\mathcal{P}(\boldsymbol{\varepsilon}, \boldsymbol{P})$ of probabilities $\mathbf{P}^* = \left(\mathbf{p}_1^*, \ldots, \mathbf{p}_C^*\right)$ such that $\mathbf{p}_i^* = (1 - \boldsymbol{\varepsilon})\mathbf{p_i} + \boldsymbol{\varepsilon} h_i$, where $\mathbf{P} = (\mathbf{p_1}, \ldots, \mathbf{p_C})$ is an elicited probability distribution (in the considered case of SHAP, these distributions are $\mathbf{P_S}$, $\mathbf{P_{S,i}}$, $\mathbf{P_N}$); $\mathbf{h_i} \geq 0$ is arbitrary with $\mathbf{h}_1 + \ldots + \mathbf{h_C} = 1$; $0 \leq \boldsymbol{\varepsilon} \leq 1$. Parameter $\boldsymbol{\varepsilon}$ controls the size of set $\mathcal{P}(\boldsymbol{\varepsilon}, \mathbf{P})$. The greater the number of training examples, the less the uncertainty for probability distribution and the less the value of parameter $\boldsymbol{\varepsilon}$. The set $\mathcal{P}(\boldsymbol{\varepsilon}, \mathbf{P})$ is a subset of the unit simplex. It coincides with the unit simplex when $\boldsymbol{\varepsilon} = 1$. $\mathcal{P}(\boldsymbol{\varepsilon}, \mathbf{P})$ is the convex set of probabilities with lower bound $(1 - \boldsymbol{\varepsilon})\mathbf{p_i}$ and upper bound $(1 - \boldsymbol{\varepsilon})\mathbf{p_i} + \boldsymbol{\varepsilon}, \mathbf{i} = 1, \ldots, \mathbf{C}$. The convex set has $\mathbf{C}$ extreme points which are of the form: the $\mathbf{k}$-th element is $(1 - \varepsilon)\mathbf{p_i} + \boldsymbol{\varepsilon}$ and the other $\mathbf{C} - 1$ elements are equal to $(1 - \boldsymbol{\varepsilon})\mathbf{p_i}$.

Figure 4 illustrates sets $\mathcal{P}(\varepsilon, \mathbf{P_S})$, $\mathcal{P}\left(\varepsilon, \mathbf{P_{S,i}}\right)$, $\mathcal{P}(\varepsilon, \mathbf{P_N})$ for precise distributions $\mathbf{P_S}$, $\mathbf{P_{S,i}}$, $\mathbf{P_N}$, respectively, in the form of large triangles around the corresponding distributions. To take into account the imprecision, we introduce the distance between probability distributions and define a way for computing the largest and the smallest distances between all pairs of probability distributions which belong to the small subsets of distributions $\mathcal{P}(\varepsilon, \mathbf{P_S})$, $\mathcal{P}\left(\varepsilon, \mathbf{P_{S,i}}\right)$, $\mathcal{P}(\varepsilon, \mathbf{P_N})$. The use of the KL divergence leads to complex optimization problems for computing bounds for distances between subsets of probability distributions. However, Shapley values can be computed by means of generating many points in subsets $\mathcal{P}(\varepsilon, \mathbf{P_S})$, $\mathcal{P}\left(\varepsilon, \mathbf{P_{S,i}}\right)$, $\mathcal{P}(\varepsilon, \mathbf{P_N})$ and computing the smallest and the largest distances or computing the smallest and the largest Shapley values.

One of the well-known and simple algorithms for generating uniformly distributed random points in the unit simplex is of the form [32]:

1. Generate k unit-exponential random variables $Y_1, \ldots, Y_k$ and compute $T_k = \sum_{i=1}^{k} Y_i$.
2. Define $E_i = Y_i / T_k$ and return vector $E = (E_1, \ldots, E_k)$ which is uniformly distributed in the unit simplex.

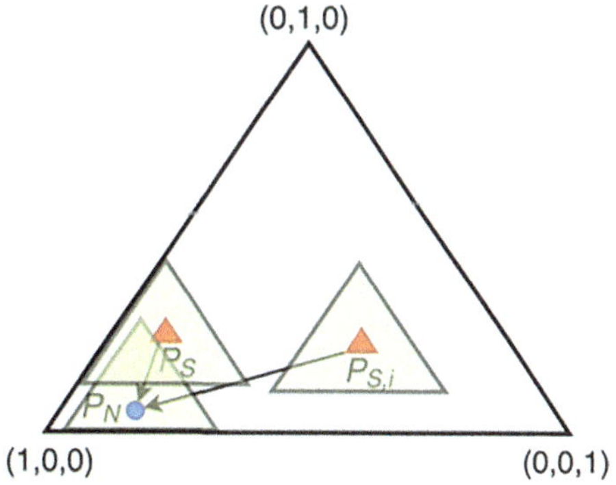

Fig. 4 Illustration of distances between subsets produced by the ε-contamination model

Convexity of $\mathcal{P}$ implies that $\mathcal{P}$ is defined by its extreme points $\mathcal{E}(\mathcal{P})$. Suppose that we have a set of r extreme points $Q_k = \left(q_1^{(k)}, \ldots, q_C^{(k)}\right)$ of $\mathcal{P}$, i.e., $q_i \in \mathcal{E}(\mathcal{P}), i = 1, \ldots, r$. For the imprecise ε-contamination model, there holds $r = C$. Then every distribution $P = (p_1, \ldots, p_C)$ from $\mathcal{P}$ can be represented as a linear combination of the extreme points $P = \sum_{k=1}^{r} \lambda_k \cdot Q_k$, where $\lambda_1 + \ldots + \lambda_r = 1$. If to uniformly generate vectors $(\lambda_1, \ldots, \lambda_r)$ from the unit simplex by using the above generation algorithm, then random points P from $\mathcal{P}$ can be obtained. From all generated points in $\mathcal{P}(\varepsilon, P_S)$, $\mathcal{P}(\varepsilon, P_{S,i})$, $\mathcal{P}(\varepsilon, P_N)$, three points $P_1 \in \mathcal{P}(\varepsilon, P_S)$, $R_1 \in \mathcal{P}(\varepsilon, P_{S,i})$, $Q_1 \in \mathcal{P}(\varepsilon, P_N)$ are selected such that $D(P, Q) - D(R, Q)$ achieves its minimum. These points contribute into ϕ_i^L. In the same way, three points P_2, R_2, Q_2 are selected from the same sets, respectively, such that the difference $D(P, Q) - D(R, Q)$ achieves its maximum. These points contribute into ϕ_i^U. The procedure is repeated for all $\phi_i, i = 1, \ldots, m$.

An advantage of the approach is that an arbitrary distance between distributions can be used for implementing the approach. However, it is extremely complex in case of a high dimension. Moreover, it is difficult to control accuracy of the obtained Shapley values because there is a chance that the selected points P_1, R_1, Q_1 and P_2, R_2, Q_2 may be not optimal, i.e., the differences $D(P, Q) - D(R, Q)$ do not achieve the minimum and the maximum. To overcome this difficulty, another approach is proposed, which is based on applying the KSD leading to a set of simple linear programming problems.

7 The Kolmogorov–Smirnov Distance and the Imprecise SHAP

The KSD between two probability distributions $P = (p_1, \ldots, p_C)$ and $Q = (q_1, \ldots, q_C)$ is defined as:

$$D_{KS}(P, Q) = \max_{i=1,\ldots,C-1} |\pi_i - \alpha_i|, \tag{14}$$

where $\pi_i = \sum_{j=1}^{i} p_j$, $\alpha_i = \sum_{j=1}^{i} q_j$ are cumulative distributions and $\pi_C = \alpha_C = 1$.

Let us consider an arbitrary event A. If $P(A)$ is an elicited probability of event A, then, according to the ε-contamination model, bounds for $P^*(A)$ are defined as:

$$(1 - \varepsilon)P(A) \leq P^*(A) \leq (1 - \varepsilon)P(A) + \varepsilon, \tag{15}$$

If $A = \{1, 2, \ldots, i\}$, then $P^*(A)$ is the i-th cumulative probability. Cumulative probability distributions π, τ, α produced by the ε-contamination model for distributions P_S, $P_{S,i}$, P_N have the following lower and upper bounds for $i = 1, \ldots, C - 1$:

$$\pi_i^L \leq \pi_i \leq \pi_i^U,\ \tau_i^L \leq \tau_i \leq \tau_i^U,\ \alpha_i^L \leq \alpha_i \leq \alpha_i^U, \pi_C = \tau_C = \alpha_C = 1. \quad (16)$$

Let us consider the lower bound L for $D_{KS}(P, Q) - D_{KS}(R, Q)$ derived from the following optimization problem:

$$L = \min_{\pi,\tau,\alpha} \left(\max_{i=1,\dots,C-1} |\pi_i - \alpha_i| - \max_{i=1,\dots,C-1} |\tau_i - \alpha_i| \right), \quad (17)$$

subject to (16).

This problem is non-convex. However, it can be represented as a set of $2(C-1)$ simple linear programming problems.

Proposition 2. *The lower bound L for $D_{KS}(P, Q) - D_{KS}(R, Q)$ is determined by solving $2(C-1)$ linear programming problems:*

$$L_1(k) = \min_{B,\pi,\alpha} \left(B - \tau_k^U + \alpha_k\right), k = 1, \dots, C-1, \quad (18)$$

$$L_2(k) = \min_{B,\pi,\alpha} \left(B - \alpha_k + \tau_k^L\right), k = 1, \dots, C-1, \quad (19)$$

subject to $\pi_i^L \leq \pi_i \leq \pi_i^U, \alpha_i^L \leq \alpha_i \leq \alpha_i^U, i = 1, \dots, C-1, B \geq \pi_i - \alpha_i, B \geq \alpha_i - \pi_i$, $i = 1, \dots, C-1$, $\pi_i \leq \pi_{i+1}$, $\alpha_i \leq \alpha_{i+1}$, $i = 1, \dots, C-2$.

For $L_1(k)$, constraints are added: $\alpha_k \leq \tau_k^U$ and $\tau_k^U - \alpha_k \geq \tau_i^U - \alpha_i$, $i = 1, \dots, C-1, i \neq k$.

For $L_2(k)$, constraints are added: $\alpha_k \geq \tau_k^L$ and $\alpha_k - \tau_k^L \geq \alpha_i - \tau_i^L$, $i = 1, \dots, C-1, i \neq k$.

If to find $k_1 = \arg\max_k(\tau_k^U - \alpha_k)$ for $L_1(k)$ and $k_2 = \arg\max_k(\alpha_k - \tau_k^L)$ for $L_2(k)$, then $L = L_1(k_1)$ if $\tau_{k_1}^U - \alpha_{k_1} \geq \alpha_{k_2} - \tau_{k_2}^L$, otherwise $L = L_2(k_2)$.

Problems (18) do not have solutions for some k when $\tau_k^U < \alpha_k^L$, problems (19) do not have solutions when $\tau_k^L > \alpha_k^U$.

The upper bound U for $D_{KS}(P, Q) - D_{KS}(R, Q)$ is derived from the problem:

$$U = \max_{\pi,\tau,\alpha} \left(\max_{i=1,\dots,C-1} |\pi_i - \alpha_i| - \max_{i=1,\dots,C-1} |\tau_i - \alpha_i| \right), \quad (20)$$

subject to (16).

Proposition 3. *The upper bound U for $D_{KS}(P, Q) - D_{KS}(R, Q)$ is determined by solving $2(C-1)$ linear programming problems:*

$$U_1(k) = \max_{B,\tau,\alpha} \left(\pi_k^U - \alpha_k - B\right), k = 1, \dots, C-1, \quad (21)$$

$$U_2(k) = \max_{B,\tau,\alpha} \left(\alpha_k - \pi_k^L - B\right), k = 1, \dots, C-1, \quad (22)$$

subject to $\tau_i^L \leq \tau_i \leq \tau_i^U, \alpha_i^L \leq \alpha_i \leq \alpha_i^U, i = 1, \ldots, C-1, B \geq \tau_i - \alpha_i, B \geq \alpha_i - \tau_i,$ $i = 1, \ldots, C-1, \tau_i \leq \tau_{i+1}, \alpha_i \leq \alpha_{i+1}, i = 1, \ldots, C-2.$

For $U_1(k)$*, constraints are added* : $\alpha_k \leq \pi_k^U, \pi_k^U - \alpha_k \geq \pi_i^U - \alpha_i, i = 1, \ldots, C-1, i \neq k.$

For $U_2(k)$*, constraints are added:* $\alpha_k \geq \pi_k^L, \alpha_k - \pi_k^L \geq \alpha_i - \pi_i^L, i = 1, \ldots, C-1, i \neq k.$

If to find $k_1 = argmax_k\left(\pi_k^U - \alpha_k\right)$ *for* $U_1(k)$ *and* $k_2 = argmax_k\left(\alpha_k - \pi_k^L\right)$ *for* $U_2(k)$*, then* $U = U_1(k_1)$ *if* $\pi_{k_1}^U - \alpha_{k_1} \geq \alpha_{k_2} - \pi_{k_2}^L$*, otherwise* $U = U_2(k_2)$.

Problems (21) do not have solutions for some k when $\pi_k^U < \alpha_k^L$. Problems (22) do not have solutions when $\pi_k^L > \alpha_k^U$.

The KSD for the binary classification model is of the form: $D_{KS}(P, Q) = |\pi_1 - \alpha_1|$, $L = \min_{\pi,\tau,\alpha}(|\pi_1 - \alpha_1| - |\tau_1 - \alpha_1|)$. The index 1 will be omitted below for short.

Corollary 4. *For the binary classification black-box model with* $C = 2$*, the lower* L *bound for* $D_{KS}(P, Q) - D_{KS}(R, Q)$ *is* $2\alpha^L - \pi^U - \tau^U$ if $\pi^U \leq \alpha^L \leq \tau^U$, $\pi^L - \tau^U$ if $\alpha^L \leq \min\left(\tau^U, \pi^L\right)$, $\tau^L - \pi^U$ if $\alpha^U \geq \max\left(\tau^L, \pi^U\right)$, $\pi^L + \tau^L - 2\alpha^U$ if $\tau^L \leq \alpha^U \leq \pi^L$.

The upper U *is* $2\alpha^U - \pi^L - \tau^L$ if $\pi^L \leq \alpha^U \leq \tau^L$, $\pi^U - \tau^L$ if $\alpha^U \leq \min\left(\tau^L, \pi^U\right)$, $\tau^U - \pi^L$ if $\alpha^L \geq \max\left(\tau^U, \pi^L\right)$, $\pi^U + \tau^U - 2\alpha^L$ if $\tau^U \leq \alpha^L \leq \pi^U$.

It follows from the above that there holds for the case of precise probabilities $D_{KS}(P, Q) - D_{KS}(R, Q) = |\pi - \alpha| - |\tau - \alpha|$.

The next question is how to compute D^L and D^U in (9) to reduce intervals of Shapley values. Let π and α be cumulative distributions corresponding to $P_\emptyset$ and P_N from (5), respectively, i.e., $\pi \in P_\emptyset$, $\alpha \in P_N$. Then the lower bound is $D^L = \min_{\pi,\alpha}\max_{i=1,\ldots,C-1}|\pi_i - \alpha_i|$. The upper bound is $D^U = \max_{\pi,\alpha}\max_{i=1,\ldots,C-1}|\pi_i - \alpha_i|$.

Proposition 5. *The lower bound* D^L *is determined from the following linear programming problem:* $D^L = \min_{\pi,\alpha} B$*, subject to constraints to (18 and 19). The upper bound* D^U *is determined as* $D^U = \max_{k=1,\ldots,C-1}\max\left(\pi_k^U - \alpha_k^L, \alpha_k^U - \pi_k^L\right)$.

Finally, we have obtained basic tools for dealing with imprecise Shapley values, which include Propositions 2, 3, and 5.

8 Numerical Experiments

In numerical experiments, the black-box model is the random forest consisting of 100 decision trees with largest depth 8. The choice of the random forest is due to simplicity of getting the class probabilities.

To study properties of the imprecise SHAP and to compare its predictions with the original SHAP, three datasets of synthetic data are generated. Every dataset consists of 1000 training examples and 250 testing examples with 2 features (x and y) uniformly generated from $[0, 5] \times [0, 5]$. The datasets are described in Table 1 where the second

Table 1 A brief introduction about datasets

Dataset	Classes	Training	Testing
1	2	879, 121	21, 535
2	3	340, 336, 324	928, 078
3	4	249, 263, 238, 250	64, 507, 462

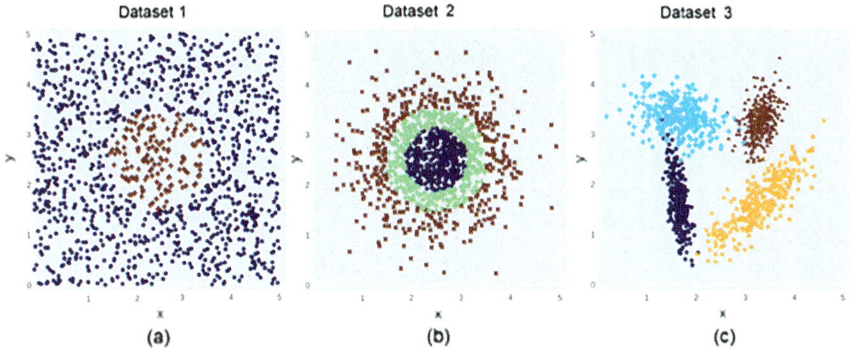

Fig. 5 Three synthetic datasets for numerical experiments

column shows the number of classes, numbers of training and testing examples in each class are given in the third and fourth columns, respectively.

Points of the first dataset are shown in Fig. 5a. Points inside a circle with the unit radius and center (2.5, 2.5) belongs to the first class, other points belong to the second class.

Points of the dataset are shown in Fig. 5b. Examples are randomly generated from the normal distribution with the expectations (2.5, 2.5) and the covariance matrix $0.5 \cdot \mathrm{I}$, where I is the unit matrix. Points inside a circle with the unit radius and center (2.5, 2.5) belong to the first class. Points located outside the circle with radius 2 and center (2.5, 2.5) belong to the third class. Other points belong to the second class.

The third dataset is depicted in Fig. 5c and consists of clusters of points normally distributed with the variance 1 about vertices of the square with sides of length 2.

An example of interval-valued Shapley values as functions of the contamination parameter ε is shown in Fig. 6. The testing example with features (1.5, 2.5) is chosen because this point is very close to the boundary between classes and the corresponding class distribution is non-trivial. Moreover, the location of the point simply shows that feature x is important because its small perturbation changes the class (see Fig. 5a) in contrast to feature y which does not change the class. One can see from Fig. 6 that intervals of Shapley values for feature x are clearly larger than the intervals for y.

Similar examples for the second and the third synthetic datasets are shown in Figs. 7 and 8. The results illustrate that imprecise Shapley values should be studied instead of precise values because intervals can reduce mistakes from the incorrect explanation.

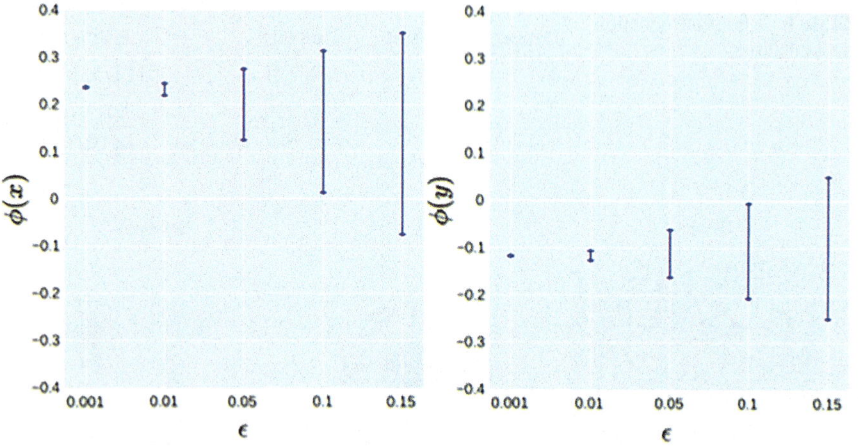

Fig. 6 Intervals of Shapley values for point (1.5, 2.5) from the first synthetic dataset with two features, x and y, by different parameters ε

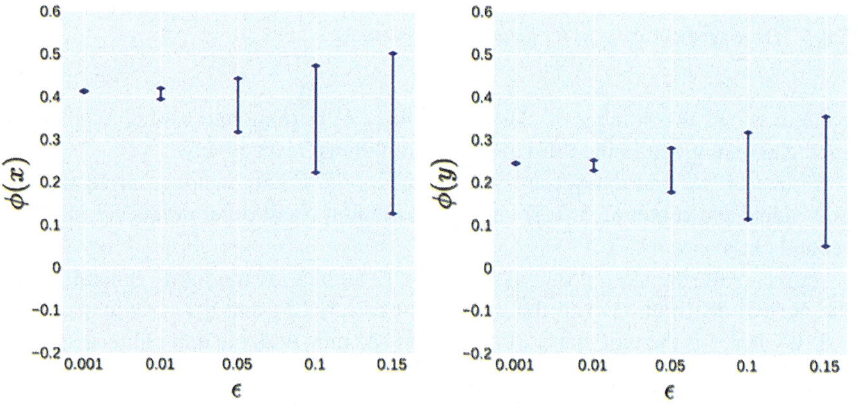

Fig. 7 Intervals of Shapley values for point (3.5, 1.5) from the second synthetic dataset with two features, x and y, by different parameters ε

Let us consider well-known real datasets from UCI Machine Learning Repository [33]: Seeds with $m = 7$, $n = 210$, $C = 3$; Ecoli with $m = 8$, $n = 336$, $C = 8$.

As an example, Fig. 8a illustrates intervals of Shapley values for the random point (18.1, 16.1, 0.88, 6.06, 3.56, 3.62, 6.0) from the Seeds dataset by $\varepsilon = 0.15$. Features are denoted as follows: area (area), perimeter (per), compactness (comp), length of kernel (len_ker), width of kernel (width), asymmetry coefficient (asym), length of kernel groove (len_gr). Red circles, which are close to centers of intervals, show precise Shapley values obtained for $\varepsilon = 0$.

Figure 9b shows intervals of Shapley values for the random point (0.8, 0.8, 0.65, 0.6, 0.17, 0.45, 0.85) from the Ecoli dataset $\varepsilon = 0.15$. Features are denoted in

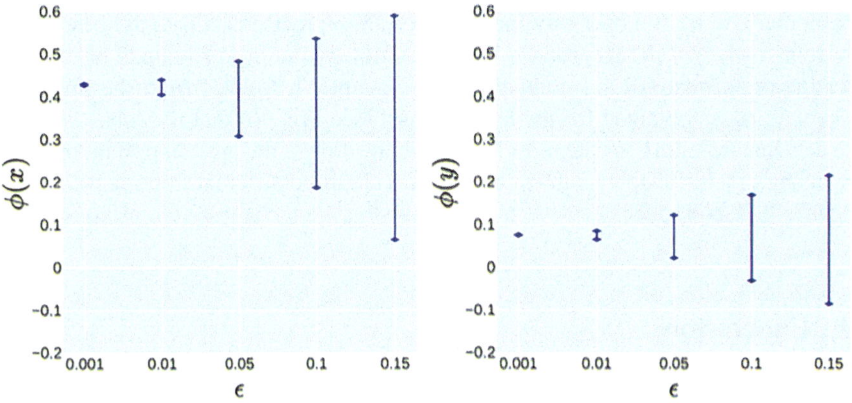

Fig. 8 Intervals of Shapley values for point (4.0, 4.0) from the third synthetic dataset with two features, x and y, by different parameters ε

accordance with the data resources as follows: mcg, gvh, lip, chg, aac, alm1, alm2. It can be seen from Fig. 8b that two features “mcg” and “gvh” provide the highest contribution to the corresponding prediction. However, it is difficult to select a single feature among features “mcg” and “gvh”. This example clearly illustrates that the use of imprecise Shapley values may change our decision about the feature contributions.

The above experiments have illustrated importance of the imprecise SHAP as a method for considering possible aleatoric uncertainty of predictions due to the limited amount of training data or other reasons. We could see from the experiments that the precise explanation may differ from the imprecise explanation. The decision making based on intervals is ambiguous. It depends on a selected decision strategy. In

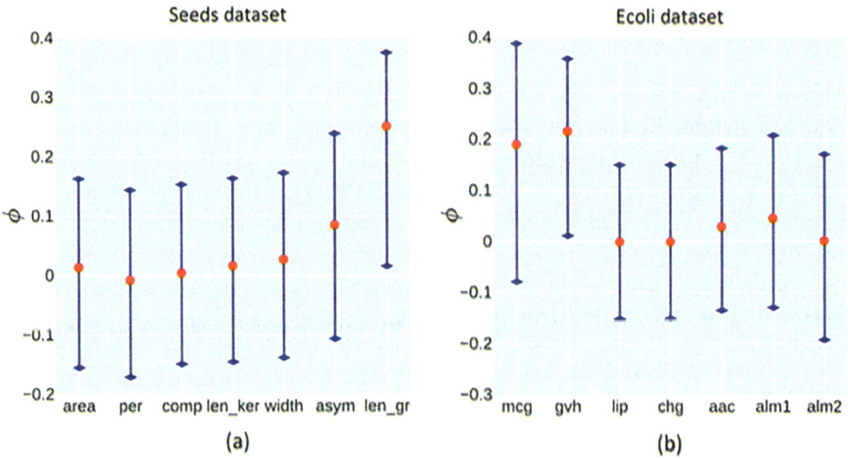

Fig. 9 Intervals of Shapley values by $\varepsilon = 0.15$ the random point from the Seeds dataset

particular, the well-known way for dealing with imprecise data is to use pessimistic or robust strategy. In accordance with this strategy, the lower bounds of intervals for important features are compared. For example, it follows from the imprecise Shapley values obtained for the Ecoli dataset (see Fig. 9b) that the robust strategy selects feature "alm1" as the most important one though the most important intervals by using precise Shapley values ($\varepsilon = 0$) provide three different important features: "gvh", "lip" and "aac".

9 Conclusion

A new approach to explanation of the black-box machine learning model probabilistic predictions has been proposed, which aims to consider the imprecision of probabilities composing the predictions. It can be applied to explanation of various machine learning models. The presented results are general and do not directly depend on the imprecise statistical model which describes imprecision of predictions.

A direction for further research is to study how to combine the obtained intervals explaining predictions on parts of intervals to reduce the SHAP complexity. This idea is mainly based on using an ensemble of random SHAPs [34] where the ensemble consists of many SHAPs such that every SHAP explains only a part of features.

Acknowledgements The research is partially funded by the Ministry of Science and Higher Education of the Russian Federation as part of state assignments "Development and research of machine learning models for solving fundamental problems of artificial intelligence for the fuel and energy complex" (topic FSEG-2024-0027).

Appendix

Proof of Proposition 1: Suppose that the sum $\sum_{i=1}^{m} \phi_i$ is precise and equals to $t \in \left[D^L, D^U\right]$. First, we consider the upper bound $\tilde{\phi}_k^U$ for ϕ_k. Then the optimization problem is of the form: $\tilde{\phi}_k^U = \max \phi_k$, subject to $\sum_{i=1}^{m} \phi_i = t$, $\phi_i^L \leq \phi_i \leq \phi_i^U$, $\phi_i \geq 0$, $i = 1, \ldots, m$. Let us write the dual problem. It is of the form:

$$\tilde{\phi}_k^U = \min \left(t \cdot w_0 + \sum_{i=1}^{m} \phi_i^U \cdot v_i - \sum_{i=1}^{m} \phi_i^L \cdot w_i \right), \tag{23}$$

subject to$w_0 + (v_i - w_i) \geq 1_k(i)$, $v_i \geq 0$, $w_i \geq 0$, $i = 1, \ldots, m$.

Here $w_0 \in \mathbb{R}$, $v_i \in \mathbb{R}_+$, $w_i \in \mathbb{R}_+$ are optimization variables; $1_k(i) = 1$ if $k = i$, otherwise $1_k(i) = 0$. The problem has $2m + 1$ variables and $3m$ constraints. It is well-known that $2m + 1$ constraints among all constraints are equalities. It is simply to prove that either $v_i = 0$ or $w_i = 0$. Suppose that the k-th constraint is one of the equalities. Suppose also that $v_k = 1$ and $w_k = 0$ (only for the k-th constraint). To minimize the objective function, w_0 should be as small as possible. The assumption $\phi_i^L \geq 0$ for all i is used here. It follows from the k-th constraint that $w_0 = 0$. Then all variables v_i, w_i are zero. Substituting the values into objective function, we get the first solution $\tilde{\phi}_k^U = \phi_k^U$. Another solution when the k-th constraint is one of the equalities is obtained if $w_0 = 1$, $v_k = 0$, $w_k = 1$. In this case, $v_i = 0$ and $w_i = 1$ for all $i \neq k$. Hence, there holds $\tilde{\phi}_k^U = t - \sum_{i=1}^{m} \phi_i^L$. Other combinations of equalities lead to a larger objective function. In sum, we get the solution $\tilde{\phi}_k^U = \min\left(\phi_k^U, t - \sum_{i=1}^{m} \phi_i^L\right)$. The solution is valid for all t from interval $\left[D^L, D^U\right]$. This implies that the smallest $\tilde{\phi}_k^U$ is achieved when $t = D^L$. The same approach can be used to derive the lower bound $\tilde{\phi}_i^L$ as was to be proved.

Proof of Proposition 2: Introduce a new variable $B = \max_{i=1,\dots,C-1} |\pi_i - \alpha_i|$. Then problem (17) becomes $\min_{\pi,\tau,\alpha} \left(B - \max_{i=1,\dots,C-1} |\tau_i - \alpha_i|\right)$, subject to (16). Constraints (16) mean that π_i and α_i are elements of the cumulative distributions. Suppose that the largest value of $|\tau_i - \alpha_i|$ is achieved for some index k, i.e., $k = \operatorname{argmax}_i |\tau_i - \alpha_i|$. Then we can rewrite the objective function as $\min_{\pi,\tau,\alpha}(B - |\tau_k - \alpha_k|)$. To achieve the minimum of the objective function, the term $|\tau_k - \alpha_k|$ must be maximized. Since, the cumulative distribution τ does not depend on π and α, then the maximum of $|\tau_k - \alpha_k|$ is reduced to two cases: $\max_\tau \tau_k - \alpha_k = \tau_k^U - \alpha_k$ and $\alpha_k - \min_\tau \tau_k = \alpha_k - \tau_k^L$. The first case requires condition $\alpha_k \leq \tau_k^U$. Condition $\alpha_k \geq \tau_k^L$ is for the second case. In sum, we get $C - 1$ linear programming problems (18) and $C - 1$ problems (19). It is obvious that the final solution L to (17) is determined by comparison of all optimal terms $\alpha_k - \tau_k^U$ from (18) and optimal terms $\tau_k^L - \alpha_k$ from (19).

Proof of Proposition 3: The proof is similar to the proof of Proposition 2.

Proof of Corollary 4: It directly follows from the proof of Propositions 2 and 3 and from considering all variants of signs of $\pi_1 - \alpha_1$ and $\tau_1 - \alpha_1$.

Proof of Proposition 5: The lower bound is derived by introducing a new variable $B = \max_{i=1,\dots,C-1} |\pi_i - \alpha_i|$. The upper bound D^U is derived similarly to problems from Propositions 2 and 3. Let $k = \operatorname{argmax}_i |\tau_i - \alpha_i|$. Introduce $D^U(k) = \max_{\pi,\alpha} |\pi_k - \alpha_k|$. Hence, we get two problems. The first one is $D_1^U(k) = \max_{\pi,\alpha}(\pi_k - \alpha_k)$ under constraints from Proposition 3 and $\pi_k \geq \alpha_k$. The second problem is $D_2^U(k) = \max_{\pi,\alpha}(\alpha_k - \pi_k)$, under the same constraints and $\alpha_k \geq \pi_k$. Solutions are trivial $D^U(k) = \pi_k^U - \alpha_k^L$ if $\pi_k^U \geq \alpha_k^L$, and $D^U(k) = \alpha_k^U - \pi_k^L$ if $\pi_k^L < \alpha_k^U$. In sum, we get $D^U = \max_{k=1,\dots,C-1} D^U(k)$, as was to be proved.

References

1. Guidotti, R., Monreale, A., Ruggieri, S., Turini, F., Giannotti, F., Pedreschi, D.: A survey of methods for explaining black box models. ACM Comput. Surv. **51**(5), 93 (2019)
2. Liang, Y., Li, S., Yan, C., Li, M., Jiang, C.: Explaining the black-box model: a survey of local interpretation methods for deep neural networks. Neurocomputing **419**, 168–182 (2021)
3. Ras, G., Xie, N., Van Gerven, M., Doran, D.: Explainable deep learning: a field guide for the uninitiated. J. Artif. Intell. Res. **73**, 329–396 (2022)
4. Ribeiro, M.T., Singh, S., Guestrin, C.: "Why should I trust You?" Explaining the predictions of any classifier. In: Proceedings of the 22nd ACM SIGKDD International Conference on Knowledge Discovery and Data Mining, pp. 1135–1144 (2016)
5. Strumbelj, E., Kononenko, I.: An efficient explanation of individual classifications using game theory. J. Mach. Learn. Res. **11**, 1–18 (2010)
6. Covert, I., Lundberg, S., Lee, S.I.: Explaining by removing: a unified framework for model explanation. J. Mach. Learn. Res. **22**(209), 1–90 (2021)
7. Senge, R., Bosner, S., Dembczynski, K., Haasenritter, J., Hirsch, O., Donner-Banzhoff, N., llermeier, E.H.: Reliable classification: learning classifiers that distinguish aleatoric and epistemic uncertainty. Inf. Sci. **255**, 16–29 (2014)
8. Walley, P.: Statistical Reasoning with Imprecise Probabilities. Chapman and Hall, London (1991)
9. Ribeiro, M.T., Singh, S., Guestrin, C.: Anchors: high-precision model agnostic explanations. In: AAAI Conference on Artificial Intelligence, pp. 1527–1535 (2018)
10. Rabold, J., Deininger, H., Siebers, M., Schmid, U.: Enriching visual with verbal explanations for relational concepts: combining LIME with Aleph. In: Machine Learning and Knowledge Discovery in Databases: International Workshops of ECML PKDD 2019, pp. 180–192 (2020)
11. Huang, Q., Yamada, M., Tian, Y., Singh, D., Yin, D., Chang, Y.: GraphLIME: local interpretable model explanations for graph neural networks. IEEE Trans. Knowl. Data Eng. **35**(7), 6968–6972 (2022)
12. Kovalev, M.S., Utkin, L.V., Kasimov, E.M.: SurvLIME: a method for explaining machine learning survival models. Knowl.-Based Syst. **203**, 106164 (2020)
13. Garreau, D., von Luxburg, U.: Explaining the explainer: a first theoretical analysis of LIME. In: International Conference on Artificial Intelligence and Statistics, pp. 1287–1296 (2020)
14. Chang, C. H., Tan, S., Lengerich, B., Goldenberg, A., Caruana, R.: How interpretable and trustworthy are GAMs? In: Proceedings of the 27th ACM SIGKDD Conference on Knowledge Discovery & Data Mining, pp. 95–105 (2021)
15. Agarwal, R., Melnick, L., Frosst, N., Zhang, X., Lengerich, B., Caruana, R., Hinton, G.E.: Neural additive models: interpretable machine learning with neural nets. Adv. Neural. Inf. Process. Syst. **34**, 4699–4711 (2021)
16. Utkin, L.V., Satyukov, E.D., Konstantinov, A.V.: SurvNAM: the machine learning survival model explanation. Neural Netw. **147**, 81–102 (2022)
17. Van den Broeck, G., Lykov, A., Schleich, M., Suciu, D.: On the tractability of SHAP explanations. J. Artif. Intell. Res. **74**, 851–886 (2022)
18. Rozemberczki, B. Sarkar, R.: The Shapley value of classifiers in ensemble games. In: Proceedings of the 30th ACM International Conference on Information & Knowledge Management, pp. 1558–1567 (2021)
19. C. Frye, D. de Mijolla, L. Cowton, M. Stanley, I. Feige.: Shapley-based explainability on the data manifold. arXiv:2006.01272. June 2020
20. Bento, J., Saleiro, P., Cruz, A.F., Figueiredo, M.A.T., Bizarro, P.: TimeSHAP: explaining recurrent models through sequence perturbations. In: Proceedings of the 27th ACM SIGKDD Conference on Knowledge Discovery & Data Mining, pp. 2565–2573 (2021)
21. Antwarg, L., Miller, R.M., Shapira, B., Rokach, L.: Explaining anomalies detected by autoencoders using shapley additive explanations. Expert Syst. Appl. **186**, 115736 (2021)
22. Aas, K., Jullum, M., Loland, A.: Explaining individual predictions when features are dependent: more accurate approximations to Shapley values. Artif. Intell. **298**, 103502 (2021)

23. Arrieta, A.B., Diaz-Rodriguez, N., Del Ser, J., Bennetot, A., Tabik, S., Barbado, A., Garcia, S., Gil-Lopez, S., Molina, D., Benjamins, R., Chatila, R., Herrera, F.: Explainable artificial intelligence (XAI): concepts, taxonomies, opportunities and challenges toward responsible AI. Inf. Fusion **58**, 82–115 (2020)
24. Rudin, C.: Stop explaining black box machine learning models for high stakes decisions and use interpretable models instead. Nat. Mach. Intell. **1**, 206–215 (2019)
25. Abellan, J., Moral, S.: Building classification trees using the building classification trees using the total uncertainty criterion. Int. J. Intell. Syst. **18**(12), 1215–1225 (2003)
26. Abellan, J., Mantas, C.J., Castellano, J.G., Moral-Garcia, S.: Increasing diversity in random forest learning algorithm via imprecise probabilities. Expert Syst. Appl. **97**, 228–243 (2018)
27. Moral-Garcia, S., Mantas, C.J., Castellano, J.G., Benitez, M.D., Abellan, J.: Bagging of credal decision trees for imprecise classification. Expert Syst. Appl. **141**(112944), 1–9 (2020)
28. Destercke, S., Antoine, V.: Combining imprecise probability masses with maximal coherent subsets: application to ensemble classification. In: Synergies of Soft Computing and Statistics for Intelligent Data Analysis, pp. 27–35 (2013)
29. Matt, P.-A.: Uses and computation of imprecise probabilities from statistical data and expert arguments. Int. J. Approximate Reasoning **81**, 63–86 (2017)
30. Utkin, L.V.: An imprecise deep forest for classification. Expert Syst. Appl. **141**, 112978 (2020)
31. Covert, I., Lundberg, S.M., Lee, S.I.: Understanding global feature contributions with additive importance measures. Adv. Neural. Inf. Process. Syst. **33**, 17212–17223 (2020)
32. Rubinstein, R.Y., Kroese, D.P.: Simulation and the Monte Carlo method, 2nd edn. Wiley, New Jersey (2008)
33. Lichman, M.: UCI machine learning repository (2013)
34. Utkin, L., Konstantinov, A.: Ensembles of random SHAPs. Algorithms **15**(11), 431 (2022)

Soft Relational Temporal Cognitive Models for Scenario Modeling of Electrical Load Distribution

Vadim Borisov

Abstract The paper presents Soft Relational Temporal Cognitive Models (SRTCM), which combine the advantages of fuzzy cognitive models and machine learning models. These models are designed to perform a set of problems of scenario modeling and predictive analytics of complex systems, processes and problem situations in conditions of inaccuracy, incompleteness, uncertainty of data and nonlinear interdependence between heterogeneous systemic and external factors. The following issues are considered: designing the structure of SRTCM; selecting models of scenario dynamics for SRTCM concepts; structural and parametric settings of SRTCM using training samples for each of the concepts; setting modeling scenarios; scenario modeling and predictive analytics based on trained SRTCM. An example of scenario modeling of electrical load distribution for consumers in the regional energy subsystem using the SRTCM is illustrated. The formulation of the inverse problem of scenario modeling is presented, which consists in determining all the retrospective values of the SRTCM concepts in the time range, leading to the required values of the concepts at the time of scenario modeling. Various approaches are considered that expand the capabilities of SRTCM for diagnosing the state of complex systems and problem situations, process planning, resource management, with a given "time depth" of retrospective diagnosis.

Keywords Soft relational temporal cognitive model · Fuzzy and soft computing · Scenario modeling · Predictive analytics · Forward and inverse problems · Electrical load distribution

V. Borisov (✉)
Branch of the National Research University, Moscow Power Engineering Institute, Energetichesky proezd 1, Smolensk, Russia
e-mail: vbor67@mail.ru

I. Ilin and M. Youzhong (eds.), *Digital Systems and Information Technologies in the Energy Sector*, Lecture Notes in Networks and Systems 1244,
https://doi.org/10.1007/978-3-031-80710-7_28

1 Introduction

Scenario modeling and forecasting of complex systems, processes and problem situations are performed, as a rule, in conditions of inaccuracy, incompleteness, uncertainty of data and nonlinear interdependence between different-quality systemic and external factors [1–5]. Such problem situations also include the electrical load distribution for consumers.

Various types of fuzzy cognitive models have proven themselves well to perform these tasks: B. Kosko's fuzzy cognitive maps and their modifications [6]; V. Silov's fuzzy cognitive maps [7]; Rule-Based Fuzzy Cognitive Maps [8], Generalized Rule-Based Fuzzy Cognitive Maps [9], Relational Fuzzy Cognitive Maps [10, 11], "Compatible" Fuzzy Cognitive Maps [12].

In the paper [13], the authors proposed a methodology for using fuzzy cognitive models for scenario modeling of poorly formalized systems and processes.

In [14], fuzzy temporal cognitive models were developed, and the method for multidimensional analysis and forecasting of the complex systems, processes and problem situations in conditions of non-stochastic uncertainty, non-linearity of mutual influence, partial inconsistency and interdependence of systemic and external factors was proposed.

In the papers [15] and [16], the authors proposed Soft Relational Cognitive Temporal Models (SRTCM), which combine the advantages of fuzzy cognitive models and machine learning models when performing tasks of scenario modeling and predictive analytics of the complex systems, processes and problem situations. Features of these models:

- assessment of the direct, indirect and aggregated interaction of systemic and external factors with their different time lags relative to each other;
- individual selection of a scenario dynamics model for all concepts;
- fast-acting adjustment of all parameters and fine tuning of models both during the operation of the system and according to the results of multidimensional forecasting, using constantly updated training samples;
- fuzzy computing of dependencies between concepts in vector-matrix form.

In the paper [17], the method of scenario modeling and forecasting of the complex systems, processes and problematic situations in conditions of uncertainty based on SRTCM was designed.

In this paper, scenario modeling using SRTCM is illustrated using the example of the electrical load distribution for consumers in a regional energy subsystem.

The paper considers the formulation of the inverse problem of scenario modeling based on SRTCM. Different approaches have been proposed to expand the capabilities of SRTCM for diagnosing the state of complex systems and problem situations, process planning, resource management, with different "time depth" of retrospective diagnostics.

2 Scenario Modeling Based on Soft Relational Temporal Cognitive Models

2.1 A Brief Description of the Soft Relational Temporal Cognitive model

The Soft Relational Temporal Cognitive Model is presented in the following form [15]:

$$SRTCM = \langle C, R \rangle,$$

$$C = \{C_i | i = 1, \ldots, I\}, R = \{R_i | i = 1, \ldots, I\},$$

$$R_i = \left\{\tilde{r}_{ij}(t-l) | l = 0, \ldots, L_j^i, j = 1, \ldots, J^i\right\},$$

$$c_i : \tilde{c}_i(t) = \tilde{F}_i\left(\begin{array}{l}\left\{\tilde{c}_i(t-k), \tilde{r}_{ii}(t-k) | k = 1, \ldots, L_i^i\right\}, \\ \left\{\tilde{c}_j(t-l), \tilde{r}_{ij}(t-l) | j = 1, \ldots, J^i, l = 1, \ldots, L_j^i\right\}\end{array}\right), i = 1, \ldots, I,$$

where C—a set of SRTCM concepts; I—the number of SRTCM concepts; R—a set of fuzzy influence relations of SRTCM concepts on each other; R_i—a subset of fuzzy influence relations of concepts that directly affect the concept C_i; J—the number of concepts that directly affect the concept C_i; $\tilde{r}_{ii}(t-k)$—the fuzzy influence relation of the concept C_i on itself at a time $(t-k)$; L_i^i—the maximum significant value of the time lag when the concept C_i affects itself; $\tilde{r}_{ij}(t-l)$—the fuzzy influence relation of the concept C_j on the concept C_i at a time $(t-l)$; L_j^i—the maximum significant value of the time lag when the concept C_j affects the concept C_i; $\tilde{c}_i(t)$, $\tilde{c}_i(t-k)$, $\tilde{c}_j(t-l)$—fuzzy values of concepts C_i and C_j at the appropriate time points.

2.2 Setting up Scenario Dynamics Models for the Soft Relational Temporal Cognitive model

In [17], various models of scenario dynamics are considered, which can be individually selected for each SRTCM concept.

Let's consider as a model of scenario dynamics for all SRTCM concepts a generalized dependence that takes into account fuzzy values and increments of concepts:

$$\tilde{c}_i(t) = \tilde{c}_i(t-1) \oplus \left(\bigoplus_{m=1}^{L_i^i} (\Delta\tilde{c}_i(t-m) \cdot \tilde{r}_{ii}(t-m))\right)$$

$$\oplus\left(\bigoplus_{j=1}^{J^i}\left(\bigoplus_{l=1}^{L_j^i}\left(\Delta\tilde{c}_j(t-l)\cdot\tilde{r}_{ij}(t-l)\right)\right)\right),$$

where $\tilde{c}_i(t-1)$—the fuzzy meaning of the concept C_i at time $(t-1)$; $\Delta\tilde{c}_i(t-m)$—the fuzzy increment of the meaning concept C_i at time $(t-m)$; $\Delta\tilde{c}_j(t-l)$—the fuzzy increment of the meaning concept C_j at time $(t-l)$; $\cdot$—fuzzy composition operation; $\bigoplus_{m=1}^{L_i^i}$, $\bigoplus_{l=1}^{L_j^i}$, $\oplus$—operations of fuzzy aggregation of impacts of concepts; L_i^i—the maximum significant value of the time lag with fuzzy aggregation of the concept's influence C_i on itself; L_j^i—the maximum significant value of the time lag in case of fuzzy influence aggregation of the C_j concept on the C_i concept.

2.3 Structural Setting of the Soft Relational Temporal Cognitive Model

The structural setting of the SRTCM consists in determining the maximum values of the time lags L_i^i and L_j^i, for which the impact of and from the corresponding subsets R_i is significant.

To determine the values L_i^i и L_j^i in the paper [17], the method of fuzzy multiple linear regression was used, which allows calculating fuzzy regression coefficients, which, after normalization, correlate with fuzzy influence relations:

$R_i=\left\{\tilde{r}_{ij}(t-l)=\tilde{a}'_{ij}(t-l)|l=0,\ldots,L_j^i,j=1,\ldots,J^i\right\}$,

where $\tilde{a}'_{ij}(t-l)$—normalized fuzzy regression coefficients.

Those fuzzy influence relations whose modal values are less than a certain set threshold $\varepsilon_{\min}$ (for example, 0.1) are excluded from the subsequent consideration, that is:

$$\forall C_j: L_j^i=\max_{j=1,\ldots,J^i} l|\tilde{r}_{ij}(t-l)<\varepsilon_{\min}.$$

2.4 Parametric Setting of the Soft Relational Temporal Cognitive Model

The parametric setting of the SRTCM consists in adjusting the values of the fuzzy influence relations all the concepts of the SRTCM if there is an appropriate training sample for each of them:

$$\forall C_i:\left(\left\{\left(\tilde{c}_i^{(k)}(t-1),\ldots,\tilde{c}_i^{(k)}\left(t-L_i^j\right)\right)|j\in 1,\ldots,I_i\right\},\tilde{c}_i^{(k)}(t)\right),k=1,\ldots,K_i,$$

where $\left\{\left(\tilde{c}_i^{(k)}(t-1), \ldots, \tilde{c}_i^{(k)}\left(t-L_i^j\right)\right) | j \in 1, \ldots, I_i\right\}$—the fuzzy input values of the concept C_i for the k_{th} example; $\tilde{c}_i^{(k)}(t)$—the fuzzy input value of the concept C_i for the kth example; K_i—the number of examples of the training sample for the concept C_i.

Note that the number of examples in the training samples for different SRTCM concepts may be different.

2.5 *The Soft Relational Temporal Cognitive Model for Scenario Modeling of Electrical Load Distribution for Consumers in the Regional Energy Subsystem*

Let's consider an example of using SRTCM for scenario modeling of electric load distribution for several consumers in the regional energy subsystem.

To assess the fuzzy influence relations R between 4 consumers of the regional energy subsystem (concepts C_1–C_4), an analysis of a multidimensional time series of electrical load consisting of 720 (30 days x 24 h) measurements was carried out (Fig. 1).

Figure 2 shows an example of the SRTCM structure (described in [17]) for scenario modeling of an electric load distribution for consumers in the regional energy subsystem, and Table 1 shows a matrix of fuzzy influence relationships between SRTCM concepts in the form of fuzzy triangular numbers.

Based on the results of scenario modeling using a trained SRTCM, the reliability of the results of the following typical scenarios for the electrical load distribution for consumers in the regional energy subsystem is justified:

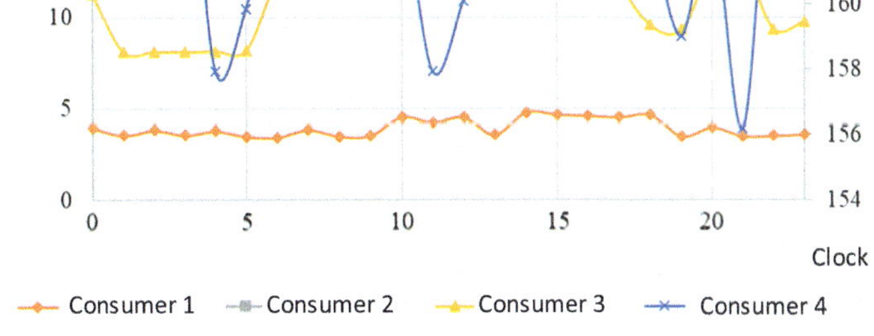

Fig. 1 Electrical load schedule for consumers in the regional energy subsystem, March 2023

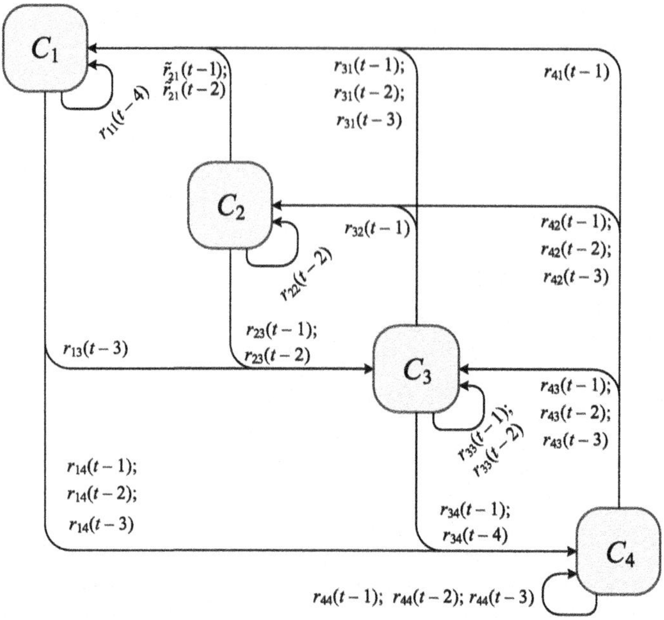

Fig. 2 SRTCM structure for scenario modeling of an electric load distribution in the regional energy subsystem [17]

Table 1 Fuzzy influence relationships between SRTCM concepts in the form of fuzzy triangular numbers [17]

R	l	C_2	C_2	C_3	C_4
C_1	1	–	–	–	[0,74; 1,00; 1,29]
	2	–	–	–	[0,53; 0,67; 0,76]
	3	–	–	[0,59; 0,62; 0,69]	[0,54; 0,71; 0,84]
	4	[0,70; 0,75; 0,78]	–	–	–
C_2	1	[0,72; 0,86; 1,00]	–	[0,48; 0,51; 0,55]	–
	2	[0,64; 0,73; 0,85]	[0,54; 0,72; 0,81]	[0,58; 0,70; 0,81]	–
C_3	1	[0,47; 0,66; 0,70]	[0,56; 0,69; 0,89]	[0,52; 0,58; 0,71]	[0,58; 0,59; 0,66]
	2	[0,48; 0,60; 0,77]	–	[0,55; 0,66; 0,81]	–
	3	[0,46; 0,65; 0,77]	–	–	–
	4	–	–	–	[0,49; 0,62; 0,75]
C_4	1	[0,87; 1,00; 1,00]	[0,44; 0,61; 0,63]	[0,36; 0,50; 0,64]	[0,45; 0,59; 0,64]
	2	–	[0,84; 1,00; 1,00]	[0,70; 0,90; 0,92]	[0,39; 0,41; 0,51]
	3	–	[0,52; 0,61; 0,66]	[0,90; 1,00; 1,00]	[0,50; 0,65; 0,71]

- changes in electricity consumption due to weather changes;
- lack of capacity as a result of damage to the electrical network;
- repair of power grid equipment;
- voltage regulation for consumers;
- ensuring the installed electrical power of consumers.

3 Formulation and Approaches to the Inverse Problem for Soft Relational Temporal Cognitive Models

The inverse problems include the problems of diagnosing the state of complex systems, process planning, and resource management.

The formulation of the inverse problem for SRTCM is presented as follows: finding all the retrospective values of the SRTCM concepts in the time range $\left[t - L_i^j, t - 1\right]$ that lead to the required values of the concepts $\tilde{c}_i(t)$ at time t of scenario modeling.

This formulation of the inverse problem for SRTCM is reduced to two successively solved subtasks:

- the 1st subtask—the computing of all retrospective values of SRTCM concepts in the time range $\left[t - L_i^j, t - 1\right]$, which correspond to the required values of the concepts $\tilde{c}_i(t)$ (in accordance with the reverse time "scan" of scenario modeling shown in Fig. 3);
- the 2nd subtask—the justification that the retrospective values of the SRTCM concepts obtained in the time range $\left[t - L_i^j, t - 1\right]$ correspond to the required values of the concepts $\tilde{c}_i(t)$ at time t of the scenario modeling.

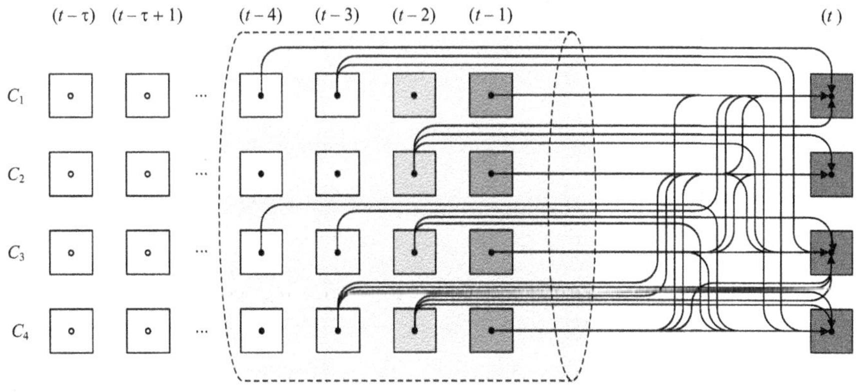

Fig. 3 Inverse time "scan" of scenario modeling for SRTCM

Note that the 1st subtask may not have a single solution, but may correspond to a family (subdomain) of solutions. In turn, 2nd subtask may not have an exact solution.

Therefore, the approach to solving the inverse problem for SRTCM in this formulation may be to search for approximate retrospective values of SRTCM concepts in the time range $\left[t - L_i^j, t - 1\right]$.

To do this, in addition to setting the required values of concepts $\tilde{c}_i(t)$, an indicator and a criterion (threshold) value of proximity to the specified values of concepts are additionally introduced. Then, a search is carried out for such retrospective values of the SRTCM concepts in the time range $\left[t - L_i^j, t - 1\right]$, which correspond (2nd subtask) to the values of concepts $\tilde{c}_i'(t)$, the degree of difference of which with a given value $\tilde{c}_i(t)$ is less than the threshold. Obviously, these values must belong to the domain of acceptable solutions (the 1st subtask). In the proposed formulation, an exhaustive algorithm can be used to find the final solution. If the search for acceptable solutions does not lead to a positive result, then you can change the threshold value of the degree of difference.

An alternative formulation of the inverse problem for SRTCM is reduced to an optimization problem with constraints imposed on the solution domain and with an objective function determining the degree of difference to the required values $\tilde{c}_i(t)$. Genetic algorithms can be used to perform the inverse problem in such a formulation.

To solve the inverse problem for SRTCM, the approach proposed in [18] is also applicable, which is focused on identifying the structure and parameters of fuzzy cognitive models based on the joint use of expert and statistical information.

Another promising variant of the formulation and solution of the inverse problem for SRTCM is possible, which is a development of the formulations of the inverse problem for fuzzy relational models proposed in [19, 20] and based on the solution of fuzzy relational equations. At the same time, the implementation of this approach requires further research aimed at substantiating the requirements and selecting fuzzy operations in order to ensure the reliability of the results of the inverse problem for SRTCM.

4 Conclusions

The paper considers Soft Relational Temporal Cognitive Models that combine the advantages of fuzzy cognitive models and machine learning models, designed for scenario modeling, predictive analytics of complex systems, processes and problematic situations under uncertainty, and provided flexible structural and parametric adaptation to changes in modeling conditions.

The description of the SRTCM is presented, the models of scenario dynamics for the SRTCM are briefly considered, the issues of structural and parametric tuning of the SRTCM are described.

An example of scenario modeling of electrical load distribution for consumers of the regional energy subsystem using a SRTCM is illustrated.

Various approaches have been proposed that expand the capabilities of SRTCM for diagnosing the state of complex systems, process planning, and resource management. These approaches are based on the considered formulations of inverse problems for SRTCM with the setting of the "time depth" of retrospective diagnosis, and consist in determining all the retrospective values of SRTCM concepts in the time range that correspond to the required values of concepts at a given moment of scenario modeling.

Acknowledgments The study was carried out as part of a state assignment, project no. FSWF-2023-0012.

References

1. Dli, M.I., Kakatunova, T.V.: A three-level fuzzy cognitive model for analyzing the processes of innovative development of the region. Prikladnaya informatika=J. Appl. Inform. **1**(43), 5–10 (2013) (in Russian)
2. Podgorskaya, S.V., Podvesovsky, A.G., Isaev, R.A., Tarasov, A.S., Bakhmatova, G.A.: Modeling the scenario development of rural areas based on a fuzzy cognitive model. Control Sci. **5**, 49–59 (2019). (in Russian)
3. Sokolov, B., Ushakov, V., Zaharov, V.: Optimal planning and scheduling of information processes during interaction among mobile objects. Int. J. Prod. Res. **62**(16), 5905–5924 (2024). https://doi.org/10.1080/00207543.2024.2302388
4. Svetunkov, S.: Elementary image of the Kolmogorov-Gabor polynomial in economic modeling. Technoeconomics 3 **2**(9), 4–21 (2024). https://doi.org/10.57809/2024.3.2.9.1
5. Sokolov, B., Ushakov, V.: Formation reachability area as a data vector using a dynamic model for controlling information processes in the automated control system for moving objects. In: Models and Methods for Researching Information Systems in Transport 2020 (MMRIST 2020), pp. 67–75 (2020)
6. Thulukkanam, K., Vasuki, R.: Two new fuzzy models using fuzzy cognitive maps model and Kosko hamming distance. Ultra Sci. **27**(1), 43–55 (2015)
7. Silov, V.B.: Strategic Decision-Making in a Fuzzy Environment. INPRO-RES, Moscow (1995). (in Russian)
8. Carvalho, J.P., Tome, J.A.B.: Rule based fuzzy cognitive maps in socio-economic systems. IFSA-EUSFLAT 2009, pp. 1821–1826 (2000)
9. Borisov, V.V., Fedulov, A.S.: Generalized rule-based fuzzy cognitive maps: structure and dynamics model. Lect. Notes Comput. Sci. **3316**, 918–922 (2004)
10. Fedulov, A.S.: Fuzzy relational cognitive maps. News of the Russian Academy of Sciences. Theory Control. Syst. **1**, 120–132 (2005) (in Russian)
11. Borisov, V., Kurilin, S., Prokimnov, N., Chernovalova, M.: Fuzzy cognitive modeling of heterogeneous electromechanical systems. Prikladnaya informatika=J. Appl. Inform. **16**, 32–39 (2021)
12. Borisov, V., Fedulov, A., Fedulov, Y.: "Compatible" fuzzy cognitive maps for direct and inverse inference. In: Proceeding of the 18th International Conference on Computer Systems and Technologies, CompSysTech'17, Ruse, Bulgaria, ACM International Conference Proceeding Series, 1369, pp. 20–27, ACM Inc., NY, USA (2017)
13. Zakharova, A.A., Podvesovsky, A.G., Isaev, R.A.: Fuzzy cognitive models in the management of poorly structured socio-economic systems. Inf. Math. Technol. Sci. Manage. **4**(20), 5–23 (2020). (in Russian)

14. Borisov, V., Luferov, V.: Neuro-fuzzy cognitive temporal models for predicting multidimensional time series with fuzzy trends. Computación y Sistemas **24**(3), 1165–1177 (2020)
15. Borisov, V.V., Kurilin, S.P., Zharkov, A.P., Sokolov, A.M.: Multidimensional prediction of heterogeneous electromechanical systems for risk management based on fuzzy temporal ontological and cognitive models. Syst. Control Commun. Secur. **4**, 83–102 (2022). (in Russian)
16. Borisov V., Kurilin S., Luferov V.: Fuzzy relational cognitive temporal models for analyzing and state prediction of complex technical systems. Prikladnaya informatika=J. Appl. Inform. **17**(1), 27–38 (2022) (in Russian)
17. Borisov, V.V., Zharkov, A.P., Luferov, V.S.: Scenario modeling based on fuzzy relational temporal cognitive models. In: Proceeding of the 21st National Conference on Artificial Intelligence, CAI-2023, vol. 2, pp. 30–41, Smolensk (2023) (in Russian)
18. Podvesovsky, A.G., Isaev, R.A.: Identification of the structure and parameters of fuzzy cognitive models: expert and statistical methods. Int. J. Open Inf. Technol. **7**(6), 35–61 (2019). (in Russian)
19. Sanchez, E.: Resolution of composite fuzzy relation equations. Inform. Contr. **30**, 38–48 (1976)
20. Li, G., Fang, S.C.: Solving interval-valued fuzzy relation equations. IEEE Trans. Fuzzy Syst. **6–2**, 321–324 (1998)

Creation of Automatic Monitoring Systems Using AutoML Technologies

Vladislav Sobolevskii

Abstract Today all branches of industrial production are being actively automated. This determines the relevance of the development of tools for creating automated systems. The article proposes the use of AutoML approach to automate the creation of machine learning models that are used for automatic monitoring. The model of object recognition in the image VGG19 was chosen as an example. On its example the automation of structural-parametric synthesis of models and optimization of hyperparameters was demonstrated. The presented system was realized as a part of the AutoGenNet software platform. This platform realizes the concept of No-Code development, which allows to hide from users the complexity of the processes of model creation and training. The use of the No-Code development concept allows to reduce the entry threshold for working with the program. Also, on the basis of AutoGenNet platform the mechanism of generation of program shells for operation of trained models was realized. All this allowed to implement AutoML approach for automation of VGG19 model generation and training processes, which in turn simplified and accelerated the process of solving automatic monitoring problems with deep learning models. The created system can be scaled and used for automated generation of other object recognition models for solving a wide variety of applied problems.

Keywords Deep learning · Object detection · AutoML · VGG19

V. Sobolevskii (✉)
St. Petersburg Federal Research Center of the Russian Academy of Science, 14 Line V.O., 39, St. Petersburg, Russia
e-mail: arguzd@yandex.ru

I. Ilin and M. Youzhong (eds.), *Digital Systems and Information Technologies in the Energy Sector*, Lecture Notes in Networks and Systems 1244,
https://doi.org/10.1007/978-3-031-80710-7_29

1 Introduction

Photo and video image recognition technologies are being more and more intensively implemented in modern industries. The main purpose of such systems is to recognize defects in production, abnormal situations, incorrect behavior of workers, etc. [1–3]. The development of this field became possible due to the emergence of convolutional neural network (CNN) architectures. This type of architecture allows to solve the tasks of recognition, segmentation and semantic analysis of objects in the image with high accuracy. The higher the accuracy and capabilities of CNN models, the more sophisticated they become. The vast majority of currently relevant CNN architectures have many heterogeneous layers [4–6]. This leads not only to higher recognition accuracy, but also to increased complexity in building and training such networks.

At the same time, the number of applied tasks that can be solved using CNN models is increasing. These tasks do not always require the most complex CNN architectures. But they are complex enough that ordinary users without deep learning (DL) knowledge will not be able to correctly create and adapt such networks. In addition, understanding the results obtained using such models may require specialized knowledge in specific application domains.

Also, one of the major challenges is the need for extensive training data sets, which are necessary for the model to learn patterns in the data and then make accurate predictions. The quality of training data can significantly affect the accuracy of CNN models. Thus, proper data collection and training are critical to the successful use of CNN models.

To overcome the described limitations, the scientific and technical branch of automated machine learning (AutoML) [7] was formed. The main tasks of AutoML include the development and application of models and methods for automated creation and training of DL models. This scientific and technical field has emerged only in the last decade, but there are already many research and design groups working in this direction [8–10].

There are already systems that allow CNNs to be created in the form of software prototypes (e.g., using MatLab [11]), and such prototypes require refinement for implementation into existing monitoring systems that are developed in specific stacks of application programming languages (C++, Java, Python, etc.). This, in turn, complicates further development and subsequent implementation of the prototypes.

Professional software based on the AutoML concept has also been created, for example auto-sklearn [12], AutoKeras [13] and other similar solutions. However, these software systems are often presented as software libraries, which does not allow them to be used as stand-alone software products. Also, such tools are used by DL specialists and are not intended for ordinary users without appropriate training.

Currently, the most large-scale and versatile AutoML systems are created by large companies. An example of such a system is Google Cloud AutoML [14], developed by Alphabet Inc. The basis of this system is a cloud software platform designed for automated creation of DL models. However, at the moment, the result of this system is a cloud service that requires modification for integration with other software. Google

Cloud AutoML has all the disadvantages of a cloud platform. It is critically dependent on an internet connection, so without a reliable connection, using the platform may be difficult or impossible. Since the data is stored on servers owned and operated by a third party, there is a risk of unauthorized access or information leakage. Users have limited control over the infrastructure and resources used to store and process data in the cloud. In the future, it may be difficult or even impossible to transfer the model to another platform or move the trained model to local servers.

The systems discussed above illustrate that the automation of the machine learning model development process can expedite the creation of software products aimed at addressing a multitude of applied tasks. The system presented in this article is distinguished by its focus on non-professional users and its capacity to automate, among other functions, the creation of software wrappers for CNN models. By adopting a No-Code development approach [15], the presented system facilitates the creation of a user interface that enables non-professionals, who may not be familiar with the intricacies of neural network configuration, to utilize it for solving standard tasks. The outcome of this system is not only a constructed model but also an executable file generated with additional REST and SOAP wrappers. This approach allows the developed CNN models to be launched as a service without prior preparation and accessed from other systems and software complexes. The described features position the system as a tool for quickly and easily addressing simple, routine tasks by ordinary users.

The VGG19 architecture [16] was chosen as the test architecture. This architecture is highly versatile and has demonstrated high accuracy in solving various object detection tasks. Automating the creation of VGG19 models will enable the application of the developed software system to a wide range of automatic monitoring tasks in the industrial and energy sectors.

2 Materials and Methods

2.1 AutoML Approach

To create the presented software complex, a decision was made to focus on the following basic AutoML tasks that would be addressed during the automation of the VGG19 model training:

- Structural-parameter synthesis of the model and hyperparameter optimization;
- Development of user interfaces for AutoML;
- Automatic selection of evaluation metrics;
- Creation of software wrappers for model deployment.

The complete list of AutoML tasks is significantly more extensive. However, in the presented work it was decided to concentrate only on those listed above. This decision is motivated by the fact that AutoML is a relatively new area in which only a limited number of research and project groups are currently active. Established methods and approaches for describing and solving problems in the domain of AutoML are presently lacking. Each AutoML task, taken individually, is inherently complex to formalize and requires a tailored approach for finding solutions. Currently, there are emerging algorithms capable of addressing these tasks separately [17–19], but no existing solution provides comprehensive automation across all stages of CNN model development. Therefore, presented work emphasizes the stages of model development that are common to most standard object detection tasks requiring the creation of CNN models. Automating these specific stages will help lower the level of expertise needed from specialists when developing software based on CNN models, leading to time and resource savings.

This work is oriented toward a unified approach to various object detection tasks. Consequently, a unified algorithm must be employed for the comprehensive automation of the generation and training processes of CNN models. A modification of the genetic algorithm [20], which has proven effective in solving a range of practical problems [21, 22], was selected as such an algorithm. In this study, the algorithm was adapted for use with CNN architectures (see Fig. 1).

The key differences between this algorithm and the classical genetic algorithm are as follows:

- The "genotype" is represented as an expandable array, with elements that can be either integer or real-valued. This is because the hyperparameters of CNN models can be either integer or real-valued.
- Mutation (gene inversion) operators are not applied to child CNN models due to the specific format of the "genotype."
- Crossover is random and multiplex (during each crossover, the gene is divided into a number of parts equal to the number of array elements). This modification is designed for future expansion of the developed system. Since different CNN architectures have different hyperparameters, an algorithm capable of working with scalable sets of hyperparameters in a unified manner was necessary.

The use of this algorithm facilitates the resolution of AutoML tasks related to structural-parameter synthesis of models and hyperparameter optimization, as well as the automatic selection of evaluation metrics, given that this algorithm can encapsulate and substitute methods for assessing model accuracy.

2.2 No-Code Development

When implementing the user interface for the presented system, the No-Code development approach was actively used. The No-Code development paradigm has emerged as a transformative approach to software creation, enabling individuals without formal programming expertise to design, develop, and deploy applications

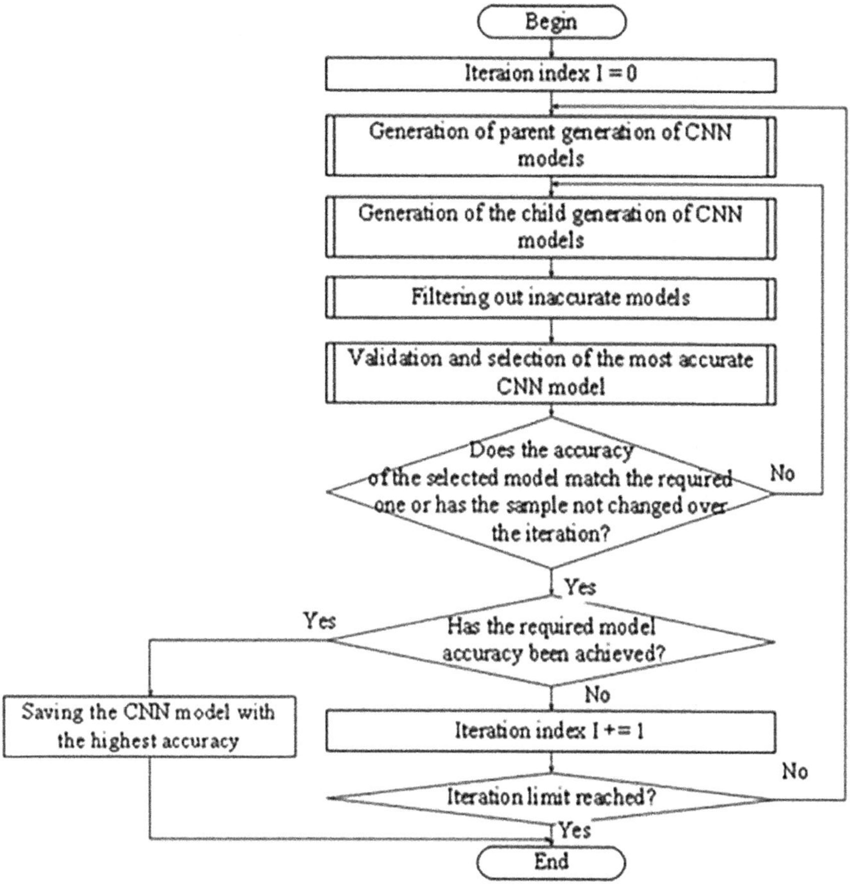

Fig. 1 Structural diagram of the modification of the genetic algorithm for automating the processes of generating and training CNN models

through intuitive graphical interfaces. This paradigm democratizes software development by allowing a broader range of users to participate in the application creation process.

At its core, No-Code development leverages visual design tools, drag-and-drop functionalities, and pre-built templates, thereby eliminating the need for traditional coding skills. By abstracting the complexities associated with programming languages and software architecture, No-Code platforms facilitate rapid prototyping and iterative design, empowering users to bring their ideas to fruition more efficiently.

Furthermore, the integration capabilities of No-Code platforms allow for seamless connectivity with existing systems and data sources, further enhancing their utility.

Despite its advantages, the No-Code approach is not without limitations. Concerns around scalability, customization, and governance must be addressed, particularly for larger program systems with complex requirements. Nevertheless, as the No-Code

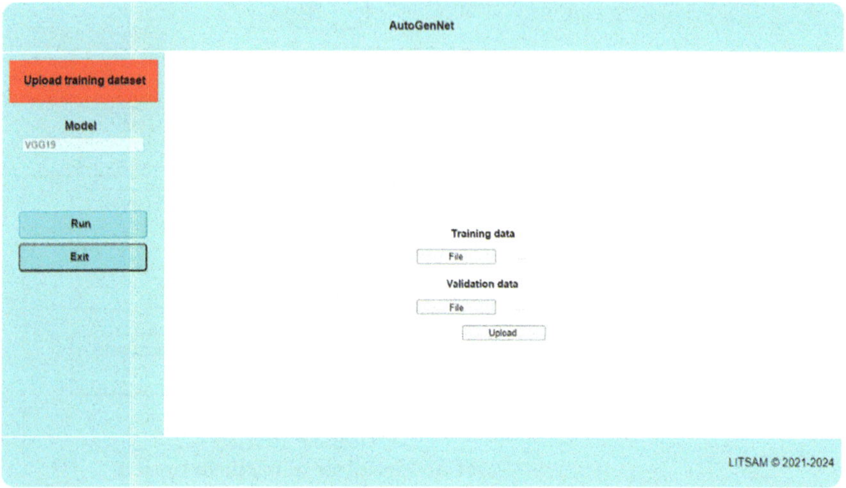

Fig. 2 Form of training CNN models of the software system for complex automation of the synthesis of artificial neural networks AutoGenNet

area continues to evolve, it holds the potential to significantly alter the landscape of software development. Ultimately, the No-Code paradigm represents a significant shift towards inclusive technology, promoting a more participatory approach to digital solution development.

Today, there are already many systems implementing No-Code development in various industries [23–25]. The result of their use shows the high efficiency of such solutions. In the presented system, the No-Code development approach is used at all stages of working with the VGG19 model. Figure 2 shows the interface for the automated creation of new VGG19 models. As can be seen, the user does not need to write program code to create models.

Also, the concept of No-Code development was extended to the automatic generation of software wrappers for trained CNN models. After training the model, the user has the opportunity to generate a software package containing a CNN trained to solve a specific problem and software wrappers. This service is cross-platform and can be launched without preliminary installation and additional configuration of software based on several operating systems (which is possible due to the cross-platform nature of the language for implementing these modules—Python [26]). Accordingly, such a module can be used in systems that support both the service-oriented architecture (SOA) paradigm [27] and the Internet of Things (IoT) [28] through REST and SOAP interfaces [29–31].

The proposed solutions enable the use of the developed software not only by specialists in the field of DL but also by professionals from various other domains who may lack sufficient knowledge in DL to manually construct models. This accessibility broadens the user base and empowers individuals without extensive technical training to leverage DL tools effectively. By simplifying the model creation process and providing user-friendly interfaces, the software allows these users to tackle DL tasks relevant to their specific fields, ultimately facilitating the integration of advanced analytical capabilities across diverse sectors.

3 Results and Discussion

The software complex was tested on the task of recognizing individual cows in a herd based on images captured with a smartphone camera. The specificity of this task lies in the fact that photographs are taken from various angles and distances, under different lighting conditions. Furthermore, due to the size of the herd, there is a need to distinguish between several hundred individual animals. The relevance of employing AutoML in this task stems from the dynamic nature of the herd, which is constantly updated, necessitating regular retraining of the model to recognize new individuals. By leveraging AutoML, the process of fine-tuning the recognition model can be streamlined, allowing for efficient adaptation to the ongoing changes in the herd composition. This capability not only enhances the accuracy of individual identification but also reduces the manual effort required for model updates, making it feasible to maintain up-to-date recognition capabilities in a real-world agricultural setting.

Training the VGG19 model for this task was conducted in an automated manner using a small dataset consisting of 444 images, which depicted 148 individual cows. Several models were trained during this process. The average accuracy of the automated models was approximately 80%, with a maximum achieved accuracy of 96%. The results of testing one of the trained models are presented in Fig. 3. While this accuracy cannot be classified as the best, it is largely dictated by the limited size of the training dataset. On the other hand, it is noteworthy that the models were created automatically without any code writing. This underscores the feasibility and effectiveness of the presented approach, which can potentially be scaled to address other tasks. The ability to achieve such performance levels through automation not only demonstrates the viability of the methodology but also opens avenues for broader applications in various domains where rapid model development and deployment are needed.

4 Conclusion

The work presented in this article addresses the task of automating the process of generating and training CNN models for automatic monitoring. The AutoML approach utilized in this research significantly simplifies and accelerates the implementation of CNN models across various industries and the energy sector. The application of the presented software system leads to a reduction in costs and an expedited

Fig. 3 Interface for interaction with CNN models

development of software complexes based on CNN models, which are designed to solve various practical tasks related to automated monitoring.

The testing results of the software demonstrate an increase in the level of automation in the model creation process, owing to a substantial reduction in the requirements for specialists who engage in building these models. Future development plans include industrial testing of the software to assess the required computational resources and to broaden the range of CNN architectures that can be applied for automated monitoring. Furthermore, there are plans to create a mobile web version of the application, enabling its use on portable devices. Such a version could prove highly relevant in industries and the energy sector when addressing tasks related to automated monitoring of the condition and fault-finding in large industrial facilities by specialists.

Thanks to the versatility of the algorithms used for automating the processes of model generation and training, as well as the flexibility and scalability of the AutoGenNet software platform, this work can serve as a foundation for the further development of a multi-model software complex that addresses a wide range of tasks in automatic and automated monitoring. This approach not only enhances efficiency but also fosters innovation in the application of machine learning and data analysis across different domains.

Acknowledgements The research conducted in this work was partially supported by the grant of Russian Science Foundation №24-16-20017, https://rscf.ru/en/project/24-16-20017/, and St. Petersburg Science Foundation.

References

1. Yoo, Y.-J., Yang, G., Park, K., Hyun, Y., Jeong, S.: Extendable machine tool wear monitoring process using image segmentation based deep learning model and automatic detection of depth of cut line. Eng. Appl. Artif. Intell. **135**, 108570 (2024)
2. Jrhilifa, I., Ouadi, H., Jilbab, A., Mounir, N., Ouaguid, A.: A VMD-deep learning approach for individual load monitoring and forecasting for residential buildings energy management. Adv. Electr. Eng. Electron. Energy **8**, 100624 (2024)
3. Pandiyan, V., Wrobel, R., Richter, R.A., Leparoux, M., Leinenbach, C., Shevchik, S.: Monitoring of laser powder bed fusion process by bridging dissimilar process maps using deep learning-based domain adaptation on acoustic emissions. Addit. Manuf. **80**, 103974 (2024)
4. Krizhevsky, A., Sutskever, I., Hinton, G.E.: ImageNet classification with deep convolutional neural networks. Commun. ACM **60**(6), 84–90 (2017)
5. He, K., Gkioxari, G., Dollar, P., Girshick, R.: Mask R-CNN. In: Proceedings of the IEEE International Conference on Computer Vision, pp. 2980–2988 (2017)
6. Zeiler, M.D., Fergus, R.: Visualizing and understanding convolutional networks. 3th Eur. Conf. Comput. Vis. **8689**(1), 818–833 (2014)
7. AutoML Homepage. https://www.automl.org/automl. Last accessed 29 Jul 2024
8. Geng, Z., Wang, Y.: Automated design of a convolutional neural network with multi-scale filters for cost-efficient seismic data classification. Nat. Commun. **11**(1), 3311 (2020)
9. Witsuba, M., Rawat, A., Pedapati, T.: Automation of deep learning. In: Proceedings of the 2020 International Conference on Multimedia Retrieval, pp. 5–6 (2020)
10. Baker, B., Gupta, O., Naik, N., Raskar, R.: Designing neural network architectures using reinforcement learning. In: 5th International Conference on Learning Representations. arXiv: 1611.02167 [cs] (2017)
11. MathWorks Automated machine learning (AutoML). https://www.mathworks.com/discovery/automl.html. Last accessed 29 Jul 2024
12. Feurer, M., Klein, A., Eggensperger, K., Springenberg, T.J., Blum, M., Hutter, F.: Auto-sklearn: efficient and robust automated machine learning. In: Automated Machine Learning, pp. 113–134 (2019)
13. Jin, H., Song, Q., Hu, X.: Auto-Keras: An Efficient Neural Architecture Search System. arXiv: 1806.10282 [cs] (2019)
14. Zeng, Y., Zhang, J.: A machine learning model for detecting invasive ductal carcinoma with Google Cloud AutoML Vision. Comput. Biol. Med. **122**, 103861 (2020)

15. Mote, D.S.: No Code App Development: Learn to Build Apps Without Code. Notion Press, Chennai, India (2022)
16. Simonyan, K., Zisserman, A.: Very deep convolutional networks for large-scale image recognition. In: 3rd International Conference on Learning Representations. arXiv:1409.1556 [cs] (2015)
17. Bello, I., Zoph, B., Vasudevan, V., Le, Q. V.: Neural optimizer search with reinforcement learning. In: 34th International Conference on Machine Learning, ICML 2017, vol. 1, pp. 712–721 (2017)
18. Cai, H., Chen, T., Zhang, W., Yu, Y., Wang, J.: Efficient architecture search by network transformation. In: 32nd AAAI Conference on Artificial Intelligence, AAAI 2018, pp. 2787–2794 (2018)
19. Dong, J.-D., Cheng, A.-C., Juan, D.-C., Wei, W., Sun, M.: DPP-Net: device-aware progressive search for pareto-optimal neural architectures. In: Lecture Notes in Computer Science (including subseries Lecture Notes in Artificial Intelligence and Lecture Notes in Bioinformatics), vol. 11215, pp. 540–555, LNCS (2018)
20. McCall, J.: Genetic algorithms for modelling and optimization. J. Comput. Appl. Math. **184**(1), 205–222 (2005)
21. Sobolevskii, V.A.: The system of convolution neural networks automated training. CEUR Work. Proc. **2803**, 100–106 (2020)
22. Mikhailov, V.V., Sobolevskii, V.A., Kolpaschikov L.A.: Mask R-CNN-based system for automated reindeer recognition and counting from aerial photographs. In: Pattern Recognition and Information Processing (PRIP'2021): Proceedings of the 15th International Conference, pp. 137–151 (2022)
23. Dasegowda, G., Sato, J.Y., Elton, D.C., Garza-Frias, E., Schultz, T., Bridge, C.P., Bizzo, B.C., Kalra, M.K., Dreyer, K.J.: No code machine learning: validating the approach on use-case for classifying clavicle fractures. Clin. Imaging **112**, 110207 (2024)
24. Sundberg, L., Holmstrom, J.: Democratizing artificial intelligence: How no-code AI can leverage machine learning operations. Bus. Horiz. **66**(6), 777–788 (2023)
25. Chhor, J., Fischer, V., Kroppel, F., Schmitt, R.H.: Rule-based decision support for no-code digitalized processes. Procedia CIRP **107**, 258–263 (2022)
26. Guttag, V.J.: Introduction to Computation and Programming Using Python: With Application to Understanding, 2nd edn. MIT Press, Cambridge, MA (2016)
27. Bell, M.: Introduction to Service-Oriented Modeling, in Service-Oriented Modeling: Service Analysis, Design and Architecture. Wiley, New York, NY (2008)
28. The Internet of Things: How the next evolution of the internet is changing everything. https://www.cisco.com/c/dam/en_us/about/ac79/docs/innov/IoT_IBSG_0411FINAL.pdf. Last accessed 29 Jul 2024
29. Mesmoudi, Y., Lamnaour, M., Khamlichi, Y.E.L., Tahiri, A., Touhafi, A., Braeken, A.: Design and implementation of a smart gateway for IoT applications using heterogeneous smart objects. In: 4th International Conference on Cloud Computing Technologies and Applications, Cloudtech (2018)
30. Hanes, D.: IoT Fundamentals: Networking Technologies, Protocols, and Use Cases for the Internet of Things. Cisco Press, Indianapolis, IN (2017)
31. Erl, T.: Service-Oriented Architecture: Analysis and Design for Services and Microservices, 2nd edn. Prentice Hall, Upper Saddle River, NJ (2016)

Automatic Systems of Time Series Analysis on the Basis of AutoML

Vladislav Sobolevskii

Abstract In many commercial sectors, the analysis of time series data has become a pressing issue. This encompasses tasks ranging from forecasting currency metrics to identifying failures in industrial equipment. Such demands underline the relevance of developing tools for the creation of automated time series analysis systems. This paper proposes the use of an AutoML approach to automate the development of machine learning models employed in the automatic analysis of time series data. The analysis is suggested to be conducted on the basis of two models: one aimed at time series forecasting and the other at time series classification. This combination of models enables the resolution of a majority of practical problems encountered in the field. The automation of structural-parametric synthesis of the models and optimization of hyperparameters is demonstrated using these models. The presented system is implemented within the AutoGenNet software platform. This platform embodies the No-Code development concept, enabling users to circumvent the complexities inherent in the processes of model creation and training. The adoption of the No-Code development paradigm contributes to lowering the entry barrier for working with the software. Additionally, a mechanism for generating software wrappers for the utilization of trained models has been implemented within the AutoGenNet platform. Collectively, these advancements facilitate the application of the AutoML approach for automating the processes of model generation and training, thereby simplifying and accelerating the resolution of tasks associated with automatic time series analysis. The developed system is scalable and can be utilized for the automated generation of additional time series analysis models to address a wide range of practical challenges.

Keywords Deep learning · Time series · AutoML · XGBoost · Linear regression · ResNet

V. Sobolevskii (✉)
St. Petersburg Federal Research Center of the Russian Academy of Science, 14 Line V.O., 39, St. Petersburg, Russia
e-mail: arguzd@yandex.ru

I. Ilin and M. Youzhong (eds.), *Digital Systems and Information Technologies in the Energy Sector*, Lecture Notes in Networks and Systems 1244,
https://doi.org/10.1007/978-3-031-80710-7_30

1 Introduction

In today's world, the volume of data generated and accumulated by various organizations is rapidly increasing. There is a growing demand for automatic forecasting and classification models for time series data across numerous domains, addressing a wide array of practical challenges. Time series forecasting is crucial in both the financial sector—where it serves to predict heterogeneous metrics—and in industry, where it is employed to forecast production line performance. Classification, on the other hand, is important in the field of information security for identifying latent attacks, as well as in industry for analyzing the condition of industrial equipment. Currently, these tasks are often addressed using machine learning (ML) models, both for forecasting [1–3] and classification [4–6].

At the same time, hybrid and ensemble MK models are gaining traction [7–10]. These models represent a combination of several algorithms, enabling the effective merger of their strengths while mitigating the weaknesses of individual methods. Their advantages manifest in various aspects of machine learning and data analytics, including time series analysis tasks. Hybrid models enhance the stability and reliability of the system compared to standalone models. For instance, one component of a hybrid model might be governed by decision trees, which perform well with categorical data, while another component could be based on the k-nearest neighbors method, which is effective for continuous data. An ensemble of multiple models can tackle a complex set of tasks, with one model potentially responsible for classification, while another addresses forecasting.

Therefore, in this work, the decision was made to employ an ensemble of specialized models for addressing the task of comprehensive time series analysis, where each model is designed to solve its specific problem. However, a significant drawback of this approach is the complexity involved in training an ensemble of models compared to the development of a single model. This complexity necessitates the use of tools that facilitate the automation of the model creation process.

Another significant issue is the need for extensive training datasets, which are essential for the model to learn patterns within the data and subsequently conduct qualitative analyses. The volume of training data can have a substantial impact on the accuracy of ML models. Consequently, proper data collection and training are vital for the successful deployment of ML models.

To address these challenges, the field of automated machine learning (AutoML) [11] has emerged. The primary objectives of AutoML involve the development and application of models and methodologies for the automated creation and training of ML models. Although this scientific and technical domain has only surfaced in the last decade, it has already seen the establishment of numerous research and design groups focused on advancing its scope and applications [12–14].

At present, the most extensive and adaptable AutoML systems are being developed by major corporations. One such instance is Google Cloud AutoML [15], which has been created by Alphabet Inc. This system is built on a cloud-based software platform intended for the automated development of ML models. However, it currently functions as a cloud service that necessitates adjustments for seamless integration with other software applications. Google Cloud AutoML exhibits all the typical limitations associated with cloud platforms. It is heavily reliant on a stable internet connection; thus, using the platform may pose challenges or become impossible in the absence of reliable connectivity. Furthermore, since data is hosted on servers managed by a third party, there exists a risk of unauthorized access or potential data breaches. Users also have restricted control over the infrastructure and resources utilized for data storage and processing in the cloud. Looking ahead, transferring the model to alternative platforms or migrating the trained model to local servers may prove to be difficult or even unfeasible.

The works presented above demonstrate that automating the machine learning model development process can significantly accelerate the creation of software products designed to tackle a diverse array of practical problems. The system discussed in this article is noteworthy for its emphasis on catering to non-professional users while also possessing the capability to automate various functions, including the creation of software wrappers for model ensembles. By implementing a No-Code development methodology [16], the proposed system simplifies the process of designing a user interface that enables non-specialists, who may lack familiarity with the nuances of neural network setup, to effectively utilize it for conventional tasks. The system not only produces the trained model but also generates an executable file embedded with additional REST and SOAP wrappers. This strategy permits the models to be deployed as a service without any preliminary setup and facilitates access from other systems and software suites. The features outlined establish the system as a valuable tool for the efficient and straightforward resolution of basic, repetitive tasks by general users.

2 Materials and Methods

2.1 Ensemble of Models

For the primary model employed to address time series forecasting tasks, a hybrid model was selected that combines linear regression and eXtreme Gradient Boosting (XGBoost) [17]. Linear regression is a statistical modeling method used to assess the relationship between a dependent variable (the target variable) and one or more independent variables (features), represented by a linear dependency. In the hybrid model, linear regression is utilized to forecast the trends of the processes under consideration. Due to its simplicity, interpretability, and low computational resource requirements, this model effectively identifies trends within the framework of the hybrid approach.

The second model utilized is XGBoost, a modified version of the gradient boosting algorithm designed to enhance the efficiency and speed of model training. This algorithm is implemented as an open-source software library, which facilitates improvements in model accuracy when tackling practical problems. XGBoost is also adept at handling regression tasks where continuous numerical values need to be predicted. In a hybrid model, XGBoost is used to forecast the divergence between the trend and the actual value of the parameter at a specific point in time.

The configuration of the hybrid model is dictated by the fact that linear regression effectively addresses the task of extrapolating trends but fails to capture nonlinear dependencies among parameters. Conversely, XGBoost excels in performing regression analysis of nonlinear relationships in data but is unable to extrapolate trends and, as a result, is poorly suited for time series analysis and forecasting. The combination of these algorithms mitigates their respective weaknesses while leveraging their strengths to enhance the accuracy of time series predictions. The mathematical representation of this hybrid model is as follows:

$$F_{hb}(x) = F_{lr}(x) + F_{xgboost}(x) \tag{1}$$

where x represents the input data vector;

F_{lr} signifies the linear regression algorithm;
$F_{xgboost}$ signifies the XGBoost algorithm;
F_{hb} signifies the hybrid machine learning algorithm.

To solve the task of time series classification, a Residual Networks (ResNet) [18] was chosen in this work. This architecture was chosen due to its proven effectiveness in solving various applied problems [19–21]. ResNet, originally developed for image classification tasks, introduces the concept of skip connections, which facilitate the training of deep networks by mitigating the vanishing gradient problem. This attribute is particularly advantageous when applied to time series data, where capturing complex temporal patterns is crucial for accurate classification.

Time series data often exhibit intricate nonlinear relationships that traditional methods may struggle to interpret effectively. The deep architecture of ResNet enables the modeling of these complexities by learning hierarchical features at multiple levels of abstraction. Each layer in the network can focus on specific characteristics of the temporal data. Furthermore, the integration of residual connections ensures the preservation of lower-level information throughout the depths of the network. This feature is vital for maintaining the integrity of temporal data as it passes through multiple layers, allowing the model to build upon previously learned representations.

Thus, the application of ResNet for time series classification is justified by its ability to effectively learn complex temporal relationships, preserve information via skip connections, and adapt to various datasets.

The models described above were selected for a comprehensive analysis of time series data. Established models excel at addressing specialized tasks, and their combination will enable solutions to a wide range of applied time series analysis problems.

However, the manual creation of such an ensemble of models is a resource-intensive process. Therefore, the development of a tool to automate their synthesis is a key objective of this work.

2.2 AutoML Approach

In the development of the proposed software system, the decision was made to concentrate on several fundamental AutoML tasks that would be addressed during the automation of model training. These tasks include:

- Structural-parameter synthesis of the model and hyperparameter optimization;
- Development of user interfaces for AutoML;
- Automatic selection of evaluation metrics;
- Creation of software wrappers for model deployment.

While the complete array of AutoML tasks is considerably broader, in this article we focus on the aforementioned areas. This choice is due to the fact that AutoML is a relatively young field, with only a limited number of research and project groups actively contributing. Currently, there is a lack of established methods and approaches for articulating and addressing challenges within the AutoML domain. Each individual AutoML task is inherently complex to formalize and necessitates customized strategies for effective resolution.

Although emerging algorithms exist that can tackle these tasks individually [22–24], no existing solutions offer comprehensive automation across all phases of model development. This paper highlights model development steps that are common in most standard time series analysis problems. Automation of these specific steps is expected to reduce the level of expert knowledge required from specialists when developing software based on ML models, resulting in significant time and resource savings.

The system presented in this paper employs several different models that vary in their creation and training processes. Consequently, a universal algorithm is required for the comprehensive automation of the generation and training processes of the ensemble models for time series analysis. A modified genetic algorithm [25] was selected as this universal approach, demonstrating its effectiveness in addressing a range of practical problems [26, 27]. In this work, the algorithm has been adapted for use with the chosen architectures (see Fig. 1).

The primary distinctions between this algorithm and traditional genetic algorithms are outlined as follows:

- The "genotype" is represented as a dynamic array, with elements that may take on either integer or real-valued forms. This design choice accommodates the fact that the hyperparameters of ML models can be either type.
- Mutation operations, specifically gene inversion, are not utilized on offspring ML models due to the unique structure of the "genotype."

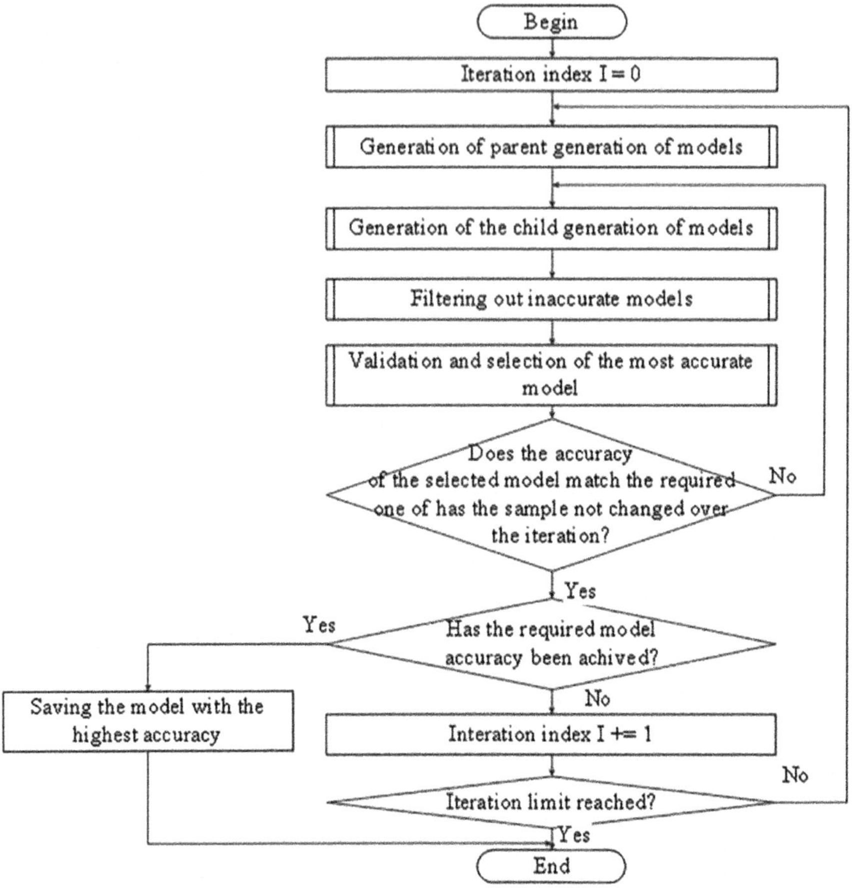

Fig. 1 Structural diagram of the modification of the genetic algorithm for automating the processes of generating and training ML models

- Crossover is executed randomly and multiplexedly; during each crossover event, the genotype is partitioned into segments corresponding to the number of elements in the array. Given that different ML architectures possess varying hyperparameters, it was essential to create an algorithm capable of managing expandable sets of hyperparameters in a cohesive manner.

The implementation of this algorithm aids in solving AutoML problems related to structural-parameter synthesis of models and hyperparameter optimization, as well as the automatic selection of evaluation metrics. This is facilitated by the algorithm's capability to encapsulate and substitute methods for evaluating model accuracy.

2.3 No-Code Development

In the implementation of the user interface for the proposed system, the No-Code development approach was prominently utilized. This paradigm has emerged as a transformative methodology for software creation, allowing individuals lacking formal programming skills to design, develop, and deploy applications through intuitive graphical user interfaces. By democratizing software development, the No-Code approach opens the application creation process to a broader of users.

Fundamentally, No-Code development employs visual design tools, drag-and-drop functionalities, and pre-built templates, thereby eliminating the necessity for traditional coding expertise. By abstracting the intricacies associated with programming languages and software architecture, No-Code platforms facilitate rapid prototyping and iterative design processes, enabling users to realize their ideas more efficiently.

However, despite its numerous advantages, the No-Code approach is not devoid of limitations. Challenges related to scalability, customization, and governance must be addressed, particularly in relation to larger systems with intricate requirements.

Today, numerous systems incorporating No-Code development are already in operation across various industries [28–30]. The outcomes of their implementation demonstrate the high efficiency of such solutions. In the presented system, the No-Code development approach is utilized at every stage of working with the ensemble of models. Figure 2 illustrates the interface for the automated creation of new models. As depicted, users are not required to write any programming code to create models, emphasizing the accessibility and ease of use inherent in this approach.

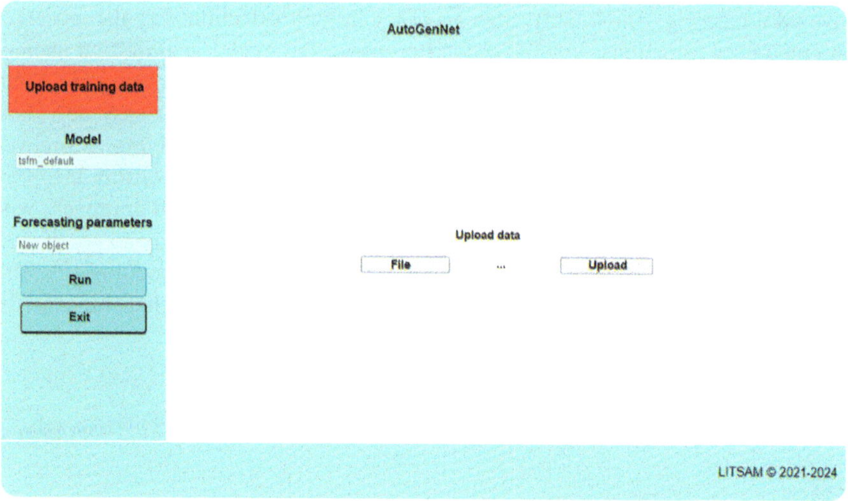

Fig. 2 Form of training models of the software system for complex automation of the synthesis of artificial neural networks AutoGenNet

The proposed solution allows for the utilization of the developed software not only by specialists in the field of ML but also by professionals from various other areas who may not possess sufficient expertise to manually construct models. This enhanced accessibility expands the user base and empowers individuals without technical training to effectively leverage ML tools. By streamlining the model creation process and offering user-friendly interfaces, the software enables these users to solves ML tasks pertinent to their specific fields, ultimately promoting the integration of advanced analytical capabilities across a wide range of areas.

3 Results and Discussion

The software package was evaluated on two distinct time series analysis tasks. The first task involved predicting equipment breakdowns in an industrial plant using sensor readings, while the second focused on classifying sales trends based on various metrics. The diversity of these test tasks highlights the flexibility and scalability of the system, making it suitable for a wide range of application areas.

The significance of utilizing AutoML lies in the fact that such tasks frequently arise in practice, necessitating the creation of new models each time. By employing AutoML, the process of configuring time series analysis models can be optimized, allowing for effective adaptation to the evolving conditions of the analysis target. While this capability does not inherently increase the accuracy of individual models, it significantly reduces the manual effort required to update them, thereby decreasing the time needed for analyzing new data and enhancing the speed of decision-making.

The average accuracy of the model developed for the first problem was an impressive 98% (see Fig. 3). Notably, high accuracy in fault detection was also observed in emergency situations. For the second problem, the model achieved an average accuracy of 96% (see Fig. 4). It is important to highlight that software shells for these models were automatically generated, allowing for their easy download and deployment on third-party servers for real-time problem-solving. Moreover, the models were created entirely automatically, without any code writing involved. This underscores the practicality and effectiveness of the proposed approach, with the potential for scalability to address other challenges. The ability to reach such performance levels through automation not only validates the methodology but also presents opportunities for broader application across various fields where rapid model development and deployment are essential.

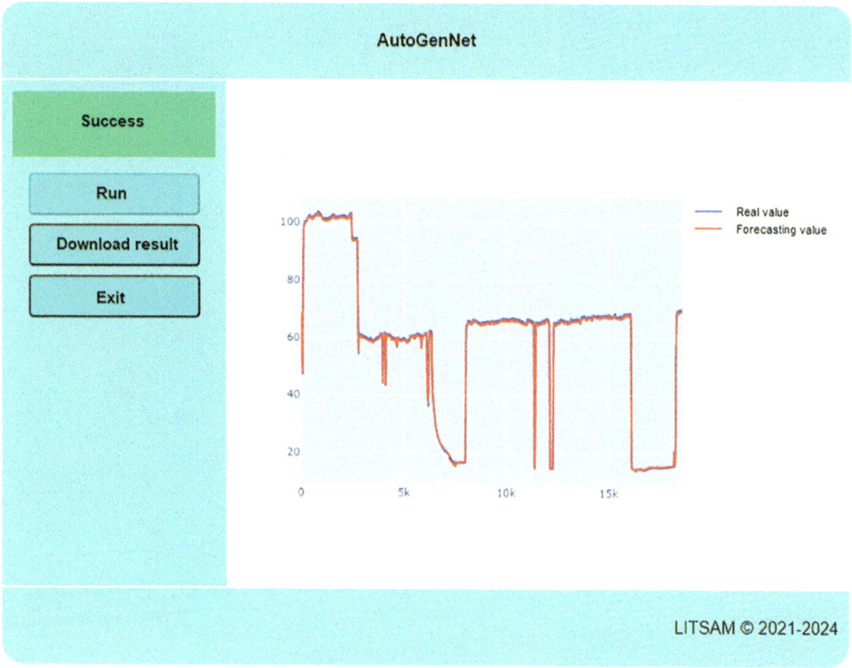

Fig. 3 Interface for time series forecasting

4 Conclusion

The work presented in this article focuses on automating the process of generating and training an ensemble of models for automatic time series analysis. The AutoML approach employed in this study greatly simplifies and accelerates the implementation of these ensemble models across various industries, particularly in the energy sector. By utilizing the software system outlined, users can achieve cost reductions and expedite the development of software packages based on ensembles of time series analysis models tailored to address diverse practical challenges.

The results from testing the developed system highlight a marked increase in the automation level of the model creation process, significantly lowering the expertise required from specialists involved in building these models. Future development plans include expanding the variety of model architectures that can be incorporated into the software, further enhancing its versatility and applicability in addressing a broader spectrum of time series analysis problems.

The versatility of the algorithms employed for automating model generation and training, combined with the flexibility and scalability of the AutoGenNet software platform, positions this work as a solid foundation for the development of a multi-model software complex. This complex can address a wide array of tasks in automatic and automated time series analysis. Such an approach not only improves efficiency

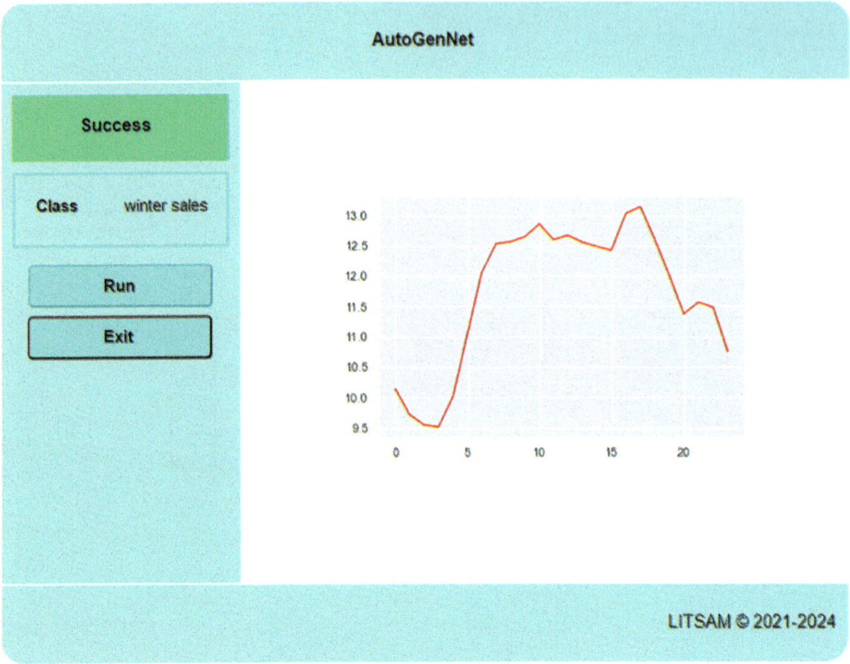

Fig. 4 Interface for time series classification

but also promotes innovation in the application of machine learning and data analysis across various domains. By leveraging these advancements, organizations can effectively harness the power of data-driven insights, further enhancing their operational capabilities and decision-making processes.

Acknowledgements The research conducted in this work was partially supported by the budgetary theme FFZF-2022-0004.

References

1. Houndekindo, F., Ouarda, T.B.M.J.: Prediction of hourly wind speed time series at unsampled locations using machine learning. Energy **299**, 131518 (2024)
2. Moreno, F.P., Rodriguez, F.I., Comendador, V.F.G., Jurado, R.D.-A., Suarez, M.Z., Valdes, R.M.A.: Prediction of air traffic complexity through a dynamic complexity indicator and machine learning models. J. Air Transp. Manag. **119**, 102632 (2024)
3. Cooper, C., Zhang, J., Ragai, I., Gao, R.X.: Multi-sensor fusion and machine learning-driven sequence-to-sequence translation for interpretable process signature prediction in machining. J. Manuf. Syst. **75**, 288–298 (2024)

4. Alveringh, D., Le, D.V., Groenesteijn, J., Schmitz, J., Lötters, J.C.: Fluid classification with integrated flow and pressure sensors using machine learning. Sens. Actuators A Phys. **363**, 114762 (2023)
5. Mariani, C., Navrotska, Y., Mancini, M.: Unsupervised machine learning for project stakeholder classification: benefits and limitations. Proj. Leadership Soc. **4**, 100093 (2023)
6. Pham, T.T.H., Luu, T.N., Nguyen, T.V., Huynh, N.T., Phan, Q.H., Le, T.H.: Polarimetric imaging combining optical parameters for classification of mice non-melanoma skin cancer tissue using machine learning. Heliyon **9**(11), e22081 (2023)
7. Castillo, A.F., Garibay, M.V., Diaz-Vazquez, D., Yebra-Montes, C., Brown, L.E., Johnson, A., Garcia-Gonzalez, A., Gradilla-Hernandez, M.S.: Improving river water quality prediction with hybrid machine learning and temporal analysis. Eco. Inform. **82**, 102655 (2024)
8. Kebede, Y.B., Yang, M.-D., Huang, C.-W.: Real-time pavement temperature prediction through ensemble machine learning. Eng. Appl. Artif. Intell. **135**, 08870 (2024)
9. Ayo, F.E., Ogundele, L.A., Olakunle, S., Awotunde, J.B., Kasali, F.A.: A hybrid correlation-based deep learning model for email spam classification using fuzzy inference system. Decis. Anal. J. **10**, 100390 (2023)
10. Zarouan, M., Mehedi, I.M., Latif, S.A., Rana, M.M.: Gradient optimizer algorithm with hybrid deep learning based failure detection and classification in the industrial environment. Comput. Model. Eng. Sci. **138**(2), 1341–1364 (2023)
11. AutoML Homepage. https://www.automl.org/automl. Last accessed 29 Jul 2024
12. Geng, Z., Wang, Y.: Automated design of a convolutional neural network with multi-scale filters for cost-efficient seismic data classification. Nat. Commun. **11**(1), 3311 (2020)
13. Witsuba, M., Rawat, A., Pedapati, T.: Automation of deep learning. In: Proceedings of the 2020 International Conference on Multimedia Retrieval, pp. 5–6 (2020)
14. Baker, B., Gupta, O., Naik, N., Raskar, R.: Designing neural network architectures using reinforcement learning. In: 5th International Conference on Learning Representations. arXiv: 1611.02167 [cs] (2017)
15. Zeng, Y., Zhang, J.: A machine learning model for detecting invasive ductal carcinoma with Google Cloud AutoML Vision. Comput. Biol. Med. **122**, 103861 (2020)
16. Mote, D.S.: No Code App Development: Learn to Build Apps Without Code. Notion Press, Chennai, India (2022)
17. Chen, T., Guestrin, C.: XGBoost: A Scalable Tree Boosting System. arXiv:1603.02754 [cs] (2016)
18. He, K., Zhang, X., Ren, S., Sun, J.: Deep Residual Learning for Image Recognition. arXiv: 1512.03385 [cs] (2015)
19. Cheng, W.X., Suganthan, P.N., Katuwal, R.: Time series classification using diversified ensemble deep random vector functional link and ResNet features. Appl. Soft Comput. **112**, 107826 (2021)
20. Wang, H., Wang, S., Sun, W., Xiang, J.: Multi-sensor signal fusion for tool wear condition monitoring using denoising transformer auto-encoder ResNet. J. Manuf. Process. **124**, 1054–1064 (2024)
21. Le-Xuan, T., Bui-Tien, T., Tran-Ngoc, H.: A novel approach model design for signal data using 1DCNN combing with LSTM and ResNet for damaged detection problem. Structures **59**:105784
22. Bello, I., Zoph, B., Vasudevan, V., Le, Q. V.: Neural optimizer search with Reinforcement learning. In: 34th International Conference on Machine Learning, ICML 2017, vol. 1, pp. 712–721 (2017)
23. Cai, H., Chen, T., Zhang, W., Yu, Y., Wang, J.: Efficient architecture search by network transformation. In: 32nd AAAI Conference on Artificial Intelligence, AAAI 2018, pp. 2787–2794 (2018)
24. Dong, J.-D., Cheng, A.-C., Juan, D.-C., Wei, W., Sun, M.: DPP-Net: device-aware progressive search for pareto-optimal neural architectures. In: Lecture Notes in Computer Science (including subseries Lecture Notes in Artificial Intelligence and Lecture Notes in Bioinformatics), vol. 11215, pp. 540–555, LNCS (2018)

25. McCall, J.: Genetic algorithms for modelling and optimization. J. Comput. Appl. Math. **184**(1), 205–222 (2005)
26. Sobolevskii, V.A.: The system of convolution neural networks automated training. CEUR Work. Proc. **2803**, 100–106 (2020)
27. Mikhailov, V.V., Sobolevskii, V.A., Kolpaschikov L.A.: Mask R-CNN-based system for automated reindeer recognition and counting from aerial photographs. In: Pattern Recognition and Information Processing (PRIP'2021): Proceedings of the 15th International Conference, pp. 137–151 (2022)
28. Dasegowda, G., Sato, J.Y., Elton, D.C., Garza-Frias, E., Schultz, T., Bridge, C.P., Bizzo, B.C., Kalra, M.K., Dreyer, K.J.: No code machine learning: validating the approach on use-case for classifying clavicle fractures. Clin. Imaging **112**, 110207 (2024)
29. Sundberg, L., Holmstrom, J.: Democratizing artificial intelligence: how no-code AI can leverage machine learning operations. Bus. Horiz. **66**(6), 777–788 (2023)
30. Chhor, J., Fischer, V., Kroppel, F., Schmitt, R.H.: Rule-based decision support for no-code digitalized processes. Procedia CIRP **107**, 258–263 (2022)

Application of Neural Network Algorithms for Monitoring the State of Objects

Maxim Mashtakov, Marina Bolsunovskaya, Svetlana Shirokova, and Olga Rostova

Abstract An important task is the possibility of controlling the impact of energy facilities on the environment. Neural networks can be of great help in automatic monitoring of the state of natural objects, in determining the best location of energy facilities, in predicting the expected amount of energy produced. The authors propose neural network algorithms for monitoring the state of objects. Emphasis is placed specifically on energy facilities. The models developed by the authors can be used in monitoring the state of natural resources, in optimizing the location of renewable energy sources, in forecasting energy production.

Keywords Neural networks · Monitoring of natural resources · Neural network algorithms · Digital projects · Renewable energy facilities

1 Introduction

The relevance of the work is due to the fact that increasingly widespread in recent years are convolutional neural networks (CNN), which are used, for example, to solve problems of object detection and image restoration, including in the digitalization of enterprises. The field of application of neural networks is wide and includes many industrial complexes.

The aim of the work is to substantiate the prospects of using convolutional neural networks for computer vision tasks in the spheres of satellite monitoring of natural objects. For what the following tasks were solved: analyzing theoretical material and technical features of existing solutions, estimation of the project efficiency and its prospects.

The object of this study is various images of natural objects, such as forests, rivers, fields and lakes. These images were acquired by satellite.

M. Mashtakov · M. Bolsunovskaya · S. Shirokova (✉) · O. Rostova
Peter the Great St. Petersburg Polytechnic University, St. Petersburg, Russia
e-mail: swchirokov@mail.ru

I. Ilin and M. Youzhong (eds.), *Digital Systems and Information Technologies in the Energy Sector*, Lecture Notes in Networks and Systems 1244,
https://doi.org/10.1007/978-3-031-80710-7_31

For the project related to satellite images, the only channel of data acquisition is Zorky satellite. The images were acquired during the SPACE-P project. The photo data were further segmented to train a neural network for recognizing natural objects and segmenting them.

It was decided to apply convolutional neural networks to solve computer vision problems using the presented images.

2 Materials and Methods

In current clause the author's analysis of existing solutions for applying neural network recognition and classification algorithms and related works is presented. One of the most popular solutions for computer vision tasks are convolution neural networks. Based on the previous statement, the following articles are devoted mainly to convolutional neural networks and related research on CNN development and related applications development.

The article "Marine Objects Recognition Using Convolutional Neural Networks" presents a convolutional neural network for tasks of recognition of marine objects. The model was trained on more than 5000 images with different conditions. In the article it is stated by authors that an accuracy of 88% was achieved during tests [1].

In the paper titled "Object Recognition for Economic Development from Daytime Satellite Imagery" authors propose a method that uses machine learning, CNN and satellite imagery to solve the limitation of the information about developing countries [2].

In the article "Building damage detection in satellite imagery using convolutional neural networks" a convolutional neural network has been created that is capable of detecting damage to buildings after natural disasters, architectures. The proposed result is considered one of the best in the field [3].

The paper "Satellite image classification with deep learning" presents an algorithm for solving problems of recognition and classification of infrastructure objects with an accuracy of 83%. The classes of objects considered in the article include ports, shipyards, residential complexes and residential buildings [4]. The paragraph reviewed various works of authors on topics related to computer vision in ecology and satellite image processing. In examining these works, it was concluded that the information and developments from the reviewed articles are suitable for application in their own works investigating computer vision problem solving and the design and training of neural networks.

One of the technical features of this and many other neural networks is the dataset, as well as tools for partitioning the dataset.

The importance of the dataset and the tools for partitioning the dataset play a key role in training neural networks. The diversity of object classes in the dataset significantly affects the accuracy of training. Unbalanced dataset can lead to undertraining and misclassification. In addition, the amount of data significantly affects the quality of the model: the more data, the more accurate the results.

However, in practice, the problem of data scarcity or limited data availability may arise. In this case, techniques such as data augmentation can be used to increase the amount of data available. Data augmentation is the process of creating new data samples by making small changes to the original data, such as changing the angle or scale of an image. This method helps to increase the variety of data and train the model on more samples [5, 6].

Another way to expand the dataset when there is a lack of data is to create synthetic data using various algorithms, such as SMOTE, the main idea of which is to increase the amount of data not by simply copying minority class data, but by generating synthetic data based on existing data [7, 8].

Note the importance of maintaining a balance in the number of data sets, since an excess of data can lead to overfitting of the model. Overfitting causes the model to become too complex to the point of "memorizing" patterns in the dataset instead of generalizing of data. To avoid overfitting, various techniques can be used, such as regularization and slower learning rates.

When choosing a dataset, we also must consider the application of the model. The examples include medical images for medical applications, car images for the automotive industry. The dataset should be appropriate for the real conditions of the model application. It is also important to take into the account the variability of external features of objects, such as hairstyles and glasses when working with face images.

The last characteristic we would like to consider in this paper is the quality of the markup. Incorrect partitioning of the dataset can have a significant impact on the quality of training of the neural network. In case of incorrect partitioning, the model will be trained on incorrect data, which will not allow to successfully apply it in the tasks for which the model was created. It is important to check the quality of data partitioning and correct errors before training the model.

Let's consider markup variants, as they are also technical features of projects in my opinion. Let's highlight 5 most popular approaches to partitioning: in-house, outsourcing, crowdsourcing, synthetic method, programmatic method.

In-house data partitioning is an effective way to ensure accuracy and quality, especially for companies with data analysts. Allows you to control the partitioning process and ensure that it meets the requirements of the task. However, can be expensive for small companies without an in-house analytics team.

Outsourcing—forming a temporary team with experienced professionals. Helps when there is a lack of resources or expertise. Reduces hardware and software costs, but requires caution in selecting service providers.

Crowdsourcing—an opportunity to obtain additional resources and accelerate data partitioning. However, may not be effective for sensitive data and requires careful platform selection.

Synthetic method—an important tool for producing high quality data. Allows to create new data with required attributes without manual partitioning. Requires computational power and risk analysis.

Software partitioning method—automates the task of data partitioning. Reduces the number of performers, but is error-prone and requires quality control. Can be effective for simple and repetitive tasks.

Consider features specific to satellite image analysis.

The application of CNNs for object classification and detection has already been mentioned, but in the study of computer vision problems in a different application area. The article "Marine Objects Recognition Using Convolutional Neural Networks" considers, among other things, the application of convolutional neural networks for solving problems of object detection in satellite images. In the study, the authors conclude that CNN is a suitable algorithm for solving such problems, which again confirms my justification for the use of convolutional neural networks in further paragraphs of this paper. In the paper we consider three activation functions for the future CNN: Tanh, ReLU and logistic sigmoid. Tanh ReLU proved to be the most suitable, but the expediency of using these functions may change due to the specifics of the project considered in this paper [1].

In the article "Object Recognition for Economic Development from Daytime Satellite Imagery" the U-Net was used as the basis for the neural network, however the project "Light UNet for Satellite Image Segmentation features the Feature Pyramid Network architecture. Both architectures are convolutional neural network architectures with a "pyramid" shape. Experts highlight the similarity in the performance of those two architectures. The U-Net was developed for solving problems that include usage of medical images, however the above-mentioned paper proves the possibility of using this architecture for segmentation in satellite images [2].

The rest of the technical features are similar to those of any other CNN for object classification development.

For the implementation of CNN we decided to use the following features of other projects: convolutional type of neural networks, FPN/U-Net architecture [9–12] and special dataset developed for this specific task.

After careful analysis of various data markup methods, it was decided that each of them has its own advantages and disadvantages. For example, using third-party companies to partition data can be an effective solution, but can lead to the risk of data leakage and high costs for the services.

In the case of current paper, we planned to use in-house data markup based on the fact that we had staff with necessary skills and experience in image tagging and a possibility to provide them with the necessary software, which allowed us to have full control over the data markup process.

FPN was preferred as it is considered more user-friendly by some experts due to the quality of features. However we also conducted some experiments with U-Net during the research.

3 Results

Project management was examined from the perspective of the PRINCE2 methodology. This standard was developed to ensure the successful completion of projects with defined objectives and within a given timeframe, as well as to manage costs and risks. The PRINCE2 standard offers a flexible and adaptive approach to project management that allows the individual characteristics of each project to be taken into account. PRINCE2 implies constant interaction between project participants and defines specific roles and responsibilities of each participant, which allows to minimize risks and increase efficiency [13–15].

Let us consider a project aimed at developing a neural network to identify rivers and lakes on satellite images. One of the main characteristics of modern technologies is the ability to acquire and process Earth images obtained from satellites, including access to current images and archives. Both local and global environmental problems can be detected using satellite images of different resolutions. High-resolution images are used to monitor localized problems such as deforestation near cities, while low-resolution images cover large areas such as fire zones. Images can be taken in different frequency bands, which allows to see different details of objects in the monitored areas.

Satellites provide information on various natural and man-made phenomena such as weather, floods, fires, precipitation, volcanic eruptions, changing seasons, movement of sand in deserts, melting glaciers and movement of icebergs. By comparing images from different years, we can track changes in urban areas, such as the construction of new industrial plants and roads, changes in river channels and the emergence of new islands, as well as the disappearance of forests and the emergence of new fields, the drying up of lakes and the filling of reservoirs.

Existing modern space monitoring technologies can be applied to solve various tasks, but their use requires careful analysis and adaptation to specific conditions, such as climatic conditions and natural-industrial factors. Space imagery processing allows analyzing various phenomena such as smoke plume, fire spread, burning zones and forest degradation. Air and water pollution data can also be analyzed, and the area of water area pollution can be determined. All these tasks require the use of different data processing methods and adaptation to each specific situation.

The project of developing a neural network for analyzing satellite images is almost identical in the planning of project stages with few exceptions. For example, there is no stage of image markup software development, as the images were marked up in Photoshop by creating layer masks for the objects to be searched for: rivers and lakes.

However, this markup is much more complex due to the size of the images, making it take longer. Due to this, the project is completed in about the same time frame as the one mentioned above.

As mentioned earlier, FPN was chosen for the neural network architecture because of the large amount of research on this topic.

Fig. 1 Initial image for processing by neural network

As a result of the project to develop a neural network for segmentation of objects of water character on satellite images, a neural network capable of coping with its tasks was obtained.

The result of processing the satellite image by the developed neural network will be presented below. Figure 1 will show the original image and Fig. 2 will show the processed image.

The image shows that in some places the neural network detects non-existent objects, which reduces the accuracy of its work.

Next, we will consider economic and qualitative indicators of the projects [16]. Economic performance will be considered from the perspective of the neural network development business, both turnkey and projects in general. Data on the cost of acquiring a neural network will be aggregated from various public sources from the websites of software development companies of the type under consideration. Since the systems considered in this project are similar in purpose (object detection) and development cost, but differ only in the field of application, the considered indicators can be considered common for both projects.

As economic indicators we will choose simple rate of return and simple payback period (formula 1, respectively).

$$SRR = \frac{SV}{TIC}, \tag{1}$$

where SRR—simple rate of return;

SV—cost of product sales, rub;
TIC—investment costs of the project, rubles.

Fig. 2 Neural network processing result for processing satellite images

There are 2 options for selling neural networks: turnkey and conventional. Let's consider the efficiency of both options. As data on the cost of sale I will take data from open sources. Within the framework of this work we have considered a project with the preparation of a training sample, which is not standard for such projects, on the basis of which the cost of sale can be set at an average of 8,000,000 rubles. A normal sale costs less - about 2,000,000 rubles. These are the data I will take for my calculation for these projects.

To begin with, let's consider a turnkey sale. From the Gantt chart we can see that the cost of the project amounted to 5,570,196 rubles. Let's substitute the known values into formulas 1 and 2 to calculate the relevant indicators.

$$SRR = \frac{8,000,000}{5,570,196} \approx 1.4$$

Based on this result, we conclude that a turnkey sale will bring the company 1.44 rubles per each ruble of investment costs per month. The turnkey project pays for itself when sold in less than one month due to the profit covering the investment costs.

Turnkey development can be used for normal sales. It is difficult to accurately guess the number of neural network sales per month due to many factors. The price of development can affect demand, but customers are willing to pay for quality products.

Advertising, marketing, product quality and service are important for the exact number of sales. Assuming a good reputation and service level, 15 neural nets per month is a realistic assumption.

Keep in mind, the number of sales can fluctuate due to price, quality, and competition. Active promotion, marketing and product improvement can increase sales significantly. Let us substitute the new data into the formulas.

$$SRR = \frac{2,000,000}{5,570,196} \approx 5.4$$

Based on the obtained result, we conclude that the sale of 15 nets per month will bring the company 5.39 rubles for each ruble of investment costs, provided that the number of sales per month does not change in the future.

However, based on the simple payback period, we can conclude that even if the number of sales changes, the project will remain effective, because its payback period is extremely short.

Let's consider possible qualitative indicators of efficiency from the application of neural network algorithms considered in this paper. To begin with, let's consider the performance from neural network for processing satellite images.

Increase in accuracy and speed of analyzing satellite images. This will reduce decision-making time in the fields of forest management, geologic research, and environmental monitoring.

Reduced cost of data analysis. The use of neural networks will speed up and simplify this process, reducing time and cost.

More accurate and objective decisions. Neural networks automatically detect natural objects, increasing the reliability of results.

Faster response to changes in the environment. Neural networks process large amounts of data faster, which allows you to react quickly to changes.

Increased competitiveness of the company. The use of modern technologies, including neural networks, improves the quality of products and services, as well as the efficiency of business processes.

4 Discussion

The neural network project dealing with the processing of satellite images and identification of rivers and lakes on them has the potential for further development and expansion in several directions. In particular, neural networks of this type are actively used in ecology [17–19].

First, it is possible to work on improving the algorithms of image processing and object classification on them, which are used by the neural network. This will improve the accuracy of identifying rivers and lakes in satellite images, which in turn can be useful for a number of applications such as cartography, ecology, agriculture, and so on. For example, key points of features can be identified, which will give a clearer understanding of their boundaries and locations that may be of interest to users of the software in question.

Secondly, one can work on expanding the range of objects to be classified. For example, one can try to identify other important natural features in satellite images, such as forests, fields, mountains, and so on. This can also be useful for many applications, such as monitoring forest fires, assessing the condition of agricultural land, and so on.

In addition, expanding the range of classified objects in satellite images can be useful not only for natural applications, but also for urban planning and infrastructure. For example, a neural network can be used to identify areas with a high concentration of buildings and infrastructure, to monitor densely populated areas, or even to determine the level of air pollution.

An additional prospect for a neural network project engaged in processing satellite imagery could be to add the ability to analyze marine objects. For example, the neural network can be trained to recognize ships, underwater objects, and other marine vehicles [20]. This could be useful for monitoring maritime traffic, controlling the movement of cargo, and detecting illegal activities such as illegal fishing or maritime trespassing.

Another example could be the use of a neural network to monitor the state of marine ecosystems, for example, to detect algae, determine the level of water pollution and so on. In addition, it is also possible to develop a system to automatically steer ships based on data obtained from satellite images. This could be useful for improving the safety and efficiency of maritime transportation as well as reducing environmental impact. In general, adding the ability to analyze marine objects can expand the scope of the project and increase its relevance for various industries.

The application of such neural networks is also possible in environmental tasks. For example, to monitor "dead zones" that appear in the oceans due to waste dumping, as well as in areas where phytoplankton accumulate.

Small animals such as zooplankton abundantly ingest phytoplankton, and small fish then consume the zooplankton. As a result, more animal waste such as feces and plankton carcasses are produced. The increase in animal waste sinking to the bottom provides an impetus for the growth of the bacteria that feed on it. In the process of waste decomposition, the bacteria suck oxygen from the water, thereby reducing its amount [21, 22].

With the help of a neural network, it is possible to monitor plankton accumulation sites or waste accumulation sites, which will allow timely action to prevent a disaster. In addition, it is possible to use such a neural network to create an automatic flight control system for unmanned aerial vehicles (UAVs) based on data obtained from satellite images. This could be particularly useful in areas where the availability and safety of aircraft is limited, such as monitoring regions after natural disasters.

Finally, extending the range of classified objects in satellite images can also be applied to security applications. For example, a neural network can be used for monitoring to monitor areas of terrorist activity.

Thus, expanding the spectrum of classifiable objects in satellite images can lead to new and improved existing applications in different domains.

Third, work can be done to improve the training of the neural network so that it can process large amounts of data faster and more accurately. This will speed up the processing of satellite images and improve their accuracy.

Neural networks play an important role in satellite image analysis, especially in the context of environmental studies. They provide a unique opportunity to process huge volumes of data received from satellites and extract key information about the state of the natural environment. This technology allows us to automate the processes of monitoring and analyzing various parameters, such as drought, clouds, changes in vegetation and even water pollution. Thanks to the accuracy and efficiency of neural networks, researchers and environmentalists can quickly obtain critical data needed to make decisions in the field of sustainable development and conservation. Thus, the use of neural networks for the analysis of satellite images helps improve our understanding of ecosystems and contributes to the development of effective strategies for environmental conservation.

Overall, the neural network project, which deals with processing satellite images and identifying rivers and lakes on them, has a lot of prospects for further development and improvement, which can lead to more accurate and efficient applications in the field of ecology.

5 Conclusion

Developing the ability to control the impact of energy facilities on the environment is an important task. The purpose of the study was to substantiate the prospects for using neural networks for computer vision tasks in the field of satellite monitoring of natural objects. Neural networks can be of great help in automatically monitoring the state of natural objects, determining the best location of energy facilities and predicting the expected amount of energy produced.

The authors carried out an analysis of the technical features of solutions to computer vision problems in the studied and similar areas. The analysis confirmed the feasibility of using the proposed class of solutions to accomplish the assigned tasks. To implement neural networks, it was decided to use a number of features and developments from other projects, for example, the FPN/U-Net architecture, specialized datasets developed taking into account the specifics of tasks, as well as the markup method.

An assessment was made of the effectiveness of the project for implementing the developed neural network from the point of view of the business and the customer. Qualitative performance indicators from the use of this class of neural networks in the presented problems have been identified, which have confirmed the feasibility of developing and applying the proposed solutions.

Within the framework of the work, possible directions for the development of such projects were also presented. Development prospects include not only increasing the efficiency of monitoring, but also expanding the functionality of neural networks.

The models developed by the authors can be used in monitoring the state of natural objects, optimizing the placement of renewable energy sources, monitoring the safety and environmental friendliness of their operation, and more accurately forecasting energy production.

References

1. Lorencin, I., Anđelic, N., Mrzljak, V., Car, Z.: Marine objects recognition using convolutional neural networks. Naše more **66**(3), 112–119 (2019)
2. Ackermann, K., et al.: Object Recognition for Economic Development from Daytime Satellite Imagery. Monash University, SoDa Laboratories 2020-02 (2020)
3. Xu, J.Z., et al.: Building damage detection in satellite imagery using convolutional neural networks. ArXiv preprint arXiv:1910.06444 (2019)
4. Pritt, M., Chern G.: Satellite image classification with deep learning. In: IEEE Applied Imagery Pattern Recognition Workshop (AIPR), pp. 1–7 (2017)
5. Adedeji, O., Owoade, P., Ajayi O., Arowolo, O.: Image augmentation for satellite images. ArXiv preprint arXiv:2207.14580 (2022)
6. Illarionova, S., et al.: MixChannel: advanced augmentation for multispectral satellite images. Remote Sens. **13**(11), 2181 (2021)
7. Lemaître, G., Nogueira, F., Aridas, C.K.: Imbalanced-learn: a python toolbox to tackle the curse of imbalanced datasets in machine learning. J. Mach. Learn. Res. **1**, 559–563 (2017)
8. Ushakov, V.: Models and algorithms for controlling information processes in the moving objects interaction. Mar. Intellect. Technol. **3**(1), 235–247 (2022)
9. Ghiasi, G., Lin, T., Le, Q.V.: Nas-fpn: learning scalable feature pyramid architecture for object detection. In: Proceedings of the IEEE, CVF Conference on Computer Vision and Pattern Recognition, pp. 7036–7045 (2019)
10. Adelson, E.H., et al.: Pyramid methods in image processing. RCA Engineer **29–6**, 33–41 (1984)
11. Pereira, E.O., et al.: Mapping threatened canga ecosystems in the Brazilian savanna using U-Net deep learning segmentation and Sentinel-2 images: a first step toward conservation planning. Biota. Neotrop. **23**(1), e20221384 (2023)
12. Singh, N.J., Nongmeikapam, K.: Semantic segmentation of satellite images using deep-UNet. Arab. J. Sci. Eng. **48**(2), 1193–1205 (2023)
13. Shirokova, S., Kislova, E., Rostova, O., Shmeleva, A., Tolstrup, L.: Company efficiency improvement using agile methodologies for managing IT projects. In: ACM International Conference Proceeding Series (DTMIS) (2020)
14. Matos, S., Lopesa, E.: Prince2 or PMBOK—a question of choice. Procedia Technol. **9**, 787–794 (2013)
15. Shirokova, S.V., Rostova, O.V., Prosvirnina, A., Odainic, A.: Key trends in the digital transformation of business and their impact on the business processes. Lect. Notes Netw. Syst. **684**, 885–895 (2023)
16. Grozdova, A., Shirokova, S., Rostova, O., Shirokova, A., Shmeleva, A.: Rationale for information and technological support for the enterprise investment management. Lect. Notes Netw. Syst. **387**, 181–190 (2022)
17. Wang, B., et al.: Incorporating fast and intelligent control technique into ecology: a Chebyshev neural network-based terminal sliding mode approach for fractional chaotic ecological systems. Ecol. Complex. **47**, 100943 (2021)
18. Christin, S., Hervet, E., Lecomte, N.: Applications for deep learning in ecology. Methods Ecol. Evol. **10**, 1632–1644 (2019)
19. Gazya, G.V., et al.: Artificial intelligence systems based on artificial neural networks in ecology. In: Computer Science On-line Conference, pp. 149–158. Springer International Publishing (2022)

20. Rainey, K., Reedera, J.D., Corellia, A.G.: Convolution neural networks for ship type recognition. In: Automatic Target Recognition XXVI, vol. 9844, pp. 56–66. SPIE (2016)
21. Karstensen, J., et al.: Open ocean dead zones in the tropical North Atlantic Ocean. Biogeosciences **8**, 2597–2605 (2015)
22. Altieri, A.H., Gedan, K.B.: Climate change and dead zones. Glob. Change Biol. **4**, 1395–1406 (2015)

Application of Machine Learning Methods in Human Resource Management at Enterprises of the Energy Industry

Anastasiia Kondakova, Svetlana Shirokova, Olga Rostova, Anastasiia Shmeleva, and Vladislav Shpagin

Abstract The article considers the methods of machine learning and their application in the field of human resource management in the company. The reasons for the necessity of using digital technologies in the field of recruiting at energy enterprises are identified. The main tasks in personnel management that can be solved through the application of artificial intelligence technologies and machine learning algorithms are highlighted. Possible impacts of AI tools on human resource management and factors of successful implementation of machine learning technologies are analyzed. A comparative analysis of machine learning methods is made, in particular those methods that relate to learning with a teacher (supervised learning). The choice of methods that can be used to solve problems in the considered area is justified. A practical example of solving a problem related to recruiting and personnel selection is considered. The algorithm for designing a web-service for recruiting based on machine learning models for the energy industry enterprise is developed. The ways for further research in the field of application of machine learning methods for solving problems of personnel management are indicated.

Keywords Human resources · Energy industry enterprises · Machine learning methods · Artificial intelligence technologies · Learning algorithm · Neural network

A. Kondakova · S. Shirokova (✉) · O. Rostova · V. Shpagin
Peter the Great St. Petersburg Polytechnic University, St. Petersburg, Russia
e-mail: swchirokov@mail.ru

A. Shmeleva
Marriott International Yerevan, Yerevan, Armenia

I. Ilin and M. Youzhong (eds.), *Digital Systems and Information Technologies in the Energy Sector*, Lecture Notes in Networks and Systems 1244,
https://doi.org/10.1007/978-3-031-80710-7_32

1 Introduction

In today's dynamic world, where competition for talent is a key element of success, effective recruiting processes are becoming an integral part of HR strategy. With the constant growth of data and the need for rapid decision-making, traditional recruiting methods are often ineffective. In this context, the application of machine learning (ML) techniques represents not only a relevant but also a very promising paradigm for automating and optimizing recruitment processes [1, 2].

In the presented study we considered the specifics of using the capabilities of artificial intelligence technologies in the energy industry. About 2.6 million people work at the enterprises of this sector. More than 88% of jobs in the energy sector are provided by large energy companies and medium-sized businesses [3]. Human resources are one of the most important factors of a company's competitiveness in the market. Accordingly, the company's management should pay great attention to the functioning of the HR management system.

The following are the main reasons for the need to use digital technologies in the field of recruiting in energy companies:

1. Due to the rapid development of technologies used in the energy sector, the level of training and qualification of specialists must meet the highest requirements.
2. The specifics of work in this industry are difficult working conditions, so there are difficulties with the selection of qualified personnel willing to work in such conditions.
3. The complexity and variety of equipment used in the oil and gas industry require a wide range of highly specialized specialists to service it.

In this regard, one of the key problems associated with increasing the level of costs of enterprises in the energy industry is the long-term preservation of open vacancies.

New technologies in the field of human resource management will allow organizations to solve several major problems:

- Increased efficiency: artificial intelligence technologies will automate repetitive and time-consuming HR tasks such as searching and screening resumes, scheduling interviews, and evaluating candidates. This allows HR professionals to focus on more important tasks such as employee engagement, training and development [4, 5];
- Increased level of accuracy: AI can analyze large amounts of data quickly and accurately, reducing the risk of human error in HR processes [6, 7];
- Cost savings: AI can reduce the need for manual labor in HR processes, thereby reducing costs associated with recruitment and performance management [8, 9].

In this case, machine learning is considered as a method of analyzing incoming data, which allows automating the process of building an analytical model. Application of machine learning methods in human resource management at the energy sector enterprises is necessary for automation of certain HR-processes. It will allow certain algorithms to perform routine work, thereby freeing up the time of HR employees to perform strategic and current tasks [10].

The purpose of this paper is to assess the current capabilities of artificial intelligence and machine learning in the field of HRM at enterprises in the energy industry, based on the analysis of existing solutions, as well as on demonstrating the application of machine learning methods to solve problems in the field of HR management. In accordance with the set goal, the following research objectives were formed:

- Identification of problems in the sphere of recruiting at the enterprises of the energy industry, requiring the use of artificial intelligence technology;
- Study of the possible impact of artificial intelligence on HRM;
- analyzing machine learning methods and selecting the most preferable ones for the problem to be solved;
- Development of a web service related to recruiting using machine learning algorithms;
- Description of key success factors in the implementation of AI technologies.

2 Materials and Methods

Human resource management is a practical activity, which, first of all, is aimed at providing an enterprise with qualified personnel capable of properly performing their assigned duties.

The application of machine learning methods to solve problems in the field of human resource management does not have a very long history. This is due to the fact that the tools and technologies of artificial intelligence have appeared relatively recently. However, it is still possible to note a number of works that consider such tasks and their solutions [11–14].

The article by Blinnikova A.V. et al. is devoted to substantiation of the usefulness of introduction and use of machine learning methods and artificial intelligence technologies by companies in the field of personnel management. The paper investigated the current state of human capital management and highlighted the main problems. In addition, the prospects of using AI tools and possible problems that an organization may face in the process of implementing the model were presented [15].

In the paper by Wisetsri W. et al. the process of integrating artificial intelligence tools with employee management models in an organization is considered. Also, this article emphasizes the significance of the application of AI technologies in many economic areas, in particular, in the human resource management system [16].

The article by Sakka F. et al. analyzes three aspects of implementing artificial intelligence technologies in the field of human resource management [17]. The first aspect is to facilitate the decision-making process through the use of AI tools, which in turn will reduce the workload of HR employees, freeing them from routine simple tasks in favor of strategic assignments. The next aspect is to change the position of the HR department in organizations. In other words, the human resource management unit in the organization will become the center of strategic decision making concerning organizational problems. The last aspect involves the financial side of implementing AI tools. AI can help automate routine tasks, improve employee recruitment and retention, and inform data-driven decision-making. However, the authors also caution that implementing AI requires careful planning and consideration of ethical and legal issues. Organizations must ensure ethical and transparent use of AI technologies to avoid bias and discrimination.

The literature review concluded that recruiting, motivation and training are the key human resource management challenges that require the use of machine learning and AI tools [18].

Recruitment refers to the process of identifying, attracting and hiring qualified individuals to fill vacancies in an organization. It involves a series of activities that range from posting job openings to selecting and interviewing candidates, making job offers and onboarding new employees. The goal of recruiting is to find the best candidates who meet the employer's requirements and fit the organization's culture and values. Effective recruiting is critical to an organization's success because it ensures that it has the talent it needs to achieve its goals and compete in the marketplace [19].

When hiring, an employer should follow a structured recruiting process that includes the following steps:

1. Defining the job requirements: the employer should identify the skills, qualifications, experience and other job-related criteria needed to succeed in the role.
2. Source candidates: the employer should utilize various channels to attract candidates such as job boards, social media, referrals and recruitment agencies.
3. Reviewing resumes and applications: the employer should review the resumes and applications received to identify candidates who best match the job requirements.
4. Interviewing: the employer should interview selected candidates to assess their skills, experience, qualifications and cultural fit.
5. Check references: the employer should contact the candidate's references to check the candidate's work experience, skills and qualifications.
6. Make an offer: the employer should make a job offer to the selected candidate, including details such as salary, benefits and start date.
7. Adaptation of the new employee: the employer should provide the new employee with the necessary information, training and resources to ensure a smooth transition to the new position.

By following a structured hiring process, employers can increase the likelihood of finding the right candidate for the job and reduce the risk of costly hiring mistakes.

Employee motivation refers to the process of creating a work environment that inspires and motivates employees to perform to the best of their abilities. It involves the use of various methods to increase employee morale, engagement and job satisfaction, which can ultimately lead to increased employee productivity and efficiency and improved retention rates. Employee motivation methods may include:

1. Providing opportunities for career growth and development.
2. Offering compensation and benefits packages to employees.

3. Creating a positive work culture that promotes teamwork and collaboration.
4. Recognizing and rewarding good performance and achievements.
5. Providing regular feedback and training to help employees improve their skills.
6. Empowering employees by giving them more autonomy and responsibility.
7. Creating a healthy work-life balance by offering flexible work schedules.

By adopting these methods, organizations can increase the motivation level of their employees, which will further lead to increased productivity, higher job satisfaction, and higher overall performance.

Employee training is the process of providing employees with the knowledge, skills and competencies they need to perform their job responsibilities effectively. It involves a systematic and structured approach to training aimed at improving employee performance, efficiency and job satisfaction. Employee training can be provided in a variety of ways including on-the-job training, classroom training, online training, workshops and seminars. The benefits of staff training include:

1. Increasing productivity and efficiency. Staff training helps employees acquire new skills and knowledge [20].
2. Increased job satisfaction: staff training can help employees feel more confident in their abilities and more engaged in their work.
3. Reduce employee turnover: employees who receive appropriate training are more likely to stay with the organization and contribute to its success.
4. Increased organizational effectiveness: employee training can lead to increased organizational effectiveness by improving employee skills and knowledge.
5. Regulatory Compliance. Staff training can help to ensure that employees comply with legal and regulatory requirements, thereby reducing the risk of legal and financial penalties.

In general, employee training is an important component of any successful organization that seeks to develop and retain a skilled and competent workforce.

Machine learning, based on algorithms and models capable of quickly adapting to new information, provides an opportunity to create intelligent systems capable of handling multiple tasks in the field of recruiting [2]. The paper analyzes various methods and tools of machine learning (KNN method, logistic regression, decision tree method¸ support vector method¸ neural network method), which can be used to improve human resource management processes at enterprises in the energy industry.

3 Results

Stage 1. Analysis of machine learning methods and selection of the most preferable ones for the problem to be solved

Earlier it was stated that the key tasks of human resource management, which require the application of machine learning methods and AI tools for their solution, are recruiting, motivation and training of personnel. Further we will consider several machine learning algorithms and choose the most appropriate one for solving the task related to recruiting.

Machine learning algorithms can be categorized into three types. Each approach has its own characteristics, advantages and disadvantages, and may be more or less suitable for a particular type of task [21].

Supervised Learning or Supervised Learning is the process of training a model with labeled data that already contains correct answers. The model then learns to predict correct answers for new previously unlabeled data.

Unsupervised Learning or unsupervised learning is the process of training a model on unlabeled data that does not have the correct answers. The model then learns to identify patterns and relationships in the data on its own.

Reinforcement Learning is the process of teaching a machine learning model to interact with its environment and receive feedback in the form of rewards or penalties based on its actions. The model then learns to make decisions that maximize its rewards over time.

Each of these approaches provides different tools and techniques for working with data and building models to effectively solve a variety of machine learning problems.

In order to solve the task at hand—design and development of a job candidate matching service, the learning-with-teacher approach will be used. The main idea of this approach is that the model can identify dependencies between input features (parameters) and the target variable, so that it can then predict the value of the target variable for new data. In learning with a teacher, algorithms take as input feature-label pairs and, having learned from such data, seek to predict labels for new examples.

The goal of learning with a teacher is to build a model that can predict the target variable for new data with high accuracy and generalizability.

The main objectives of the learning with teacher approach are:

- Classification. The task of classification is to assign each object to one of the given categories or classes;
- Regression. In a regression task, the model predicts a numerical value or continuous variable based on the input data;
- Object detection. The object detection task is to detect and classify objects in images or videos. The model is trained on the labeled data to detect and classify objects in new images;
- Time series prediction. In time series prediction task, the model is trained on historical time series data to predict its future values.

The projected recruiting service will solve exactly the task of object classification, as the system will need to predict the object class "fit" (candidate is suitable) or "unfit" (candidate is not suitable) based on the value of the analyzed attributes (gender, age, work experience, education, marital status, etc.). Thus, the main purpose of the classification task in the context of the recruiting service is to help recruiters quickly and efficiently select candidates who best meet the requirements of the vacancy. The following methods can be used to solve the task under consideration:

- Logistic regression;

- Nearest neighbor method (kNN);
- Decision tree method;
- Support vector method (SVM);
- Neural networks.

Each of these methods has its own features and applicability depending on the type of data, the size of the dataset and the required problem. Let's take a closer look at each of them.

Logistic regression is a machine learning technique that is used to solve binary classification problems, that is, classifying objects into two classes. However, it can also be adapted to solve multiclass classification problems.

The basic idea of logistic regression is to predict the probability of an object belonging to a certain class. To do this, a logistic function is used that takes input values and returns the probability that an object belongs to a positive class. If the probability obtained is greater than a certain threshold value, the object belongs to a positive class, otherwise it belongs to a negative class.

The nearest neighbor method (kNN) is a machine learning method used to solve classification and regression problems. It is based on the nearest neighbor similarity principle, where data objects are considered similar if their feature descriptions are close in the feature space. The parameter k represents the number of nearest neighbors that are used for classification or regression. The choice of the optimal value of k may depend on the specific task and data structure. A small value of k may lead to overtraining, while a large value of k may lead to undertraining.

The kNN method is sensitive to the scale of the features and the choice of the distance function. Therefore, it is important to scale the features and experiment with different values of the k parameter and distance functions before using the method. The kNN is often used when there is little data or the data structure is complex and cannot be modeled using analytical methods.

Decision tree method is a type of machine learning algorithm used for both classification and regression tasks. The main essence of the decision tree method is to build a tree model of decisions and their possible consequences based on the features of the analyzed data [22].

The decision tree algorithm works by recursively dividing the feature space into smaller subsets based on the most distinguishing features. At each node in the tree, the algorithm selects the feature that best separates the data points into different classes. This feature becomes the decision node and the data is partitioned based on its value. This process is recursively repeated for each subset of the data until the leaves of the tree represent pure classes or regression values.

The decision tree method can be used for feature selection because it identifies the most important features for prediction.

However, the decision tree method can suffer from overfitting if the tree is too deep and complex, which can lead to poor performance when dealing with new data. To overcome this limitation, methods such as reduction, ensemble methods and random forests can be used to improve the accuracy and generalization of the decision tree model.

Support vector method (SVM) is a powerful machine learning technique that is used to solve both classification and regression problems. SVM is particularly well suited for problems with data partitioning in feature space [23].

The main idea of the SVM method is to find an optimal hyperplane that separates objects of different classes in the feature space. This hyperplane is chosen in such a way as to maximize the distance (gap) between the closest objects of different classes, called support vectors.

The SVM algorithm has several advantages over other classification algorithms. It can handle both linearly and nonlinearly separable data and is efficient in multi-dimensional spaces with a small number of data points. In addition, SVM has a regularization parameter that helps to avoid overfitting and improves model generalization.

However, SVM can be sensitive to the choice of kernel function and parameters used to define the hyperplane. It can also be computationally expensive for large datasets. To overcome these limitations, SVM variants such as support vector regression (SVR) and multiple kernel learning (MKL) have been developed to improve the accuracy and scalability of the algorithm.

Neural networks are a large class of machine learning algorithms inspired by the biological neural networks of the human brain. They consist of many interconnected artificial neurons organized into layers and are capable of adapting to complex datasets to perform various tasks such as classification, regression, image segmentation, text generation, and more [24].

However, training neural networks can be computationally expensive and can suffer from overtraining if the network is too complex or the dataset is too small. In addition, neural networks can be difficult to interpret and may require subject matter expertise to optimize their performance.

In order to understand which of the methods are the most suitable for solving the set task of recruitment service development, it is necessary to analyze the advantages and disadvantages of each algorithm. Table 1 presents the advantages and disadvantages of the previously described methods of training with the teacher.

Analyzing the advantages and disadvantages of each of the proposed methods allowed us to conclude that the most appropriate methods for designing a recruitment service would be decision tree, support vector machine (SVM) and neural network methods. Subsequently, candidate class prediction models based on these methods were developed.

Stage 2. Development of a web service related to recruiting utilizing machine learning algorithms

In this paper we solved the problem of developing a web service related to recruiting (selection of candidates for an open vacancy) using the previously described machine learning algorithms.

The web service will be able to perform the following functions:

1. Data aggregation—the service will collect data from various recruiting sources, which will allow viewing candidate resumes from different sites on one resource.

Table 1 Advantages and disadvantages of machine learning methods

Method	Advantages	Disadvantages
Logistic regression	• Ease of implementation and interpretation • Works well with linearly separable classes • Ability to estimate probabilities of class membership	• Is less effective for complex and non-linear dependencies in the data • Requires linear class separability
Nearest neighbors method (KNN)	• Easy to implement and understand • Does not require training as it is a "lazy" algorithm • Can be used for non-linear dependencies in the data • Is effective on small datasets	• Is computationally expensive with a large number of objects • Is sensitive to outliers and noise in the data • Does not work well in high dimensional spaces due to the "curse of dimensionality"
Decision tree method	• Ease of interpretation and visualization • Is able to handle both numeric and categorical features • Can handle non-linear dependencies in the data	• Is prone to overtraining, especially on large and complex data • Is unstable to changes in the data small changes can lead to strong changes in the constructed tree
Support vector method (SVM)	• Works well in high dimensional spaces • Is effective when a small number of training examples are available; • Can handle linear and nonlinear separating surfaces using kernels	• Requires proper kernel and parameter selection, which can be difficult • Difficult to interpret the results
Neural networks	• Can model complex nonlinear dependencies in data • Can process both numerical and categorical features • Can automatically extract features from data	• Require a large amount of data for training and tuning • Difficulty in interpreting results

2. Automated matching of candidates' resumes with the requirements for a specific vacancy.
3. Analyzing candidate profiles, which will allow to quickly highlight key skills, work experience and other parameters that may be crucial in making a decision to hire a candidate.
4. Predicting a candidate's employment success based on data collected from various sources, including their past experience and performance in previous jobs.

5. Making recommendations to improve the recruitment process—the service will be able to analyze data on past recruitment processes in the organization and provide recommendations to improve hiring efficiency.

These functional requirements can serve as the basis for developing detailed product specifications and an implementation plan.

To visualize the functioning and interaction with users, a Use-Case diagram was constructed to understand how the system interacts with external entities (users, other systems, or external events).

In this case, user roles would include:

- Recruiter: the primary user responsible for managing recruiting;
- Candidate: the person looking for job openings and submitting applications;
- Administrator: manages access, system settings and security.

Use Cases or options for using the system are as follows:

1. Job Vacancies:
    - Recruiter can upload job requirements to the system;
    - System benchmarks the vacancy and candidate resumes using machine learning models.
2. Job recommendation: the system can recommend job openings to candidates that match their skills and experience.
3. Success prediction: the system uses data about the candidate, past experience to predict the success of employment.
4. Analytics:
    - Data collection: the system collects data about hiring processes;
    - Process analysis: the system analyzes data to optimize and improve recruiting efficiency.
5. Security and management:
    - Administrator manages user access to the system.
    - Data security and prevention of unauthorized access is ensured.

The development of a recruiting service is a complex task that requires not only the use of machine learning methods, but also integration with web scraping technologies. Figure 1 presents the stages of development of this service.

Stage 3. Solving the recruiting problem using machine learning methods

Next, the stage of developing machine learning models is considered in more detail. Suppose that a potential candidate is evaluated by the following attributes:

1. The candidate's age.
2. Gender of the candidate.
3. Marital status.
4. Education.

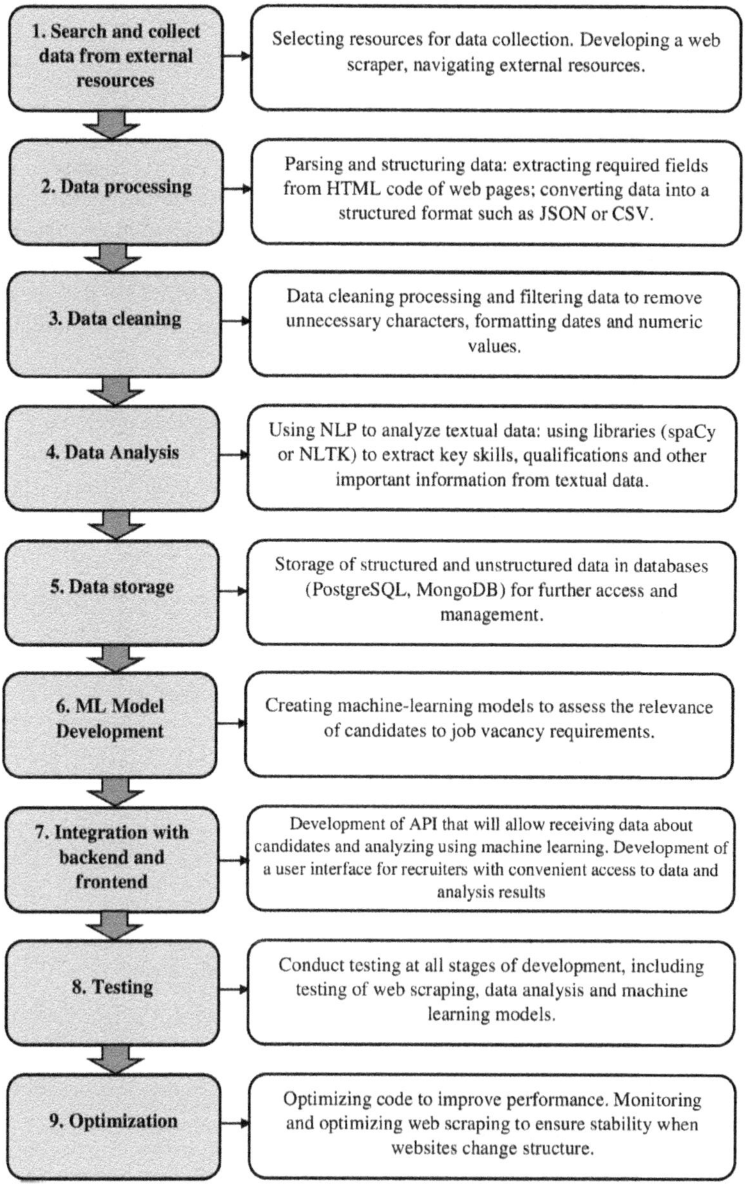

Fig. 1 Stages of development of the web-based recruiting service

5. Availability of work experience.
6. Foreign language skills.
7. Soft-skills.
8. Hard-skills.

It should be noted that in the considered task there was no assessment of the significance of features, as the main purpose is to demonstrate the possibility of applying machine learning algorithms in the field of human resource management. To solve the problem, a toy dataset (Toy Dataset) was modeled, as toy datasets play a key role in understanding the work of the algorithm. The synthetic dataset consists of 50 "candidates"; two classes: "fit"—candidate is fit and 'unfit'—candidate is not fit; 8 previously described candidate characteristics.

After loading the dataset, the data were processed—the categorical variables "candidate gender", "marital status", "education", "foreign language proficiency", "soft-skills", "hard-skills" were coded; target and factor variables were defined; further the sample was divided into training and test sets in the proportion of 70/30 and standardized. The models were built and trained as follows:

```
# Model KNN with 3 clusters (k=3)
knn = KNeighborsClassifier(n_neighbors=3)
knn.fit(X_train, y_train)
knn_y_pred = knn.predict(X_test)
# Model Decision Tree
dt = DecisionTreeClassifier()
dt.fit(X_train, y_train)
dt_y_pred = dt.predict(X_test)
# Model Support Vector Machine
svm = SVC()
svm.fit(X_train, y_train)
svm_y_pred = svm.predict(X_test)
# Model Neural Network
nn = MLPClassifier(hidden_layer_sizes=(5), max_iter=1000)
nn.fit(X_train, y_train)
nn_y_pred = nn.predict(X_test)
```

Figure 2 shows the results of calculating the Accuracy score (Accuracy) of using the above four machine learning algorithms.

Having analyzed the results, we can conclude that the neural network method showed the highest accuracy. MLPClassifier from scikit-learn library was used as

```
Model comparison results:
KNN: 0.7333333333333333
Decision Tree: 0.6666666666666666
Support Vector Machine: 0.4
Neural Network: 0.8
```

Fig. 2 Accuracy of methods

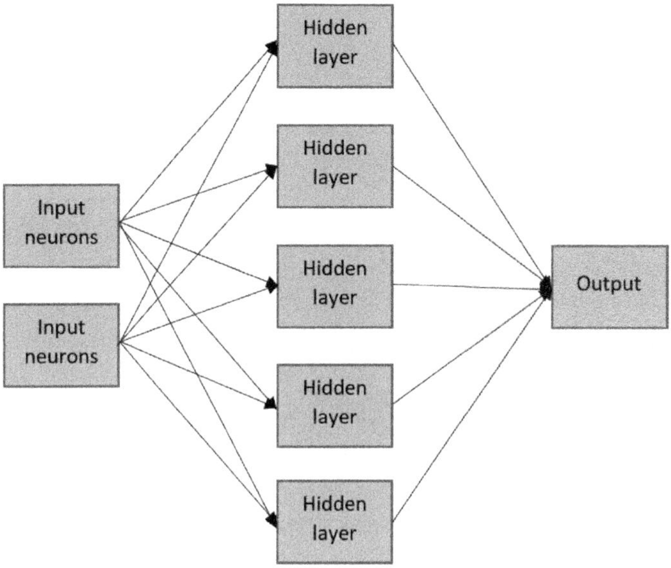

Fig. 3 Structure of neural networks

a model. MLP is a multilayer perceptron which is an artificial neural network with feedback. Where inputs and outputs are zero or one. The structure of this neural network is shown in Fig. 3. MLPClassifier from the scikit-learn library was used as the model.

As can be seen in Fig. 3, this is a classical neural network, having the first layer—the input layer, which receives data; the second layer—the hidden layer, consisting of 5 layers (hidden_layer_sizes=(5)); the third layer—the output layer, which forms the class label in the classification task. The number of iterations in the model is 1000 (max_iter=1000)—the model will go through the sample data 1000, that is, it will complete 1000 epochs.

Let's try to predict the "candidate" class by feeding to the input of the neural network the vector X, which is equal to = [31 1 0 2 5 1 2 1]:

```
nn.predict(np.array([[31,1,0,2,5,1,2,1]]))
array(['Fit']
```

The model identified this observation as belonging to class 1 "fit".

As a result of the work done, data was collected and processed, a machine learning model was built to predict the candidate class, a web service was developed to input requirements for the vacancy and output suitable candidates. After processing the entered requirements, the system outputs candidates suitable for the given vacancy.

4 Discussion

The study identified both the benefits of using artificial intelligence and the risks that can accompany such implementations [25–27].

Artificial intelligence can significantly impact the field of human resource management in several ways:

1. Hiring and selection: AI will help automate and optimize the hiring and selection process by using algorithms to identify suitable candidates based on their skills and qualifications. This can save time and reduce bias in the selection process.
2. Performance Management: AI can help track and analyze employee performance data, such as productivity and engagement levels, to identify areas for improvement and provide feedback to employees.
3. Training and development: AI can be used to personalize training and development programs according to the needs and preferences of individual employees, which can improve their learning and development.
4. Employee engagement: AI will improve employee engagement by analyzing employee sentiment and identifying areas where improvements can be made in the work environment.
5. Predictive analytics: AI can help predict employee turnover and identify factors affecting his metric, allowing organizations to take proactive measures to retain their best talent.

However, the use of AI in workforce management also raises concerns about privacy, data security and job cuts. It is important for organizations to implement AI in a responsible and ethical manner, taking into account these potential risks and ensuring that AI is used to improve, not replace, human resource management practices [28, 29].

Successfully implementing machine learning techniques in human resource management requires a number of key success factors, including:

1. Clear objectives.
2. Data quality and availability.
3. Qualified team.
4. Integration with existing systems.
5. Continuous monitoring and evaluation.
6. Change management.

Let's look at each of the factors in more detail.

Clear goals—the business goals of implementing machine learning techniques in HRM, such as improving hiring, performance management, or employee engagement, need to be clearly defined.

Data quality and availability—high quality data is essential to the success of machine learning techniques. In this case, you need to be certain that HR data is accurate, up-to-date and available in a format that can be utilized by machine learning algorithms.

Skilled team—To effectively implement and manage machine learning methods, you need a team with the right skill set. This can include data scientists, human resources and IT professionals.

Ethical considerations—there needs to be a clear understanding of the potential ethical implications of using ML in HRM, such as bias, privacy and data security. To avoid this, ethical guidelines and policies should be developed.

Integration with existing systems—Machine learning techniques should be integrated with existing HRM systems and processes to ensure their effectiveness and efficiency.

Continuous monitoring and evaluation—continuous monitoring and evaluation of the effectiveness of machine learning techniques in HR management is necessary to identify areas for improvement and ensure they are aligned with business objectives.

Change management—Implementing machine learning techniques in HRM may require changes in organizational culture and processes. Change management techniques should be utilized to ensure that employees are informed, engaged and supportive of change.

Overall, successful implementation of machine learning techniques in HRM requires a strategic approach that addresses the key success factors listed above. In this way, organizations can leverage the power of machine learning to improve HR outcomes and achieve business success.

5 Conclusion

The paper evaluated the current capabilities of artificial intelligence and machine learning in HRM, and demonstrated the application of machine learning methods to solve HRM problems in the energy industry.

To summarize, it is worth noting that the application of machine learning methods can provide numerous benefits to an enterprise's HRM system, such as improving the efficiency and accuracy of hiring, increasing employee engagement and retention, and improving the overall efficiency of the organization. Machine learning algorithms can help analyze large amounts of data, identify patterns, and predict future outcomes, helping HR professionals make more informed decisions.

Applying machine learning methods in recruiting provides an opportunity to optimize the entire recruitment cycle, from the initial screening of resumes to the final assessment of candidate fit. Effective use of machine learning in recruiting can save time, improve the quality of selection, and reduce the risk of bias in the decision-making process. Automating routine tasks such as initial candidate screening also frees recruiters for more strategic and creative aspects of their work [2].

However, it is important to note that the success of these techniques depends on the quality and relevance of the data, as well as the expertise of HR professionals in interpreting and applying the results.

As machine learning continues to evolve and become more accessible, it is likely that it will play an increasingly important role in shaping the future of HRM. Given the constant development of technologies and the growing importance of quality HR management, the application of machine learning methods in recruiting remains a relevant and promising area that will help improve the efficiency and competitiveness of energy industry companies in today's labor market.

References

1. Tambe, P., Cappelli, P., Yakubovich, V.: Artificial intelligence in human resources management: challenges and a path forward. Calif. Manage. Rev. **61**(4), 15–42 (2019)
2. Shirokova, S.V., Rostova, O.V., Prosvirnina, A., Odainic, A.: Key Trends in the digital transformation of business and their impact on the business processes. In: Lecture notes in networks and systems, vol. 684, pp. 885–895. Springer, Cham (2023). https://doi.org/10.1007/978-3-031-32719-3_67
3. Shavlinskaya, K.A.: Personnel development as a key aspect of personnel risk management in energy companies. Sci. Educ. Today **36**(29), 83–85 (2018)
4. Geetha, R., Bhanu, S.R.D.: Recruitment through artificial intelligence: a conceptual study. Int. J. Mech. Eng. Technol. **9**(7), 63–70 (2018)
5. Hmoud, B., Laszlo, V.: Will artificial intelligence take over human resources recruitment and selection? Netw. Intell. Stud. **7**(13), 21–30 (2019)
6. Jatoba, M., Gutierriz, I., Fernandes, P.O., Teixeira, J.P., Moscon, D.: Artificial intelligence in the recruitment& selection: innovation and impacts for the human resources management. In: 43rd International Scientific Conference on Economics and Social Development, pp. 96–104 (2019)
7. Pandya, B.: A competency framework for virtual HR professionals in an artificial intelligence age. In: Proceedings of the International Conference on Applied Research in Management, Business and Economics, pp. 27–48 (2019)
8. Mahmoud, A.A., Shawabkeh, T.A., Salameh, W.A., Al Amro, I.: Performance predicting in hiring process and performance appraisals using machine learning. In: 10th International Conference on Information and Communication Systems, pp. 110–115 (2019)
9. Rostova, O., Shirokova, S., Shmeleva, A.: Application of immersive technologies to improve the effectiveness of training and staff performance. In: Lecture Notes in Networks and Systems, vol. 827, pp. 197–209. Springer, Cham (2023). https://doi.org/10.1007/978-3-031-47694-5_15
10. Yano, K.: How artificial intelligence will change HR. People Strat. **40**(3), 42–47 (2017)
11. Berhil, S., Benlahmar, H., Labani, N.: A review paper on artificial intelligence at the service of human resources management. Indones. J. Electr. Eng. Comput. Sci. **18**(1), 32–40 (2020)
12. Bhardwaj, G., Singh, S.V., Kumar, V.: An empirical study of artificial intelligence and its impact on human resource functions. In: 2020 International Conference on Computation, Automation and Knowledge Management (ICCAKM), pp. 47–51 (2020)
13. Haenlein, M., Kaplan, A.: A brief history of artificial intelligence: on the past, present, and future of artificial intelligence. Calif. Manage. Rev. **61**(4), 5–14 (2019)
14. Strohmeier, S., Piazza, F.: Artificial intelligence techniques in human resource management-a conceptual exploration. In: Kahraman, C., Çevik Onar, S. (eds.) Intelligent Techniques in Engineering Management: Theory and Applications, pp. 149–172. Springer, New York (2015)
15. Blinnikova, A.V., Ying, D.K.: The use of artificial intelligence in human resource management. Bull. Univ. **7**, 14–21 (2020)
16. Wisetsri, W., Vijai, C, Chueinwittaya, K, Jirayus, P: Artificial intelligence in human resources management. J. Posit. Sch. Psychol. **6**(2), 2688–2693 (2022)
17. Sakka, F., Maknouzi, M.E., Sadok, H.: Human resource management in the era of artificial intelligence: future HR work practices, anticipated skill set, financial and legal implications. Acad. Strateg. Manag. J. **21**(S1), 1–14 (2022)

18. Cappelli, P., Tambe, P., Yakubovich, V.: Artificial intelligence in human resources management: challenges and a path forward. https://papers.ssrn.com/sol3/papers.cfm?abstract_id=3263878. Last accessed 05 Sep 2024
19. Ilchenko, S.V.: Recruiting personnel in the system of personnel management. Bull. Exp. Educ. **1**(26), 34–43 (2021)
20. Markova, T., Shirokova, S., Rostova, O., Misbakhova, C.: Possibilities of using computer-aided design systems to improve the quality of technical specialists training. In: Lecture Notes in Networks and Systems, vol. 951, pp. 93–105. Springer, Cham (2024). https://doi.org/10.1007/978-3-031-56677-6_7
21. Akzholov, R.K.: Machine learning. Sci. Bull. **6**(15), 348–351 (2019)
22. Feofanov, A.N., Egorov, M.K.: Improving the efficiency of an enterprise based on the application of the decision tree method. Autom. Model. Des. Manag. **1**(11), 29–34 (2021)
23. Zaikin, O.S., Kochemazov, S.E.: Using support vector machines to assess the usefulness of conflict clauses in CDCL inference. Int. J. Open Inf. Technol. **12**, 5–10 (2019)
24. Trishechkin, S.N.: Data Mining and the method of neural networks. Bull. Sci. Educ. **8**(62), 38–40 (2019)
25. Ernst, E., Merola, R., Samaan, D.: Economics of artificial intelligence: implications for the future of work. IZA J. Labor Policy **9**(1), 1–35 (2019)
26. Meister, J.: Ten HR trends in the age of artificial intelligence. Forbes Magazine. https://www.forbes.com/sites/jeannemeister/2019/01/08/ten-hr-trends-in-the-age-of-artificial-intelligence/. Last accessed 25 Aug 2024
27. Qiu, L., Zhao, L.: Opportunities and challenges of artificial intelligence to human resource management. Acad. J. Humanities Soc. Sci. **2**(1), 144–153 (2018)
28. Bryndin, E.: Formation of technological cognitive reason with artificial intelligence in virtual space. Br. Int. Exact Sci. (BIoEx) J. **2**(2), 450–461 (2020)
29. Johnson, K., Pasquale, F., Chapman, J.: Artificial intelligence, machine learning, and bias in finance: toward responsible innovation. Fordham Law Rev. **88**(2), 499–529 (2019)

Correction to: Arctic Zone Energy Sector Ontology

Nina Trifonova, Alexandra Borremans, Anastasia Levina, and Alissa Dubgorn

Correction to: Chapter 5 in: I. Ilin and M. Youzhong (eds.), *Digital Systems and Information Technologies in the Energy Sector*, Lecture Notes in Networks and Systems 1244, https://doi.org/10.1007/978-3-031-80710-7_5

The original version of the chapter was updated with an acknowledgement section. The chapter has been updated with the change.

The updated version of this chapter can be found at
https://doi.org/10.1007/978-3-031-80710-7_5

I. Ilin and M. Youzhong (eds.), *Digital Systems and Information Technologies in the Energy Sector*, Lecture Notes in Networks and Systems 1244,
https://doi.org/10.1007/978-3-031-80710-7_33

Author Index

I. Ilin and M. Youzhong (eds.), *Digital Systems and Information Technologies in the Energy Sector*, Lecture Notes in Networks and Systems 1244,
https://doi.org/10.1007/978-3-031-80710-7

The manufacturer's authorised representative in the EU is Springer Nature Customer Service Centre GmbH, Europaplatz 3, 69115 Heidelberg, Germany. If you have any concerns regarding our products, please contact ProductSafety@springernature.com

Printed and bound by CPI Group (UK) Ltd, Croydon, CR0 4YY
30/09/2026
02225435-0012